Energy Production, Conversion and Storage

Energy Production, Conversion and Storage

Editor: Lucas Collins

CALLISTO
REFERENCE
www.callistoreference.com

Callisto Reference,
118-35 Queens Blvd., Suite 400,
Forest Hills, NY 11375, USA

Visit us on the World Wide Web at:
www.callistoreference.com

ISBN: 978-1-64116-052-0 (Hardback)

Cataloging-in-Publication Data

Energy production, conversion and storage / edited by Lucas Collins.
     p. cm.
Includes bibliographical references and index.
ISBN 978-1-64116-052-0
1. Power resources. 2. Energy development. 3. Energy conversion. 4. Energy storage.
5. Energy facilities. I. Collins, Lucas.
TJ163.2 .E54 2019
621.042--dc23

Table of Contents

Preface

Energy development refers to the set of activities aimed at the extraction of energy from natural resources. It is also concerned with the conversion of energy for industrial usage and its storage. It encompasses the processes related to conventional, renewable as well as alternative sources of energy. This book presents innovative techniques for the production, conversion and storage of energy. It compiles researches of international experts which will help the readers develop a comprehensive understanding of the subject. This book will serve as a resource guide for engineers, scientists, experts, researchers and students alike.

This book has been the outcome of endless efforts put in by authors and researchers on various issues and topics within the field. The book is a comprehensive collection of significant researches that are addressed in a variety of chapters. It will surely enhance the knowledge of the field among readers across the globe.

It gives us an immense pleasure to thank our researchers and authors for their efforts to submit their piece of writing before the deadlines. Finally in the end, I would like to thank my family and colleagues who have been a great source of inspiration and support.

Editor

Torque Distribution Characteristics of a Novel Double-Stator Permanent Magnet Generator Integrated with a Magnetic Gear

Shehu Salihu Mustafa [1], Norhisam Misron [1,2,*], Norman Mariun [1], Mohammad Lutfi Othman [1] and Tsuyoshi Hanamoto [3]

[1] Department of Electrical & Electronic, Faculty of Engineering, Universiti Putra Malaysia, 43400 Serdang, Selangor, Malaysia; shehums@gmail.com (S.S.M.); norman@upm.edu.my (N.M.); lutfi@upm.edu.my (M.L.O.)

[2] Institute of Advanced Technology, Faculty of Engineering, Universiti Putra Malaysia, 43400 Serdang, Selangor, Malaysia

[3] Department of Biological Functions Engineering, Graduate School of Life Science and Systems Engineering, Kyushu Institute of Technology, Kitakyushu 808-0916, Japan; hanamoto@life.kyutech.ac.jp

* Correspondence: norhisam@upm.edu.my

Academic Editor: K.T. Chau

Abstract: This paper presents a novel double-stator permanent-magnet machine integrated with a triple rotor magnetic gear structure, which is proposed to address problems of mechanical geared generators for low-speed applications. Torque transmission is based on three rotors consisting of prime permanent-magnet (PM) poles in the middle rotor and field PM poles in the inner and outer rotors. The proposed machine combines the functions of magnetic gearing and electrical power generation. The operating principles of the magnetic gear and generator are discussed and the torque distribution characteristics of the integrated machine are analysed using the 2D finite-element method (2D FEM). Also the power, torque, and speed characteristics are reported. A prototype is fabricated and tested experimentally. The predicted and measured results validate the proposed machine design.

Keywords: magnetic gear (MG); permanent-magnet (PM); double-stator magnetic geared machine (DS-MGM); prime magnet; field magnet; torque distribution; transmission torque; cogging torque; finite element method; harmonic analysis; flux density

1. Introduction

Electrical power generation from renewable energy is the focus of current research in the academia and industry. In a conventional power generation system, input mechanical power is converted to electrical power energy by the rotation of a prime mover in a generator. Wind turbine generators and hydro turbines are some examples of renewable energy sources that provide clean sources of energy. In the wind industry a high-speed power generator is usually coupled to a mechanical gearbox for converting the shaft's low-speed to high-speed in low-wind regions. However, this mechanical geared configuration has problems of reduced efficiency due to maintenance, lubrication, and noise, which constrain the operation of mechanical geared generator machines. For low-speed operation, a direct drive generator operates without a mechanical gearbox. Although it eliminates the use of a gearbox, it still has issues with size and cost due to a greater number of poles, including heavier weight and lower power density. With the introduction of magnetic gears, research and development have progressed rapidly [1]. A study [2] on magnetic gearing was conducted and the authors reported that magnetic gears could be more efficient than mechanical gears because of the advantages of inherent overload protection, greater transmission torque, and oil-free maintenance. Researchers have proposed various types of magnetic gear topologies, including radial flux magnetic gears, linear magnetic gears, and axial magnetic gears [3–7] with a reported calculated torque density greater than 100 kNm/m^3. The concentric radial flux topology appears to be the most favourable concept because of its efficient use of PMs and greater torque density, but this topology results in mechanical difficulties in fabrication. Various designs of magnetic gears have been proposed [8–15] and their torque characteristics demonstrated with measured data. Chen [16] compared different topologies of magnetic gears and concluded that the radial-flux configuration produces the greatest torque density, although measured data were not reported in the study. The operating principle of a magnetic gear is similar to a mechanical gear as torque is transferred from a low-speed shaft to a high-speed shaft with permanent magnets. Studies [17–19] have shown that integrating a magnetic gear with a permanent magnet machine could realise a compact and cost-effective electromechanical machine.

The purpose of this paper is to present the torque distribution characteristics of a novel magnetic geared double-stator permanent magnet generator. The aim is to integrate a magnetic gear designed with two modulating iron rings and three permanent magnet rotors with a double-stator permanent magnet machine to address problems with mechanical gears. In Section 2, the proposed structure and machine design will be introduced. Section 3 will present the machine operating principle. Section 4 discusses the results. Finally, the conclusion is presented in Section 5.

2. Proposed Structure and Machine Operating Principle

Figure 1 shows the structure of the proposed double-stator magnetic geared machine while the parameters are listed in Table 1. It is comprised of twelve parts. The outer field PMs, the outer modulating iron ring, the prime PMs, the inner modulating iron ring, and the inner field PMs are the active components that function as a magnetic gear. In the double-stator topology the two airgaps in the centre of the electrical machine are replaced with a magnetic gear. The proposed machine is designed in a magnetically coupled configuration with three bone rotors and three layers of mutual PMs. The advantage of the coupled configuration is the combined effect of the three PM rotors to the total flux-linkage in both outer and inner stators.

As illustrated in Figure 2, the prime PM rotor, which is the prime mover shaft for the magnetic gear, rotates at low-speed and is magnetically coupled simultaneously to both inner and outer field PM rotors. The high-speed field PM rotors and stators are interconnected through magnetic field excitation and this combination is equivalent to a conventional PM electrical machine. The output torque that is produced from the high-speed field PM rotor by the magnetic gear is due to the torque applied on the low-speed prime PM rotor. The material properties of the components used for the prototype machine are listed in Table 2, while the coil winding properties are listed in Table 3. A coaxial magnetic gear with three rotors is integrated with a double-stator PM machine to form a compact double-stator magnetic geared PM machine.

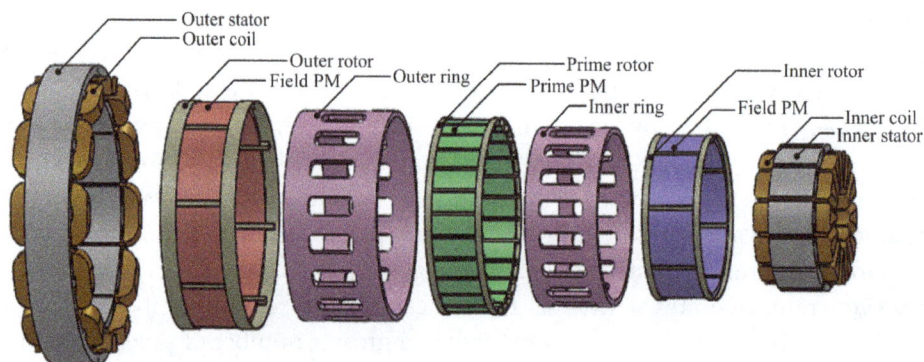

(a)

Figure 1. *Cont.*

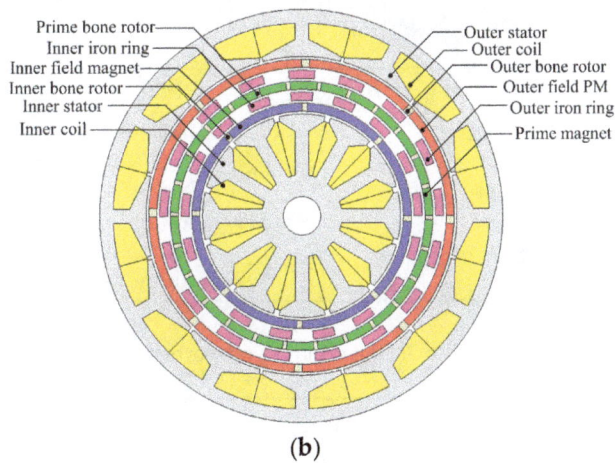

(b)

Figure 1. Proposed structure of double-stator permanent magnet machine integrated with a triple rotor magnetic gear: (**a**) Exploded view of machine components; (**b**) Structural view of integrated machine.

Table 1. Design parameters of double-stator magnetic geared machine.

Parameter	Value
Pole-pair number outer and inner field PMs	4
Pole-pair number prime PMs	13
Pole number outer and inner iron ring pieces	17
Number of outer stator slots	12
Number of inner stator slots	12
Outer and inner airgap	1 mm
Thickness of PMs	3 mm
Thickness of iron ring pieces	3 mm
Axial length	30 mm
Outer diameter	151 mm

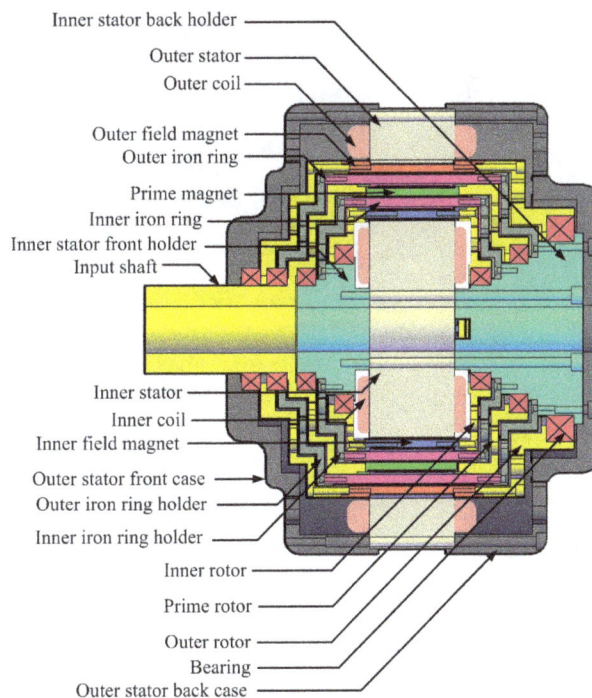

Figure 2. Axial cross-sectional diagram of the magnetic geared double-stator machine.

Table 2. Material properties of components.

Component	Material
Magnets	Nd-Fe-B-38H
Iron rings	SS400
Rotors	SS400
Stators	50H800 Laminated steel sheet
Iron ring end rings	Aluminium
Shaft	Aluminium

Table 3. Coil winding properties.

Parameter	Value
Diameter of coil wire	0.80 mm
Outer coil number of turns	75
Inner coil number of turns	31
Outer coil resistance per phase	1.20 Ω
Inner coil resistance per phase	0.40 Ω

2.1. Triple Rotor Magnetic Gear Operating Principle

The operating principle of the triple rotor magnetic gear is based on the permeance difference between the air in the slots and the two iron rings [20]. The structure shown in Figure 3a consists of two magnetic gears with one low-speed rotor rotating two high-speed rotors simultaneously. The inner permanent magnet of the inner gear generates a magnetomotive force distribution, as shown in Figure 3b, while the inner iron ring generates a permeance distribution to vary the permeance, shown in Figure 3c.

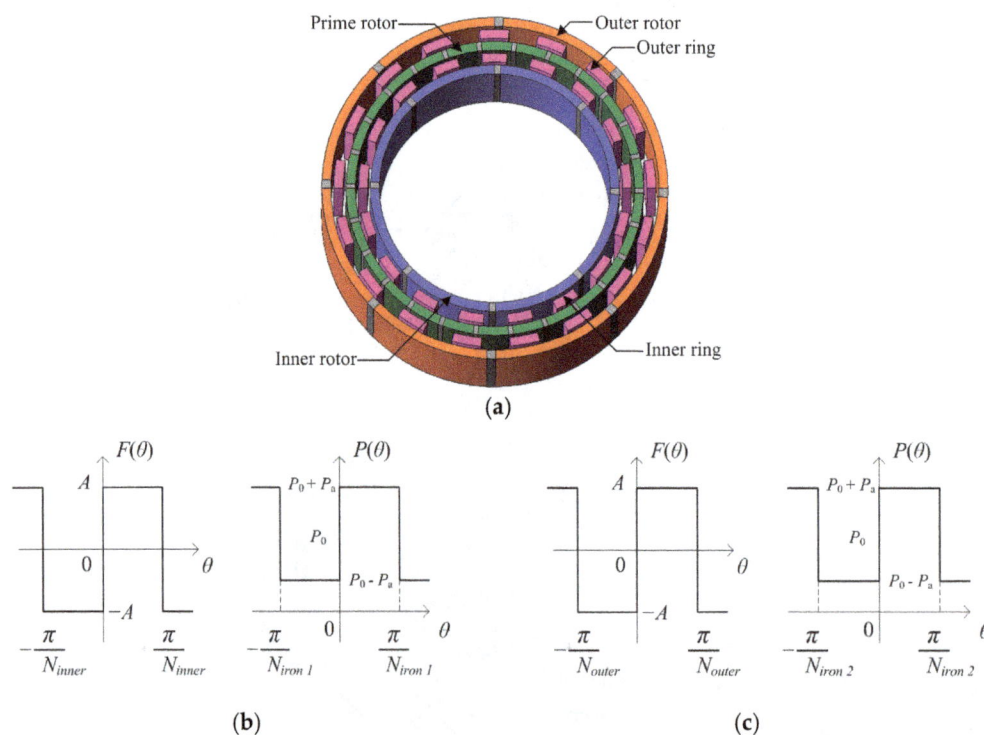

Figure 3. Structure of magnetic gear with three rotors: (**a**) Configuration of the triple-rotor magnetic gear; (**b**) Magnetomotive force and permeance distribution of the inner gear; (**c**) Magnetomotive force and permeance distribution of the outer gear.

The Fourier series expansions of $F(\theta)$ and $P(\theta)$ can be expressed by Equations (1) and (2):

$$F(\theta) = \sum_{m=1}^{\infty} a_m \, sin\{(2m-1)N_{inner}\theta\} \tag{1}$$

$$P(\theta) = P_0 + \sum_{n=1}^{\infty} a_n sin\{(2n-1)N_{iron}\theta\} \tag{2}$$

where N_{inner} is the inner rotor pole-pair number, N_{iron} is the number of stationary iron ring pieces, a_m and a_n are the amplitude of the magnetomotive force and permeance, respectively, while P_0 is the permeance of the iron ring pole pieces.

The magnetic flux distribution $\phi(\theta)$ can be derived as follows [20]:

$$
\begin{aligned}
\phi(\theta) \quad &= F(\theta){\cdot}P(\theta) \\
&= \sum_{m=1}^{\infty} P_o \, a_m sin\{(2m-1)N_{inner}\theta\} \\
&+ \sum_{m=1}^{\infty}\sum_{n=1}^{\infty} \frac{a_m a_n}{2}[cos\{(2n-1)N_{iron} - (2m-1)N_{inner}\}\theta \\
&\quad -cos\{(2n-1)N_{iron} + (2m-1)N_{inner}\}\theta]
\end{aligned}
\tag{3}
$$

The magnetic flux density distribution ϕ (θ) contains the $H_1(m)$, $H_2(n)$, $H_3(m,n)$, and $H_4(m,n)$ order constituents, which are shown in:

$$
\left\{
\begin{aligned}
H_1(m) &= (2m-1)N_{inner} \\
H_2(m) &= (2n-1)N_{iron} \\
H_3(m,n) &= (2n-1)N_{iron} - (2m-1)N_{inner} \\
H_4(m,n) &= (2n-1)N_{iron} + (2m-1)N_{inner}
\end{aligned}
\right\}
\tag{4}
$$

$$
\begin{aligned}
\phi_{\Delta t}(\theta) = F_{\Delta t}(\theta) \quad &{\cdot}P_{\Delta t}(\theta) = F(\theta - \omega_{inner}\Delta t){\cdot}P(\theta - \omega_{iron}\Delta t) \\
&= \sum_{m=1}^{\infty} P_o a_m sin\{H_1(m)(\theta - \omega_{inner}\Delta t)\} \\
&+ \sum_{m=1}^{\infty}\sum_{n=1}^{\infty} \frac{a_m a_n}{2}\left[cos\left\{H_3(m,n)(\theta - \frac{H_1(m)\omega_{iron} - H_2(m)\omega_{inner}}{H_3(m,n)}\Delta t\right\} \right.\\
&\quad \left.- cos\left\{H_4(m,n)(\theta - \frac{H_1(m)\omega_{iron} + H_2(m)\omega_{inner}}{H_4(m,n)}\Delta t\right\}\right]
\end{aligned}
\tag{5}
$$

The magnetic flux constituents $H_1(m)$, $H_3(m,n)$, and $H_4(m,n)$ each rotate ω_{inner}, $(H_1(m)\omega_{iron} - H_2(m)\omega_{inner})/H_3(m,n)$ and $(H_1(m)\omega_{iron} + H_2(m)\omega_{inner})/H_4(m,n)$, respectively. The number of pole-pairs in $H_3(m,n)$ and $H_4(m,n)$ should be equal to N_{outer} or N_{inner}. The relationship between N_{inner}, N_{iron}, and N_{outer} can be expressed as follows:

$$(2n-1)N_{iron} = N_{outer} \pm (2m-1)N_{inner} \tag{6}$$

$$(2n-1)N_{iron} \, \omega_{iron} = N_{outer}\omega_{outer} \pm (2m-1)N_{inner}\omega_{inner} \tag{7}$$

$$
\left\{
\begin{aligned}
\omega_{input} &= \omega_{inner} \, [N_{inner}] \\
\omega_{output} &= \omega_{outer1} = \omega_{outer2} \, [N_{outer1} = N_{outer2}] \\
\omega_{fixed} &= \omega_{iron\,1} = \omega_{iron\,2} = 0 \, [N_{iron\,1} = N_{iron\,2}]
\end{aligned}
\right\}
\tag{8}
$$

$$N_{iron} = N_{outer} + N_{inner} \tag{9}$$

$$G_r = -\frac{\omega_{input}}{\omega_{output\,1}} = -\frac{N_{inner}}{N_{outer1}} \tag{10}$$

$$G_r = -\frac{\omega_{input}}{\omega_{output\,2}} = -\frac{N_{inner}}{N_{outer2}} \tag{11}$$

2.2. Transmission Torque

The structure of the triple rotor magnetic gear shown in Figure 4 comprises an outer rotor with p_1 pole-pair permanent magnets (PMs) with an angular speed of ω_1; an outer iron ring consisting of n_2 ferromagnetic pole-pieces with an angular speed of ω_2; a middle rotor containing p_3 pole-pair prime PMs with an angular speed of ω_3; an inner iron ring comprising n_4 ferromagnetic pole pieces with an angular speed of ω_4; and an inner rotor comprising p_5 pole-pair PMs with an angular speed of ω_5. Both outer and inner rotors rotate in the opposite direction to the prime rotor.

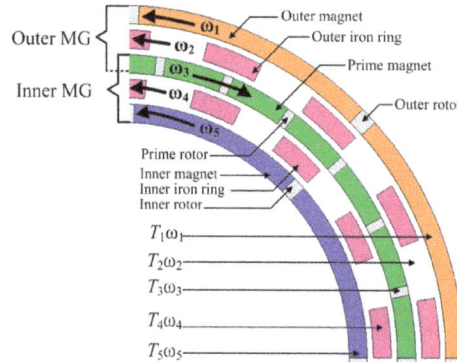

Figure 4. Mechanical power flow distribution in the triple-rotor magnetic gear.

The transmission torque of the magnetic gear is governed by:

$$N_{iron\ 1} = P_{outer} + P_{prime} \ [\text{Outer magnetic gear}] \tag{12}$$

$$N_{iron\ 2} = P_{inner} + P_{prime} \ [\text{Inner magnetic gear}] \tag{13}$$

By using the Maxwell stress tensor to calculate the average magnetic torque developed in the outer airgap between the outer rotor and prime rotor, the average magnetic torque can be expressed as [3]:

$$T_{m\_outer} = \frac{L_{ef} R_{outer}^2}{\mu_0} \int_0^{2\pi} B_{r\_outer} B_{\theta\_outer} d\theta \tag{14}$$

Similarly, the average magnetic torque developed in the inner airgap between the inner rotor and prime rotor can be expressed as [3]:

$$T_{m\_inner} = \frac{L_{ef} R_{inner}^2}{\mu_0} \int_0^{2\pi} B_{r\_inner} B_{\theta\_inner} d\theta \tag{15}$$

where L_{ef} is the active axial length of the magnetic gear, B_{r_outer} and B_{θ_outer} are the radial and tangential flux densities in the outer airgap, B_{r_inner} and B_{θ_inner} are the radial and tangential flux densities in the inner airgap, and R_{outer} and R_{inner} are the radii of the outer and inner airgaps, respectively. Applying the law of conservation of energy and neglecting power losses in the magnetic gear shown in Figure 4, this yields:

$$T_1\omega_1 + T_2\omega_2 + T_3\omega_3 + T_4\omega_4 + T_5\omega_5 = 0 \tag{16}$$

where T_1, T_2, T_3, T_4, and T_5 are the torques for the outer rotor, outer iron ring, prime rotor, inner iron ring, and inner rotor, respectively. If the two iron rings are stationary ($\omega_2 = \omega_4 = 0$), this yields:

$$T_1\omega_1 + T_3\omega_3 + T_5\omega_5 = 0 \tag{17}$$

If the outer and inner PM rotors rotate at a similar angular speed due to their equal gear ratio, i.e., $\omega_1 = \omega_5$, then substituting into Equation (17) gives:

$$T_1\omega_5 + T_3\omega_3 + T_5\omega_5 = 0 \tag{18}$$

$$T_3 = -\frac{\omega_5}{\omega_3}(T_1 + T_5) \tag{19}$$

The gear ratio G_r obtained from the ratio of the inner rotor speed and prime rotor speed is expressed as:

$$G_r = -\frac{\omega_5}{\omega_3} \tag{20}$$

The gear ratio is equal to the torque ratio, and the equation that relates the magnetic torques on the three PM rotors for the magnetic gear is expressed as:

$$T_r = \frac{T_3}{T_1 + T_5} \tag{21}$$

where T_r is the transmission torque ratio for the three PM rotors, which is equal to G_r neglecting power losses.

If both outer and inner rotors are stationary, i.e., $\omega_1 = \omega_5 = 0$, applying the law of conservation of energy it yields:

$$T_2\omega_2 + T_3\omega_3 + T_4\omega_4 = 0 \tag{22}$$

$$T_3 = -\frac{\omega_1}{\omega_3}\left(\frac{n_2 - p_3}{n_2}\right)(T_2 + T_4) \tag{23}$$

The magnetic torque relationship between the prime PM rotor, outer iron ring, and inner iron ring can be expressed as:

$$T_3 = \left(-\frac{p_3}{n_2}\right)(T_2 + T_4) \tag{24}$$

where p_3 is the pole number of prime PMs, n_2 is the outer iron ring, T_2, T_3 and T_4 are the torques for the outer iron ring, prime rotor and inner iron ring respectively.

2.3. Double-Stator Permanent Magnet Machine Operating Principle

The output power of a double-stator permanent magnet machine (DS-PM machine) [21] produced by the inner stator can be expressed as:

$$P_1 = \frac{\pi}{2}\eta K_w A_1 B_{g1} D_{g1}^2 L_a \omega_r \tag{25}$$

while the output power produced by the outer stator can be expressed as:

$$P_2 = \frac{\pi}{2}\eta K_w A_2 B_{g2} D_{g2}^2 L_a \omega_r \tag{26}$$

The output torque produced from both the inner and outer airgap of a DS-PM machine [21] shown in Figure 5 can be calculated for the inner and outer stators. The torque produced by the inner stator in the inner airgap is derived as:

$$T_1 = \frac{\pi}{2}\eta K_w A_1 B_{g1} D_{g1}^2 L_a \tag{27}$$

Similarly the torque produced by the outer stator in the outer airgap is also derived as:

$$T_2 = \frac{\pi}{2}\eta K_w A_2 B_{g2} D_{g2}^2 L_a \tag{28}$$

The total output torque T_{Total} generated by the DS-PM machine is obtained by combining Equations (27) and (28) and can be expressed as:

$$T_{Total} = \frac{\pi}{2}\eta K_w B_g(A_1 D_{g1}^2 + A_2 D_{g2}^2)L_a \tag{29}$$

where η is the efficiency, K_w is the winding factor, Bg is the magnetic loading, A_1 and A_2 are the electric loading of both outer and inner airgaps, respectively, D_{g1} is the inner stator outer diameter, D_{g2} is the outer stator inner diameter, L_a is the active axial length, and ω_r is the rotor speed. In Figure 5 the flux flows radially from the magnet across the airgap, then circumferentially through the outer stator back yoke and returns across the airgap to the magnet to form a closed loop through the inner stator back yoke.

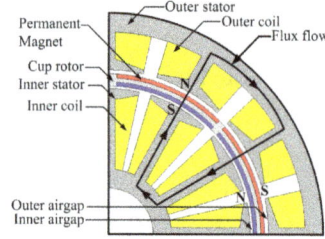

Figure 5. A quarter section of the double-stator permanent magnet machine.

2.4. Cogging Torque Characteristics

The torque ripple results from the interaction of both field and prime PMs with the iron ring pole-pieces because there is variation in the reluctance of the air-gap between the iron rings and magnets. The torque ripple on the inner rotor varies with time and can be expressed as [3]:

$$T_{ripple\_inner} = \sum_{n=1}^{\infty} s_{n\_inner} sin(nA(\omega_1 t - \theta_1)) \tag{30}$$

where S_{n_inner} is the Fourier coefficient, A is the least common multiple of N_s and $2p_1$, N_s is the number of stationary pole-pieces, and p_1 is the PM pole number. Also, the torque ripple on the outer rotor can be expressed as:

$$T_{ripple\_outer} = \sum_{n=1}^{\infty} s_{n\_outer} sin(nB(\omega_2 t - \theta_2)) \tag{31}$$

where S_{n_outer} is the Fourier coefficient, B is the least common multiple of N_s and $2p_2$, and p_2 is the number of PM poles. Cogging torque in brushless PM machines is produced from the relationship between the magnets and stator tooth [22], which is expressed as:

$$T_{cog} = -\frac{1}{2} \phi_g^2 \frac{dR}{d\theta} \tag{32}$$

where ϕ_g is the air-gap flux, R is the reluctance of the air-gap, and θ is the mechanical angle of the rotor. The reluctance of the air-gap changes with periodicity; hence, it can be expressed in Fourier series as:

$$T_{cog} = \sum_{k=1}^{\infty} T_{mk} sin(mk\theta) \tag{33}$$

where m is the least common multiple of stator slots number N_s and the pole number N_p, T_{mk} is a Fourier coefficient, and k is an integer. For a magnetic gear [23], the smallest common multiple N_c between the number of PMs on the high-speed rotor and the number of iron pole pieces is expressed as:

$$C_T = \frac{2pn_s}{Nc} \tag{34}$$

where C_T is the cogging torque factor and p is the pole number of PMs on the high-speed rotor. For PM brushless machines [24], the period of the cogging torque is expressed as:

$$N_p = \frac{2p}{HCF(Q,2p)} \tag{35}$$

where N_p is the number of periods of the cogging torque, Q is the number of stator slots, $2p$ is the number of poles, and HCF is the highest common factor between Q and $2p$. The cogging torque factor for permanent magnet brushless machines [25], in general form, is given by:

$$C_T = \frac{2pQ_s}{N_c} \tag{36}$$

where Q_s is the number of slots, $2p$ is the number of poles, and N_c is the smallest common multiple between Q_s and $2p$.

2.5. Torque Distribution Map and Analysis

The distribution of electromagnetic torque in the double-stator magnetic geared permanent magnet machine is difficult to investigate using analytical methods because of the machine's complex structure. There are six airgaps with seven independent components; also, the stator iron yokes and bone rotors have nonlinear properties. The stationary iron ring pole pieces produce modulation effects, while the stator slots produce slot effects. In conventional magnetic gears, both outer and inner rotors consist of ferromagnetic iron yokes, while in the double-stator magnetic geared permanent magnet machine a stator replaces both outer and inner field PM rotors. We can assume the iron ring pole pieces to be equivalent to imaginary stator slots and the transmission torque on both outer and inner field PM rotors are affected simultaneously by the stators and stationary iron rings. The DS-MGM machine is analysed as a five-port machine consisting of three mechanical ports and two electrical ports. This section discusses the electromagnetic torque distribution on the whole machine. The electromagnetic torques generated by the outer stator, outer field PM rotor, outer iron ring, prime PM rotor, inner iron ring, inner field PM rotor, and inner stator are labelled T_{os}, T_{ofm}, T_{oi}, T_{pm}, T_{ii}, T_{ifm}, and T_{is}, respectively. When the machine operates in generator mode, both outer and inner stators output electrical power and can be expressed as:

$$P_{os} = T_{os}\omega_{os} \tag{37}$$

$$P_{is} = T_{is}\omega_{is} \tag{38}$$

where P_{os} and ω_{os} are the mechanical power and angular speed of the outer stator, while P_{is} and ω_{is} are the mechanical power and angular speed of the inner stator. The mechanical power generated by the inner field PM, outer field PM, and prime PM rotors, respectively, can be expressed as:

$$P_{ofm} = T_{ofm}\omega_{ofm} \tag{39}$$

$$P_{pm} = T_{pm}\omega_{pm} \tag{40}$$

$$P_{ifm} = T_{ifm}\omega_{ifm} \tag{41}$$

where P_{ofm}, P_{pm}, and P_{ifm} are the mechanical power of the outer field PM, prime PM, and inner field PM rotors, respectively, while ω_{ofm}, ω_{pm}, and ω_{ifm} are the angular speeds of the outer field PM, prime PM, and inner field PM rotors, respectively. The mechanical power generated by the inner and outer modulating iron rings can be expressed as:

$$P_{ii} = T_{ii}\omega_{ii} \tag{42}$$

$$P_{oi} = T_{oi}\omega_{oi} \tag{43}$$

where P_{ii} and ω_{ii} are the mechanical power and angular speed of the inner iron ring, while P_{oi} and ω_{oi} are the mechanical power and angular speed of the outer iron ring. Neglecting all power losses and applying the law of conservation of energy, the torque sum of all five ports should be zero:

$$T_{os}\omega_{os} + T_{ofm}\omega_{ofm} + T_{ii}\omega_{ii} + T_{pm}\omega_{pm} + T_{ifm}\omega_{ifm} + T_{oi}\omega_{oi} + T_{is}\omega_{is} = 0 \tag{44}$$

$$T_{os} + T_{ofm} + T_{ii} + T_{pm} + T_{ifm} + T_{oi} + T_{is} = 0 \tag{45}$$

To illustrate the torque relations of T_{os}, T_{ofm}, T_{oi}, T_{pm}, T_{ii}, T_{ifm}, and T_{is}, a torque distribution map method is proposed and illustrated in Figure 6. Points A, B, C, D, E, F, and G, as shown in Figure 7, symbolize the outer stator, outer field PM rotor, outer iron ring, prime PM rotor, inner iron ring, inner field PM rotor, and inner stator, respectively. The distance line AB represents the torque relation between the outer stator and the outer field PM rotor, while the distance line BC indicates the torque relation between the outer field PM rotor and outer iron ring. According to the torque equilibrium principle, the relation between T_{os} and T_{ofm} can be expressed as:

$$T_{os} = -T_{ofm} \tag{46}$$

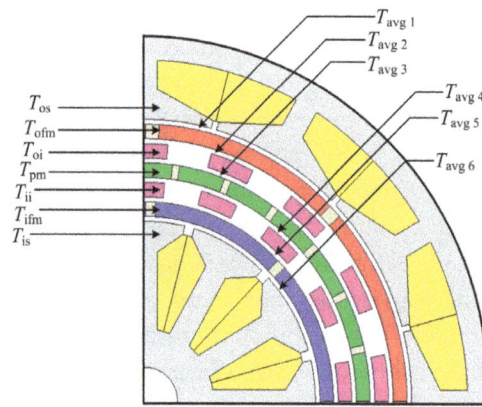

Figure 6. Torque distribution points of double-stator magnetic geared permanent magnet machine.

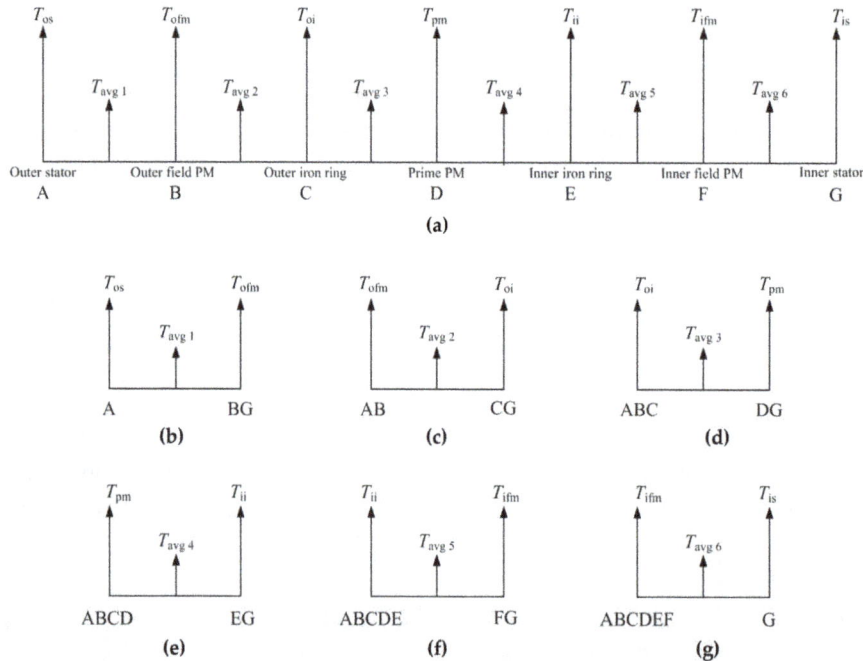

Figure 7. Torque analysis of DS-MGM: (**a**) Torque distribution between parts; (**b**) Outer stator and outer field PM; (**c**) Outer field PM and outer iron ring; (**d**) Outer iron ring and prime PM; (**e**) Prime PM and outer iron ring; (**f**) Inner iron ring and inner field PM; (**g**) Inner field PM and inner stator.

Based on Newton's third Law, T_{ofm} acts as counter-torque equal in value but in the opposite direction to T_{os}. This implies that for rotation of the outer field PM to occur the condition line $BC > AB$ should be satisfied and the midpoints $T_{avg\ 1}$, $T_{avg\ 2}$ are assumed as pivot points representing the average electromagnetic torques for lines AB and BC, respectively. Line $ABCDEFG$ acts as a lever balancing torque and the product sum of the electromagnetic torques is equal to zero, as stated in Equation (45). If midpoints $T_{avg\ 1}$, $T_{avg\ 2}$, $T_{avg\ 3}$, $T_{avg\ 4}$, $T_{avg\ 5}$, and $T_{avg\ 6}$ are selected as pivot points, a set of equations can be realized for each point to obtain:

$$T_{avg\ 1} = \frac{A + BG}{2} \tag{47}$$

$$T_{avg\ 2} = \frac{AB + CG}{2} \tag{48}$$

$$T_{avg\ 3} = \frac{ABC + DG}{2} \tag{49}$$

$$T_{avg\ 4} = \frac{ABCD + EG}{2} \tag{50}$$

$$T_{avg\ 5} = \frac{ABCDE + FG}{2} \tag{51}$$

$$T_{avg\ 6} = \frac{ABCDEF + G}{2} \tag{52}$$

where point A is the outer stator torque and point BG is the sum torques of the outer field PM rotor, outer iron ring, prime PM rotor, inner iron ring, inner field PM rotor, and inner stator, respectively. Similarly, the electromagnetic torque on the other points B, C, D, E, and F are the sum torques on each side of the pivot points. However, for the three PM rotors to rotate the following conditions should be satisfied:

$$\left\{ \begin{array}{l} T_{avg\ 2} > T_{avg\ 1}\ [\text{Torque between line BC greater than line } AB] \\ T_{avg\ 3} > T_{avg\ 2}\ [\text{Torque between line } CD \text{ greater than line } BC] \\ T_{avg\ 4} > T_{avg\ 5}\ [\text{Torque between line } DE \text{ greater than line } EF] \\ T_{avg\ 5} > T_{avg\ 6}\ [\text{Torque between line } EF \text{ greater than line } FG] \end{array} \right\} \tag{53}$$

2.6. Stator Slots and Winding Design

The outer and inner stators are both designed with fractional slots and concentrated windings. Fractional slot windings have advantages over distributed windings because they reduce the end-winding lengths, torque ripple [26], coil volume and copper losses, machine axial length, and increased slot fill factor [27,28]. The calculated slot-per-phase-per-pole $q = 1/2$ and $q < 1$. Therefore, a double-layer concentrated winding scheme is adopted for both inner and outer stators, which have two coil sides per slot and separate phases. The slot-vector star method is used to determine the correct coil winding connections shown in Figure 8a,b. The key is to achieve the largest amplitude of the main harmonics in the induced EMF and achieve equal EMF waveform in the three phases. The machine periodicity t can be expressed as:

$$t = GCD\{Q, p\} \tag{54}$$

where GCD is the greatest common divisor, Q is the number of slots, and p is the number of poles. The number of spokes is Q/t, with t number of phasors in each spoke and the angle between two spokes is given as:

$$\alpha_{ph} = \frac{2\pi}{(Q/t)} \tag{55}$$

where α_{ph} is the electrical angle between two slots, which results in $Q = 12$, $p = 4$, $t = 4$, number of spokes = 3, and $\alpha_{ph} = 120°$. There are four phasors per spoke, as shown in Figure 8a,b.

(a)

(b)

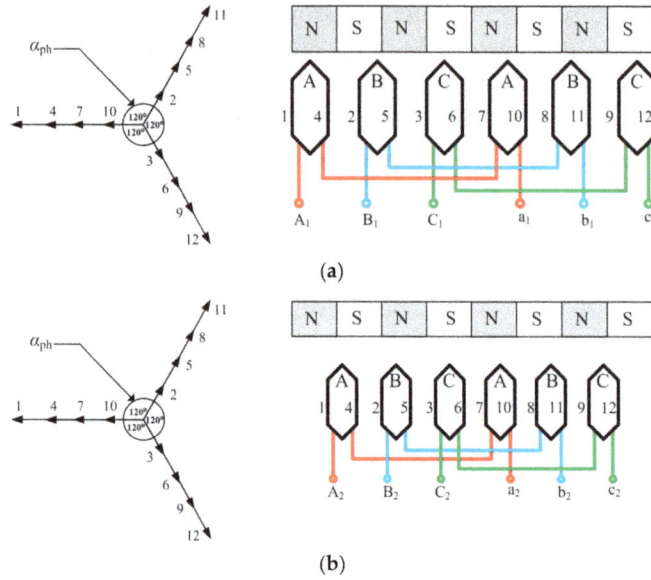

Figure 8. Star of slots and winding connections: (**a**) Outer stator machine; (**b**) Inner stator machine.

3. Finite Element Method

A two-dimensional finite element method is used to evaluate the electromagnetic characteristics of the double-stator magnetic geared PM machine. The two-dimensional electromagnetic field equation [29] of the proposed machine governed by Maxwell's equation is given as:

$$\Omega : \frac{\partial}{\partial x}\left(v\frac{\partial y}{\partial x}\right) + \frac{\partial}{\partial y}\left(v\frac{\partial y}{\partial y}\right) = -J - v\left(\frac{\partial B_{ry}}{\partial x} - \frac{\partial B_{rx}}{\partial y}\right) + \sigma\frac{\partial A}{\partial t} \tag{56}$$

where Ω is the field solution region of calculation, A is the magnetic vector potential, J is the current density, v the is reluctivity, σ is the electrical conductivity, and B_{rx} and B_{ry} are the remanent flux density components. The back-electromagnetic force (Back-EMF) of the stator coils can be expressed as:

$$e = -\frac{L}{S}\left(\iint \Omega + \frac{\partial A}{\partial t}d\Omega - \iint \Omega + \frac{\partial A}{\partial t}d\Omega\right) \tag{57}$$

where e is the Back-EMF generated by one coil, L the axial length of the machine, S is the area of conductor of each phase winding, and $\Omega+$ and $\Omega-$ are the cross-sectional areas of the input and output conductor coil, respectively. The circuit equation of the machine in motoring mode is given by:

$$R_i + L_e\frac{di}{dt} + \frac{l}{S}\iint_{\Omega} \frac{\partial A}{\partial t}d\Omega = u \tag{58}$$

where R is the resistance per phase winding, L_e is the inductance of the coil end windings, i is the phase current, l is the axial length, S is the area of each phase winding conductor, and Ω is the total cross-sectional area of conductors of each phase winding. The circuit equation in generating mode is given by:

$$(R_0 + R')i + (L_0 + L')\frac{di}{dt} - \frac{l}{S}\iint_{\Omega} \frac{\partial A}{\partial t}d\Omega = 0 \tag{59}$$

where R_0 is the resistance per phase winding, R' is the resistance of the load, L_0 is the inductance of the coil end windings, and L' is the inductance of the load. The general motion equation is given by:

$$J_m\frac{d\omega}{dt} + B\omega = T_{em} - T_L \tag{60}$$

where J_m is the moment of inertia, B is the coefficient of friction, ω is the mechanical speed, T_L is the load torque, and T_{em} is the electromagnetic torque. The rotational motions of the prime PM rotor, inner PM rotor, and outer PM rotor are governed by:

$$J_1 \frac{d\omega_1}{dt} + B_1\omega_1 = T_1 - T_L \tag{61}$$

$$J_2 \frac{d\omega_2}{dt} + B_2\omega_2 = T_{em} - T_2 \tag{62}$$

$$J_3 \frac{d\omega_3}{dt} + B_3\omega_3 = T_3 - T_L \tag{63}$$

where J_1, J_2, and J_3 are the moments of inertia of the inner rotor, prime rotor, and outer rotor, respectively; T_1, T_2, and T_3 are the torque of the inner rotor, prime rotor, and outer rotor, respectively; and ω_1, ω_2, and ω_3 are the mechanical speeds of the inner PM rotor, prime PM rotor, and outer PM rotor, respectively.

4. Results and Discussion

4.1. Flux Density Distribution

In Figure 9a there is reduced magnetic flux at no load and the flux flows across the air-gap with minimal flux fringing, while in Figure 9b the magnetic flux increases with load. Also, both the inner and outer field PMs have a pole-arc angle of 42° because in practice the full magnet pole-arc is rarely used due to manufacturing considerations and its effect on the cogging torque. In Figure 10b it is observed that the magnetic flux density is more saturated at the outer stator back iron and inner stator pole shoe compared to Figure 10a since the magnetic flux flows through the shortest back iron path of the magnetic circuit.

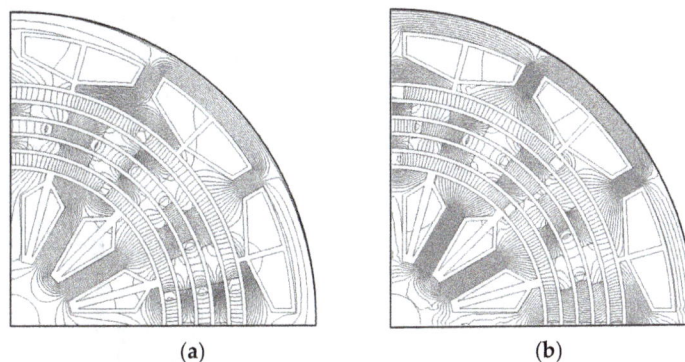

(a)   (b)

Figure 9. Magnetic flux lines distribution: (**a**) Without load; (**b**) With load.

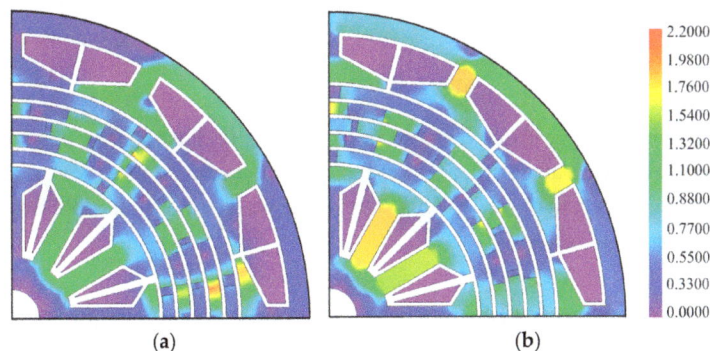

(a)   (b)

Figure 10. Magnetic flux density distribution: (**a**) Without load; (**b**) With load.

4.2. Flux Density Characteristics and Harmonic Analysis

The Fast Fourier Transform (FFT) was used for the harmonic spectrum analysis of the magnetic flux density at the centre of the magnetic gear's three inner air-gaps. Figure 11a illustrates the radial component of the magnetic flux density between the inner field PMs and inner stator, while Figure 11b shows its harmonic order. The number of flux density pulsations is eight, which corresponds to the number of poles of field PMs. Also, the main harmonic number is four, equal to the field PMs pole-pair number. Figure 11c shows the radial component of the flux density at the centre of the air-gap between the prime PMs and inner iron ring, while Figure 11f shows the harmonic order.

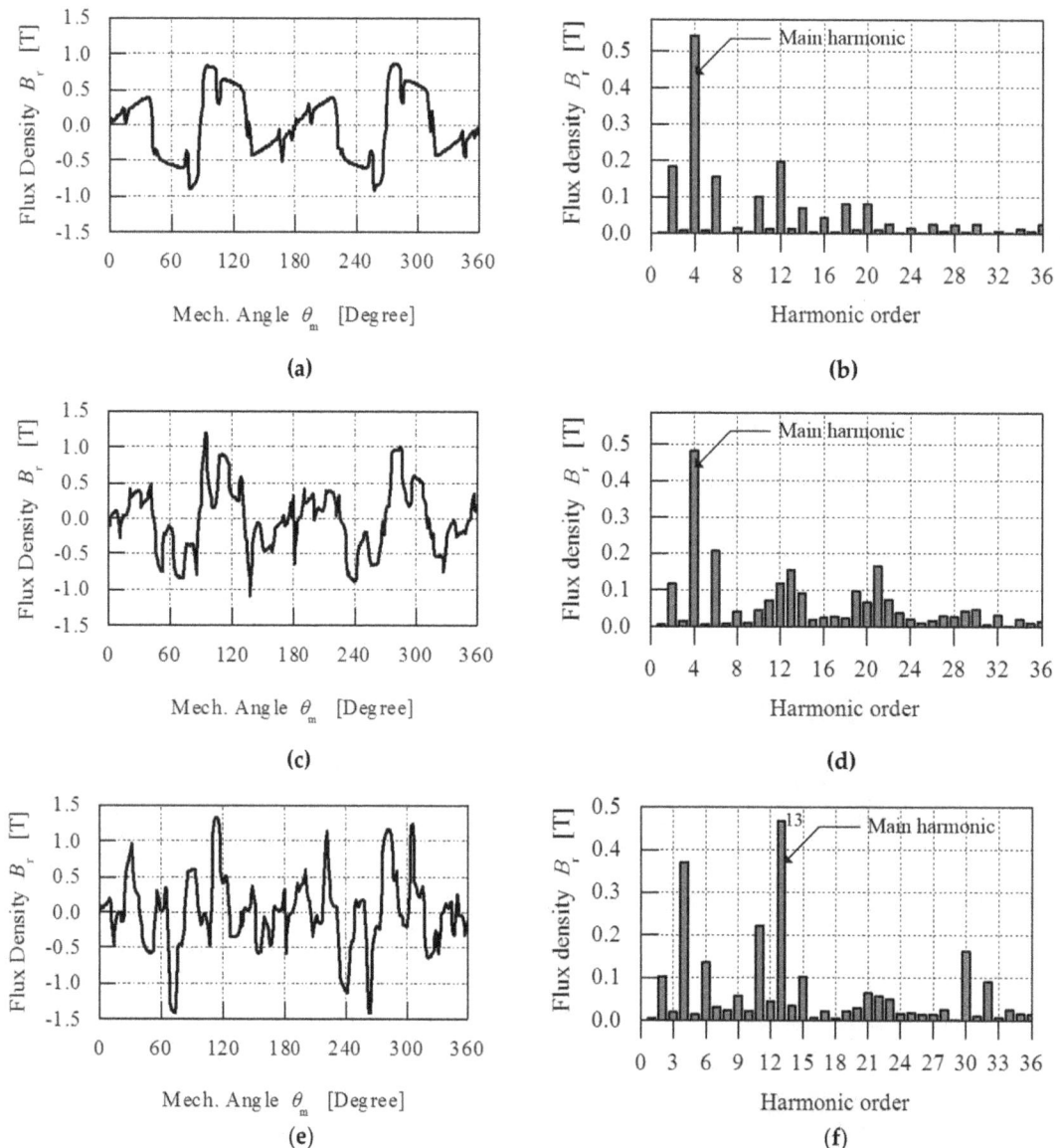

Figure 11. The inner magnetic gear flux density characteristics: (**a**) Flux density between inner PM and inner stator; (**b**) Fast Fourier Transform of (**a**); (**c**) Flux density between inner PM and inner pole-pole ring; (**d**) Fast Fourier Transform of (**c**); (**e**) Flux density between inner pole-pole ring and prime PM; (**f**) Fast Fourier Transform of (**e**).

The magnetic flux density at the centre of the three outer air-gaps is illustrated in Figure 12a,c,e, while their main harmonics are illustrated in Figure 12b,d,f. The effect of flux concentration between the prime PMs and outer iron ring shown in Figure 12c is dominant at 116°, 281°, and 306°, with

26 pulsations equal to 26 prime PMs (26 poles). In Figure 12c the flux density waveform between the outer field PMs and outer iron ring is concentrated at 94° and 279°, respectively. The ratio of main harmonics 4 and 13 is equal to 3.25, which confirms the calculated magnetic gear ratio of 3.25:1.

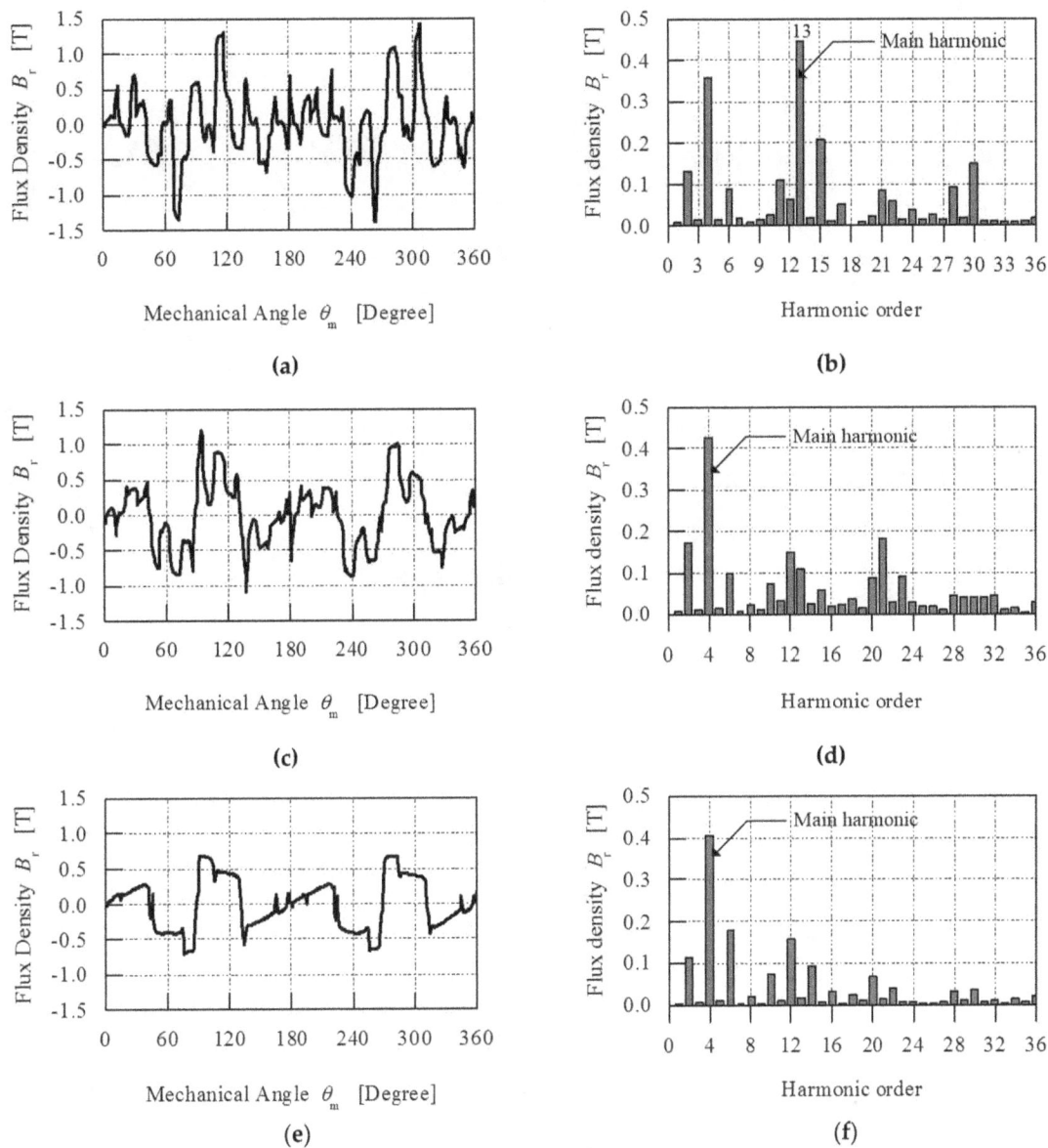

Figure 12. The outer magnetic gear flux density characteristics: (**a**) Flux density between outer pole-pole ring and prime PM; (**b**) Fast Fourier Transform of (**a**); (**c**) Flux density between outer PM and outer pole-pole ring; (**d**) Fast Fourier Transform of (**c**); (**e**) Flux density between outer PM and outer stator; (**f**) Fast Fourier Transform of (**e**).

4.3. Cogging Torque and Transmission Torque Characteristics

Figure 13a illustrates the cogging torque of the inner field, outer field, and prime PM rotors in the magnetic gear; their calculated results are shown in Table 4. Also, Figure 13b shows the cogging torque from the stators. In Table 4 it is observed that the peak-to-peak cogging torque on the inner stator is 64% less than that on the outer stator. Figure 13c illustrates the average transmission torque of the outer field, inner field, and prime PM rotors in the magnetic gear; their computed results are shown in Table 5. The simulated transmission torque ratio is 3.25: 1, which is equal to the magnetic gear ratio. The maximum torque-angle curves of the three PM rotors shown in Figure 13d are calculated by holding the prime rotor static while both inner and outer rotors are rotated step by step. It can be observed that the torque-angle curve is sinusoidal and the maximum torque value indicates the pull-out torques. The maximum torque value is shown in Table 5 and the torque ratio calculated as 3.16 is 97% accurate with the calculated gear ratio of 3.25.

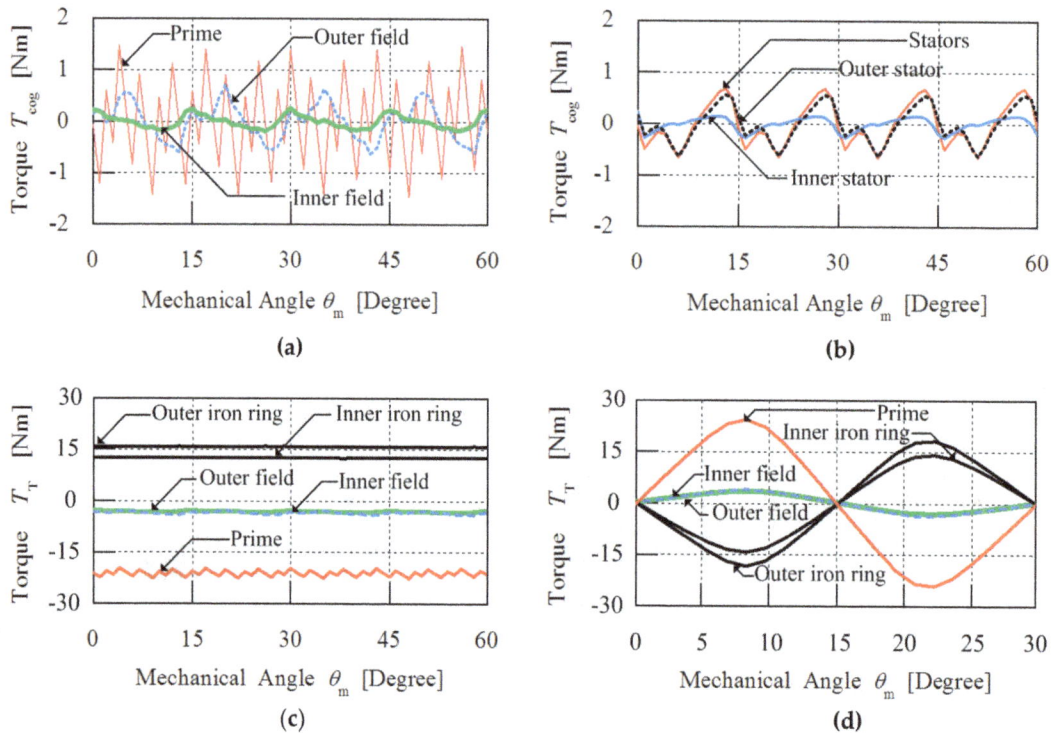

Figure 13. The cogging torque and transmission torque characteristics of the DS-MGM: (**a**) Cogging torque generated by iron rings of the magnetic gear; (**b**) Cogging torque generated by stator slots of the machine; (**c**) Transmission torque of the magnetic gear; (**d**) Pull-out torque angle curves.

Table 4. Cogging torque characteristics of the magnetic gear.

Quantity	Outer Field	Inner Field	Prime	Outer Stator	Inner Stator	Stators
Peak-to-Peak	1.35 Nm	0.46 Nm	3.00 Nm	1.21 Nm	0.43 Nm	1.35 Nm
Cogging torque	0.67 Nm	0.23 Nm	1.50 Nm	0.60 Nm	0.21 Nm	0.67 Nm

Table 5. Transmission torque characteristics of the magnetic gear.

Quantity	Outer Field	Inner Field	Prime	Inner Ring	Outer Ring
Average torque	3.43 Nm	3.05 Nm	21.08 Nm	12.55 Nm	15.76 Nm
Pull-out torque	4.05 Nm	3.65 Nm	24.37 Nm	14.14 Nm	18.21 Nm
Torque ripple	39.27%	15.09%	14.25%	1.70%	1.66%

For the 3.25 : 1 magnetic geared machine, the lowest common multiple between the number of field PM poles $2p_1 = 8$, and the number of iron ring pole-pieces $n_s = 17$, is 136 which is the value of the fundamental order of the ripple torque waveform. By substituting into Equation (34), a cogging torque factor of 1 is obtained. This means that if the lowest common multiple is greater than 1, a transmission torque with high torque ripple will be realised. As evident in Figure 14a,c,e, the prime rotor, inner rotor, and outer rotor exhibit low torque ripple in their transmission torque. Figure 14b,d,f shows the harmonic analysis of the ripple torque and it can be seen that the sixth-order ripple torque harmonic is prominent, which is caused by the flux-linkage of the coil winding flux with the magnetic flux generated by both outer and inner field PMs. The flux-linkage has harmonic components but the prime PMs are in a magnetically coupled configuration with both outer and inner field PMs and therefore the sixth-order ripple torque harmonic also appears in the prime rotor ripple torque. The sixth-order ripple torque harmonic can be reduced by optimisation of the stator teeth geometry. Also the 12th and 34th order ripple torque harmonics are clearly visible in Figure 14d, this demonstrates that both orders are multiples of 12 stator slots and 17 iron ring pole pieces containing harmonic components.

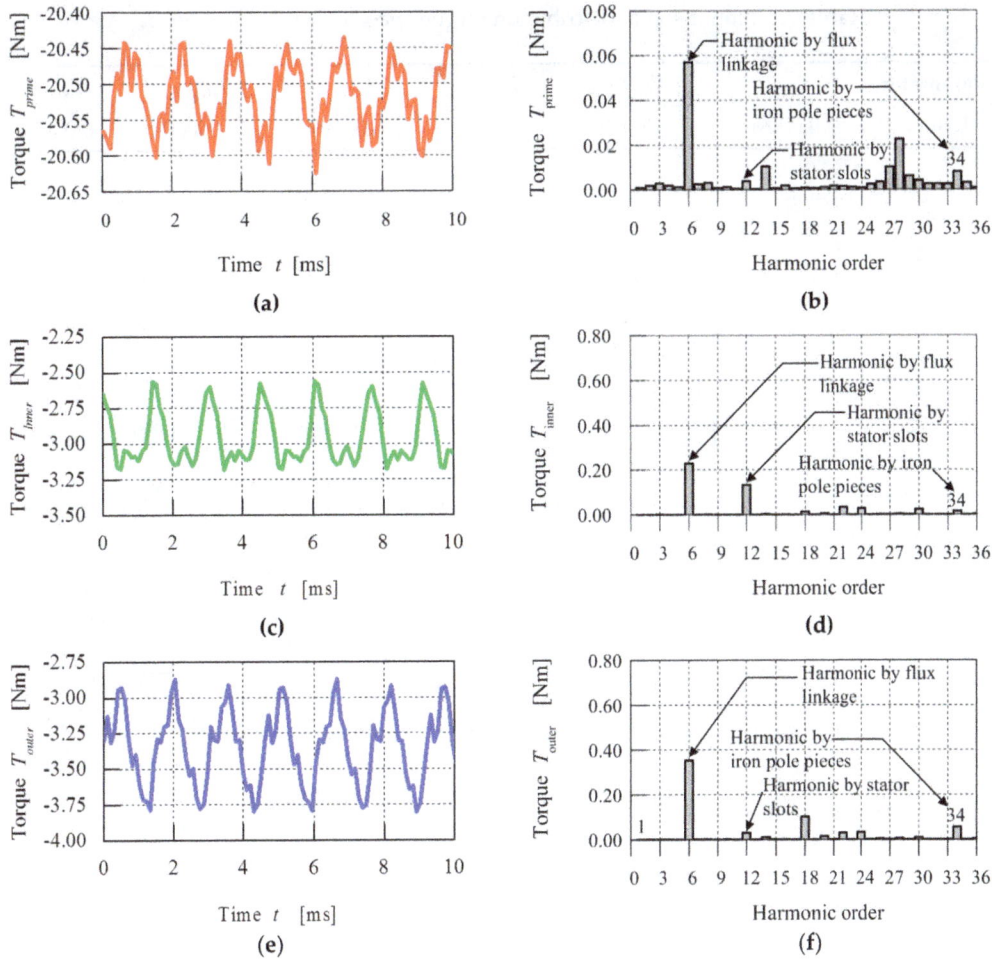

Figure 14. The ripple torque characteristics of the DS-MGM: (**a**) Prime PM rotor ripple torque; (**b**) Harmonic analysis of (**a**); (**c**) Inner PM rotor ripple torque; (**d**) Harmonic analysis of (**c**); (**e**) Outer PM rotor ripple torque; (**f**) Harmonic analysis of (**d**).

4.4. Torque Distribution Characteristics

The torque characteristics shown in Figure 15 without load and Figure 16 with load indicate that there is a linear relationship between the machine's torque and the rotational speed of the prime mover. For the analysis of the torque distribution, the slope of the line known as the torque–speed constant, k_{slope}, is chosen as the evaluating parameter, which is similar to the power–speed constant [30]. The torque-speed constant is calculated by finding the slope of the average torque distribution between two parts when operating at different speeds. The k_{slope} is calculated by Equation (64):

$$k_{slope} = \frac{T_{avg\,2} - T_{avg\,1}}{\omega_2 - \omega_1} \; (\text{Nm/rpm}). \tag{64}$$

Tables 6 and 7 shows the distribution torque speed constants at no load and with load between the seven parts of the machine.

Table 6. Comparison of distribution torque speed constants at no load.

Parameter	$T_{avg\,1}$	$T_{avg\,2}$	$T_{avg\,3}$	$T_{avg\,4}$	$T_{avg\,5}$	$T_{avg\,6}$
Kslope	0.00300	0.00040	0.00007	0.00020	0.00020	0.00070

Table 7. Comparison of distribution torque speed constants with load.

Parameter	$T_{avg\,1}$	$T_{avg\,2}$	$T_{avg\,3}$	$T_{avg\,4}$	$T_{avg\,5}$	$T_{avg\,6}$
Kslope	0.00090	0.00090	0.00350	0.00190	0.00080	0.00070

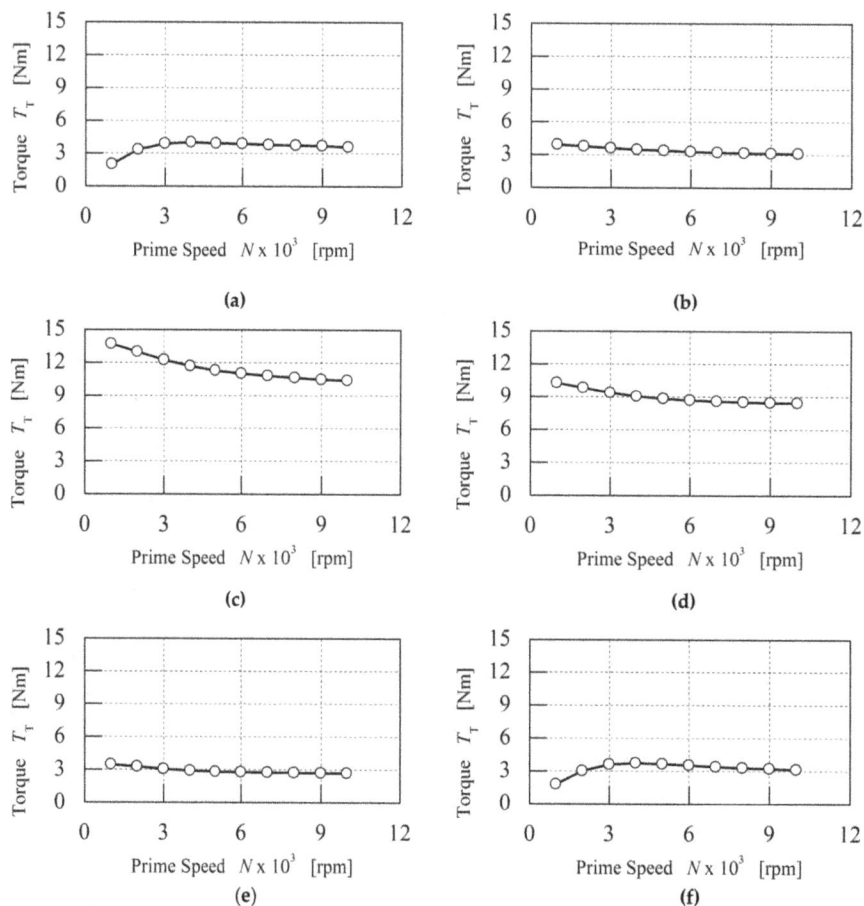

Figure 15. Torque distribution of DS-MGM without load: (**a**) Outer stator and outer field PM; (**b**) Outer field PM and outer iron ring; (**c**) Outer iron ring and prime PM; (**d**) Prime PM and outer iron ring; (**e**) Inner iron ring and inner field PM; (**f**) Inner field PM and inner stator.

The torque distribution characteristics without load between the outer stator and outer field PM shown in Figure 15a are similar to the torque distribution between the inner stator and inner field PM. At a prime rotor speed of 300 rpm maximum torque is achieved and the line curve dips slightly at speeds greater than 1000 rpm. The torque distribution between the prime PM and outer iron ring illustrated in Figure 15c,d is greater than the torque between the inner field PM and inner iron ring because the prime rotor has twenty-six poles of prime PMs compared to the eight poles of field PMs. Also, the magnetically coupled configuration confirms that the torque generated by the low-speed prime PM rotor is greater than the torque on the high-speed PM rotors. It can be observed that, as angular speed increases, the transmission torque reduces as a result of eddy current losses from the PMs and iron losses in the stationary iron rings. In Figure 15f the distribution torque between the inner stator and inner field PM uniform acts as a counter torque against the torque between inner iron ring and inner field PM shown in Figure 15e.

The torque distribution with load between the outer stator and outer field PMs as shown in Figure 16a reduces as a result of the counter torque force generated in the airgap. The fields' PMs generate excitation current in the coils, resulting in a magnetic field that produces reaction torque on the stator. By applying the principles of Lentz's law and Newton's third law of motion, the magnetic torque reacts against the torque that rotates the prime PM rotor. The torque distribution analysis shows that the torque between the outer stator and outer field PMs is greater than that between the field PMs and iron ring. Also, the magnetically coupled configuration confirms that integration of magnetic gearing with PM machines produces torque distribution in the airgaps between each component of the machine. Also, it can be observed in Figure 16c,d that the torque between the prime PMs and iron ring is reduced due to the effect of the counter torque generated in the airgap between the field PMs and iron ring.

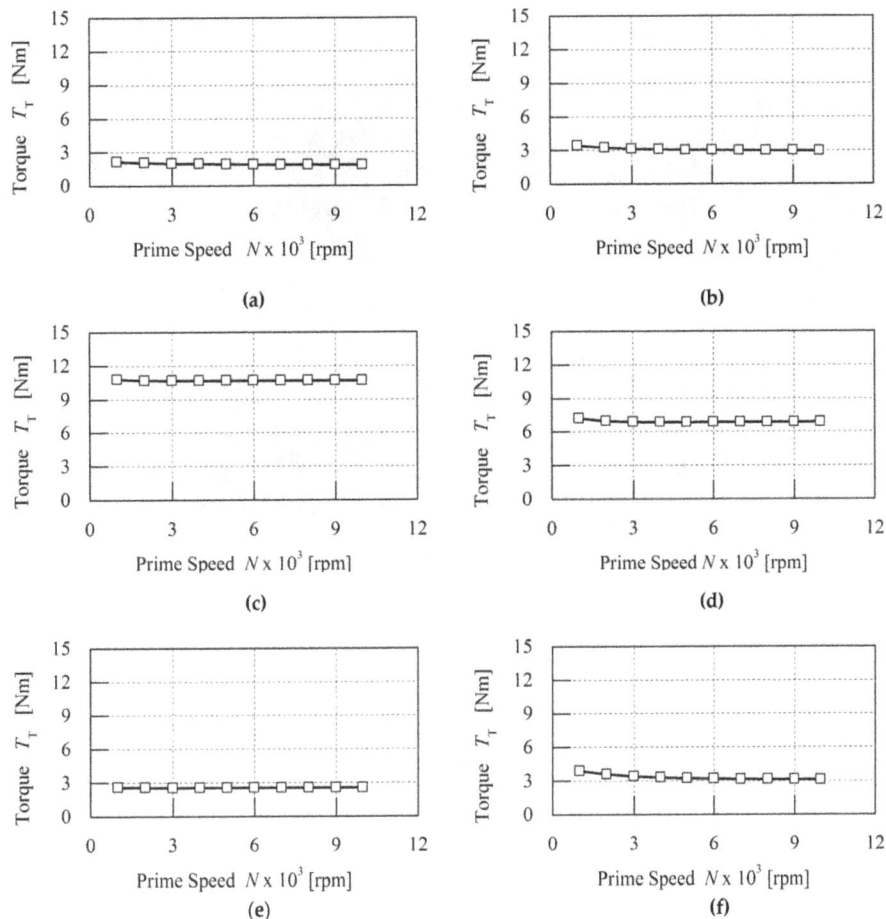

Figure 16. Torque distribution of DS-MGM with load: (**a**) Outer stator and outer field PM; (**b**) Outer field pm and outer iron ring; (**c**) Outer iron ring and prime PM; (**d**) Prime PM and outer iron ring; (**e**) Inner iron ring and inner field PM; (**f**) Inner field PM and inner stator.

4.5. Prototype Machine

Figure 17 illustrates the fabricated components of the proposed machine, including the assembled prototype. The mechanical assembly of the DS-MGM is quite complex because all three PM rotors rotate separately and both modulating iron rings have to be fixed between the three rotors with an airgap of 1 mm. The bone rotor structure is used in designing the three rotors so that a magnetically coupled configuration is achieved between the magnetic gear and double-stator PM machine. The two modulating iron ring pole pieces are fabricated from solid steel, while the connecting end rings are made from aluminium to avoid short-circuiting the magnetic flux; since the proposed machine is designed for low-speed operation, the effect of eddy current losses is ignored. Future work could investigate the effect of laminated iron rings on eddy current losses at high-speed operation. The input shaft is machined from aluminium to reduce weight and is coupled to only the prime rotor, which results in difficulties in measuring the transmission torque for both the outer PM rotor and the inner PM rotor. To address this problem, the angular speeds of the outer and inner rotors are calculated from their fundamental frequency by spectrum analysis of the back EMF voltage produced from both outer and inner stators. The power, torque, and angular speed characteristics of the prototype DS-MGM were measured with the test rig shown in Figure 18b to verify the measured results against the calculated results. An AC induction motor was coupled to the assembled prototype generator and a three-phase variable frequency drive was used to control the AC induction motor's speed and torque by varying the input frequency. The output terminals of the prototype generator were connected to a three-phase full bridge rectifier, while the output DC voltage, DC current, and DC power was measured with a single-phase programmable electronic load.

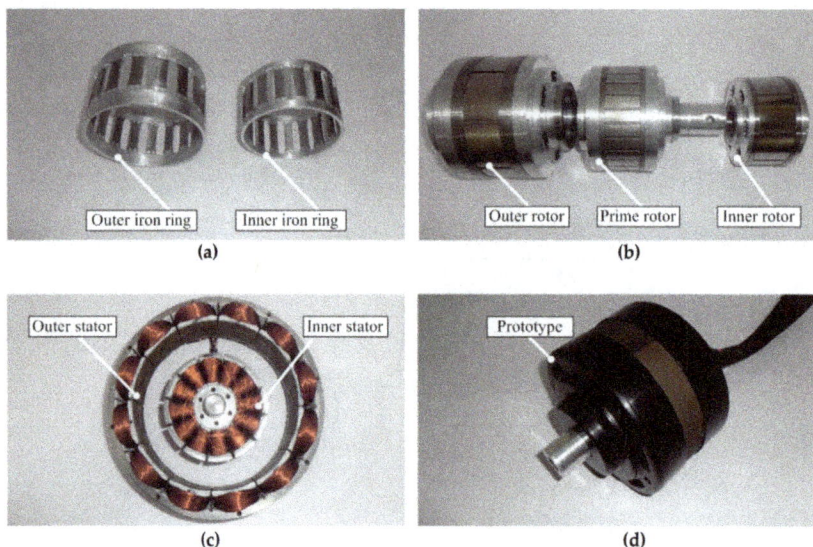

Figure 17. The fabricated components and prototype of DS-MGM: (**a**) Iron rings; (**b**) Rotors; (**c**) Stators; (**d**) Assembled prototype.

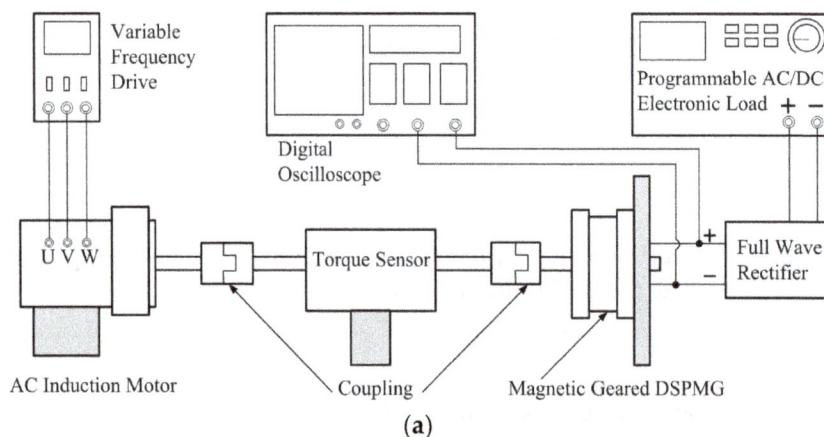

Figure 18. *Cont.*

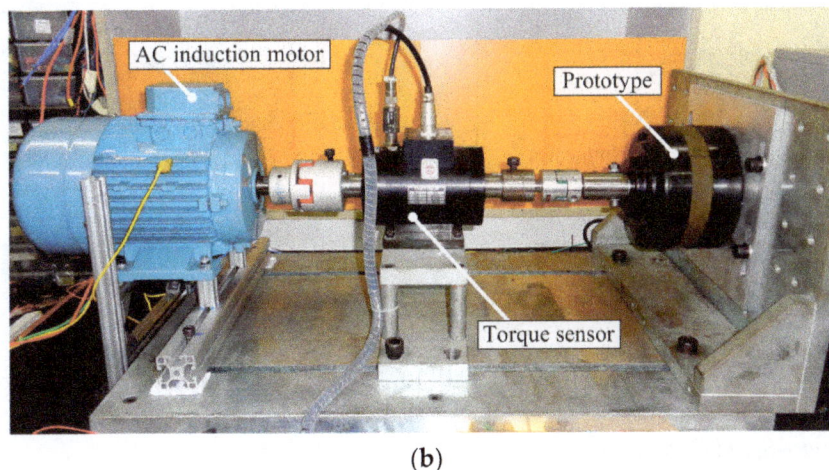

(b)

Figure 18. Experimental setup and validation: **(a)** Experimental configuration; **(b)** Test rig.

4.6. Mechanical Power–Speed Characteristics

The input mechanical power that rotates the prime rotor of the magnetic geared double-stator PM generator is calculated using Equation (65):

$$P_m = \frac{2\pi nT}{60} \ [\text{W}] \tag{65}$$

where P_m is the net mechanical power of the generator in W, T is the toque in Nm, and n is the rotational speed in rpm. The measured torque is obtained from the torque transducer shown in Figure 18b, while the calculated torque is obtained from 2D FEM. In the fabrication process of the prototype, initially the outer and inner iron rings are manufactured from solid steel to form a cage-like structure. A first experimental test was conducted on no-load and it was observed that the magnetic gear part of the generator slips when the speed of the prime PM rotor is greater than 110 rpm. An analysis of end effects indicates that magnetic flux leakage exists at the end rings due to circulating eddy currents in the pole pieces. To address this problem, the end connecting rings are designed from 3 mm thick aluminium, which is a non-ferromagnetic metal and magnetic flux barrier. Also, the aluminium end rings are connected with four steel rivet bolts per pole piece, or two rivet bolts for each end. The reason for using steel bolts was to achieve torsional strength for the iron rings against the reaction torques from both field PMs and prime PMs. Although laminated steel pole pieces bonded with epoxy and electrically insulated end rings may be an option for reducing eddy currents, but this increases the manufacturing costs and the difficulty of mechanical assembly. The second experimental test on no-load showed that the prime PM rotor achieved a speed of 700 rpm without the magnetic gear slipping. However, the effect of end connecting rings on transmission torque is not reported in the results because it is a subject for further study. Figure 19 shows the torque and mechanical power with a 100-ohm load and it is observed that the measured results are greater than the simulated values. It is likely caused by the following factors: first, the aluminium end connecting rings were secured to the pole pieces with steel bolts but the aluminium end rings are not electrically insulated from the steel bolts. This weak aspect results in induced eddy current produced by a time-varying magnetic field flowing around the iron rings and the eddy current exerts torque from the iron rings, therefore impacting the transmission torque between the prime PMs and field PMs. Secondly, end-effects are not accounted for in the two-dimensional finite element analysis of the magnetic geared machine; instead, the iron ring pole pieces are modelled as electrically insulated solid parts without end connecting rings. Ideally, the measured torque should be less than the simulated torque because friction and bearing losses are assumed to be insignificant in FEM analysis. These factors account for the large variation between the simulated torque and the measured torque.

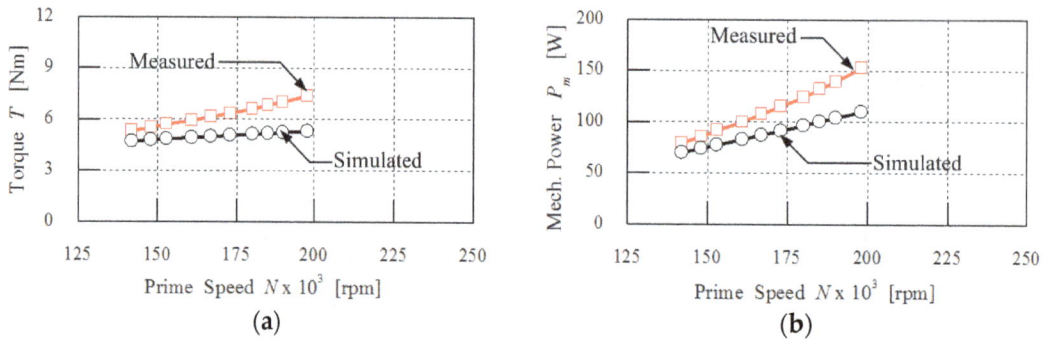

Figure 19. Torque and mechanical power with 100 Ω load: (**a**) Torque; (**b**) Mechanical power.

4.7. Electrical Power-Speed Characteristics

Figure 20 shows the calculated and measured no-load back EMF waveform at a prime rotor speed of 300 rpm. The induced peak-to-peak voltages from the calculated and measured results are 63 V and 75 V, respectively, for a double layer three-phase concentrated winding. It can be observed that both voltage waveforms match and are similar in output, which verifies the machine design. Although the induced voltage is not perfectly sinusoidal due to the presence of harmonics, these can be reduced by a proper selection of coil winding design. The measured voltage is greater than the simulated voltage by 19%; this difference could be caused by the effect of eddy current circulating between the iron ring pole pieces and aluminium end rings. The components are connected by steel rivet bolts and each iron ring pole piece would produce eddy currents; however, because the steel rivet bolts are not electrically insulated, a U-turn circuit is created by induced circulating eddy currents. This induced circulating eddy current produces a magnetic flux that can increase the back EMF. In the 2D FEM analysis these factors are not accounted for, which causes the simulated voltage to be less than the measured voltage. The use of non-electrically conductive material for the connecting end rings may be necessary to improve the accuracy of the measured results.

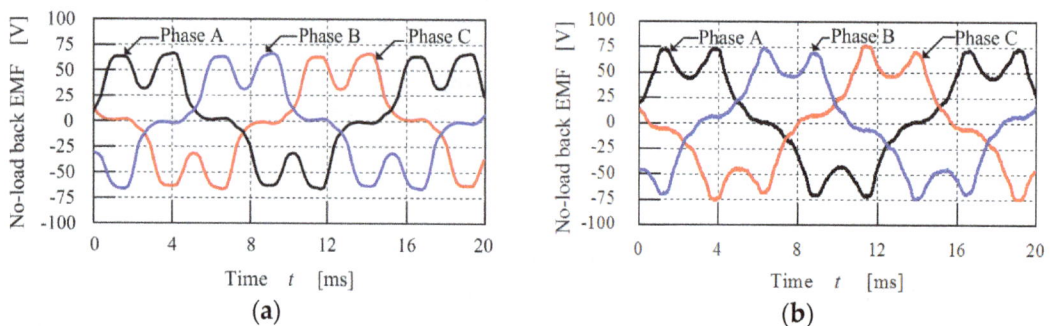

Figure 20. Comparison of no-load back-EMF waveforms at prime rotor speed = 300 rpm: (**a**) Simulated results from finite element analysis; (**b**) Measured results from the experiment.

Figure 21 compares the simulated and measured three-phase voltage waveforms at the output of the generator before DC rectification on a load of 100 ohm at prime rotor speed = 200 rpm. The field rotor speed is calculated with the formula $N = 120f/p$ by determining the frequency f from the three-phase voltage using spectrum analysis, which results in calculated and measured field rotor speeds of 650 rpm and 646 rpm, respectively. It can be seen that a good agreement exists between the simulated and measured voltage waveforms. Note that both waveforms are non-sinusoidal due to the existence of harmonic components, which is general for concentrated windings.

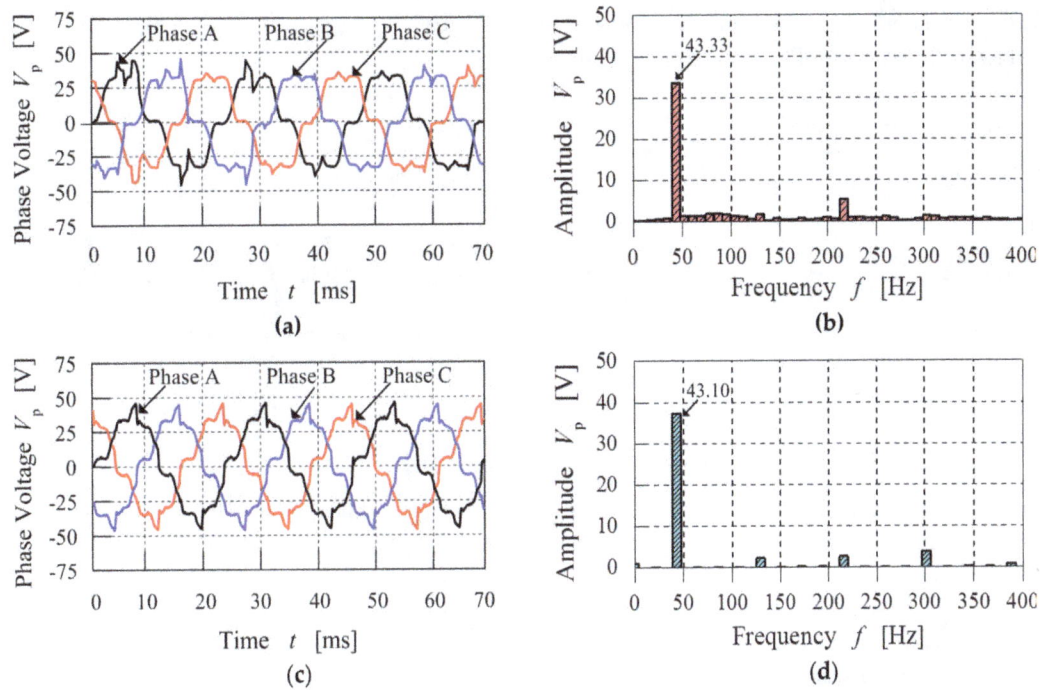

Figure 21. Comparison of output three-phase voltage from generator before DC rectification with load of 100 Ω at prime rotor speed = 200 rpm: (**a**) Simulated three-phase voltage waveform; (**b**) Spectrum analysis of (**a**); (**c**) Measured three-phase voltage waveform; (**d**) Spectrum analysis of (**c**).

The simulated three-phase output current waveform before DC rectification with a 100-ohm load at a prime rotor speed of 200 rpm is shown in Figure 22 with the measured waveforms. It can be observed that both waveforms are non-sinusoidal due to the presence of the third harmonic generated by the concentrated winding configuration of the prototype machine. Also, the amplitude of the measured phase current waveform is greater than the simulated and this could be caused by eddy currents circulating between the iron ring pole pieces and aluminium end rings as both components are not electrically insulated, which results to an additional magnetic flux that causes a difference in the induced phase current. Even though the output three-phase current waveform is non-sinusoidal, the simulated and measured three-phase current waveforms are similar. The measured three-phase current waveform shown in Figure 22b is not smooth; this is mainly caused by the noise generated by the DC rectifier when the generator is connected to a resistance load. The DC full wave rectifier used in the measurement is built in-house and its quality may not be comparable to commercially available rectifiers, which have proper shielding from noise and adequate grounding.

The programmable AC/DC electronic load has limitations and can only simulate a single-phase load; for that reason, a DC full wave rectifier was connected between the generator and the load. The DC voltage characteristics are shown in Figure 23a when the external resistance load is set at 100 ohm and it is observed that the DC voltage graph demonstrates linearity while the measured DC voltage values agree closely with the calculated results. The measured DC current and DC power characteristics of the generator shown in Figure 23b,c are consistent with this observation. A comparison of the calculated and measured performance characteristics of the prototype is shown in Table 8. By applying the law of conservation of energy, it can be observed that the measured gear ratio is 0.61% less than the calculated gear ratio as a result of losses. Also, this observation is applicable to the reduced measured rotational speed of the field PM rotor. The measured DC voltage and power characteristics are greater than the simulated values by 1.12% and 0.99%, respectively, while the measured DC current is equal to the simulated value. The effect of circulating eddy currents from the aluminium end rings and iron rings before DC rectification also affects the variation between the calculated and measured DC power characteristics.

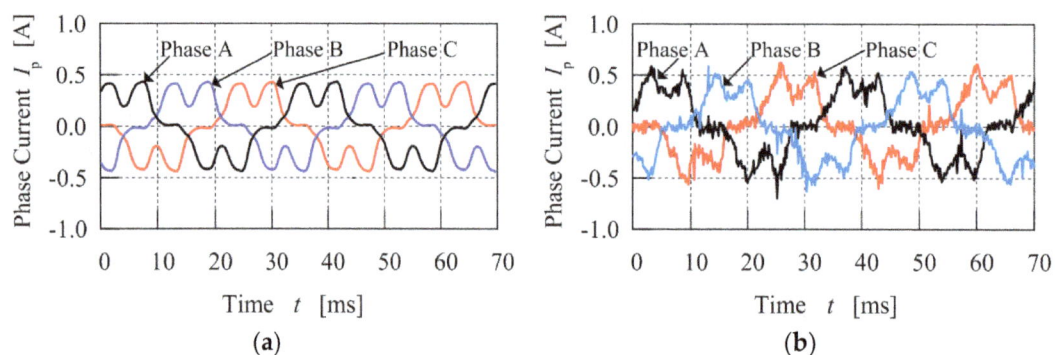

Figure 22. Comparison of output three-phase current from generator before rectification with load of 100 Ω at prime rotor speed = 200 rpm: (**a**) Simulated three-phase current waveform; (**b**) Measured three-phase current waveform.

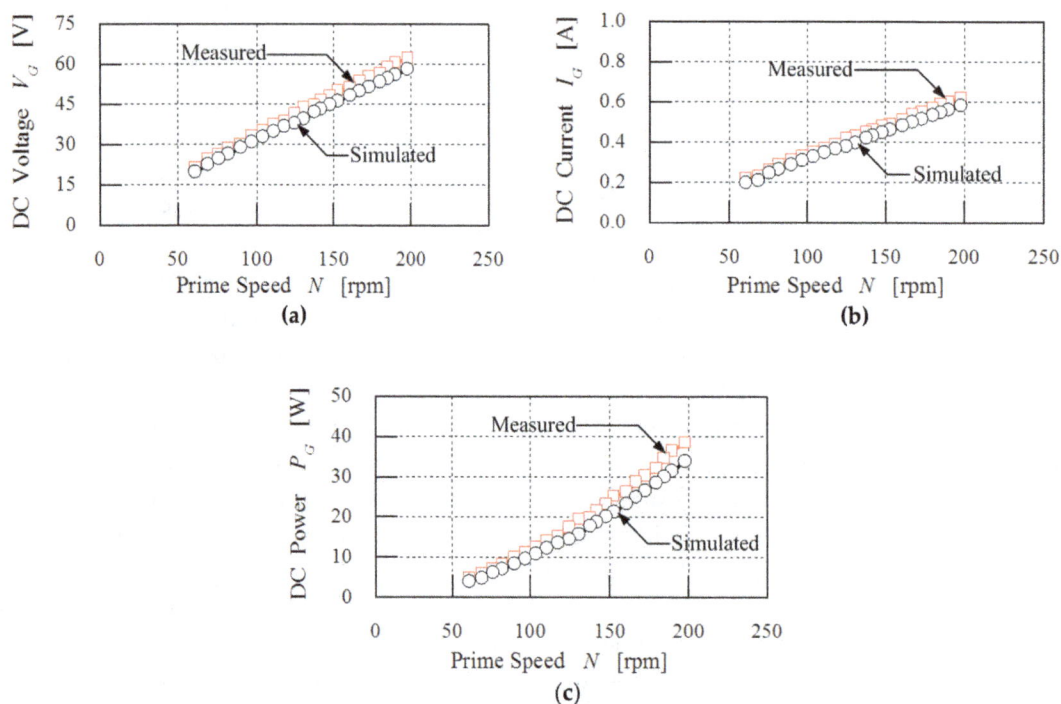

Figure 23. Electrical characteristics of the prototype DS-MGM with resistance load of 100 Ω: (**a**) DC voltage; (**b**) DC current; (**c**) DC power.

Table 8. Comparison of prototype DS-MGM performance characteristics.

Parameter	Calculated	Measured
Resistive load	100 Ω	100 Ω
Speed of prime rotor	200 rpm	200 rpm
Speed of field rotor	−650 rpm	−646 rpm
DC Voltage	62.55 V	63.25 V
DC Current	0.63 A	0.63 A
DC Power	39.41 W	39.85 W
Gear ratio	3.25	3.23

5. Conclusions

This paper has presented a novel magnetic geared double-stator PM generator and the torque distribution characteristics were studied using two-dimensional finite element analysis. Also, the power and speed characteristics were measured by the fabrication of a prototype. The torque distribution analysis shows that the magnetically coupled configuration between the three permanent magnet rotors affects the transmission torque of the magnetic gear. This can be observed in the large variation between the calculated and measured torque as a result of induced circulating eddy currents from the aluminium connecting end rings through the steel rivet bolts to the pole pieces. Although both stator slots and iron ring pole pieces equally affect the torque distribution, the torque between the field permanent magnets and the iron rings should be greater than the reaction torque between the stators and field PMs for the magnetic gear to rotate. A good agreement exists between the calculated and measured field PM rotor speed, DC voltage, DC current, and DC power characteristics. The torque distribution analysis shows that more power could be generated at a prime rotor speed greater than 1000 rpm if the transmission torque is optimised. However, the prototype could not be operated up to a prime rotor speed of 1000 rpm compared to the simulation model because of mechanical problems from the bearings, heat generation on the prime mover shaft, and manufacturing imperfections in the assembly. These problems will be addressed in future work to improve the efficiency of the generator.

Acknowledgments: The authors would like to express their gratitude to Ministry of Science, Technology and Innovation Malaysia for financial support and Universiti Putra Malaysia for the facilities provided during this research work.

Author Contributions: All authors contributed to this work by collaboration. Shehu Salihu Mustafa is the main author of this manuscript. Norhisam Misron assisted to perform the simulations and design the experiments. Norman Mariun, Mohammad Lutfi Othman and Tsuyoshi Hanamoto provided some very useful suggestions in the production of the paper. All authors revised and approved the publication of the paper.

Conflicts of Interest: The authors declare no conflict of interest.

References

1. Li, X.; Chau, K.T.; Cheng, M.; Hua, W. Comparison of magnetic-geared permanent magnet machines. *Prog. Electromagn. Res.* **2013**, *133*, 177–198. [CrossRef]
2. Rens, J.; Atallah, K.; Calverley, S.D.; Howe, D. A novel magnetic harmonic gear. *IEEE Trans. Ind. Appl.* **2010**, *46*, 206–212. [CrossRef]
3. Jian, L.N.; Chau, K.T. A coaxial magnetic gear with halbach permanent-magnet arrays. *IEEE Trans. Energy Convers.* **2010**, *25*, 319–328. [CrossRef]
4. Zhang, X.; Liu, X.; Wang, C.; Chen, Z. Analysis and design optimization of a coaxial surface-mounted permanent-magnet magnetic gear. *Energies* **2014**, *7*, 8535–8553. [CrossRef]
5. Atallah, K.; Wang, J.; Mezani, S.; Howe, D. A novel high-performance linear magnetic gear. *IEEJ Trans. Ind. Appl.* **2006**, *126*, 1352–1356. [CrossRef]
6. Mezani, S.; Atallah, K.; Howe, D. A high-performance axial-field magnetic gear. *J. Appl. Phys.* **2006**, *99*, 08R303. [CrossRef]
7. Acharya, V.M.; Bird, J.Z.; Calvin, M. Flux focusing axial magnetic gear. *IEEE Trans. Magn.* **2013**, *49*, 4092–4095. [CrossRef]
8. Jorgensen, F.; Andersen, T.; Rasmussen, P. The cycloid permanent magnetic gear. *Trans. Ind. Appl.* **2008**, *44*, 1659–1665.
9. Niguchi, N.; Hirata, K. Transmission torque analysis of a novel magnetic planetary gear employing 3-D FEM. *IEEE Trans. Magn.* **2012**, *48*, 1043–1046. [CrossRef]
10. Niguchi, N.; Hirata, K. Cogging torque analysis of magnetic gear. *IEEE Trans. Ind. Electron.* **2012**, *59*, 2189–2197. [CrossRef]
11. Tsai, M.C.; Ku, L.H. 3-D printing-based design of axial flux magnetic gear for high torque density. *IEEE Trans. Magn.* **2015**, *51*, 1–4. [CrossRef]
12. Uppalapati, K.; Bomela, W.; Bird, J.Z.; Calvin, M.; Wright, J. Experimental evaluation of low-speed flux-focusing magnetic gearboxes. *IEEE Trans. Ind. Appl.* **2014**, *50*, 3637–3643. [CrossRef]

13. Holm, R.K.; Berg, N.I.; Walkusch, M.; Rasmussen, P.O.; Hansen, R.H. Design of a magnetic lead screw for wave energy conversion. *IEEE Trans. Ind. Appl.* **2013**, *49*, 2699–2708. [CrossRef]

14. Jing, L.; Liu, L.; Xiong, M.; Feng, D. Parameters analysis and optimization design for a concentric magnetic gear based on sinusoidal magnetizations. *IEEE Trans. Appl. Supercond.* **2014**, *24*, 1–5. [CrossRef]

15. Huang, C.C.; Tsai, M.C.; Dorrell, D.G.; Lin, B.J. Development of a magnetic planetary gearbox. *IEEE Trans. Magn.* **2008**, *44*, 403–412. [CrossRef]

16. Chen, Y.; Fu, W.N.; Ho, S.L.; Liu, H. A quantitative comparison analysis of radial-flux, transverse-flux, and axial-flux magnetic gears. *IEEE Trans. Magn.* **2014**, *50*, 1–4. [CrossRef]

17. Frandsen, T.V.; Mathe, L.; Berg, N.I.; Holm, R.K.; Matzen, T.N.; Rasmussen, P.O.; Jensen, K.K. Motor integrated permanent magnet gear in a battery electrical vehicle. *IEEE Trans. Ind. Appl.* **2015**, *51*, 1516–1525. [CrossRef]

18. Liu, C.T.; Chung, H.Y.; Hwang, C.C. Design assessments of a magnetic-geared double-rotor permanent magnet generator. *IEEE Trans. Magn.* **2014**, *50*, 1–4. [CrossRef]

19. Liu, C.; Chau, K.T.; Zhang, Z. Novel design of double-stator single-rotor magnetic-geared machines. *IEEE Trans. Magn.* **2012**, *48*, 4180–4183. [CrossRef]

20. Ohno, Y.; Niguchi, N.; Hirata, K.; Morimoto, E. Radial differential magnetic harmonic gear. *JSAEM Appl. Electromagn. Mech.* **2015**, *23*, 23–28. [CrossRef]

21. Wang, Y.; Cheng, M.; Chen, M.; Du, Y.; Chau, K.T. Design of high-torque-density double-stator permanent magnet brushless motors. *IET Electr. Power Appl.* **2011**, *5*, 317–323. [CrossRef]

22. Dosiek, L.; Pillay, P. Cogging torque reduction in permanent magnet machines. *IEEE Trans. Ind. Appl.* **2007**, *43*, 1565–1571. [CrossRef]

23. Atallah, K.; Calverley, S.D.; Howe, D. Design, analysis and realization of a high performance magnetic gear. *IEE Proc. Electr. Power Appl.* **2004**, *151*, 135–143. [CrossRef]

24. Bianchi, N.; Bolognani, S. Design techniques for reducing the cogging torque in surface-mounted PM motors. *IEEE Trans. Ind. Appl.* **2002**, *38*, 1259–1265. [CrossRef]

25. Zhu, Z.; Howe, D. Influence of design parameters on cogging torque in permanent magnet machines. *IEEE Trans. Energy Convers.* **2000**, *15*, 407–412. [CrossRef]

26. Bianchi, N.; Dai, P.M. Use of the star of slots in designing fractional-slot single-layer synchronous motors. *IEE Proc. Electr. Power Appl.* **2006**, *153*, 459–466. [CrossRef]

27. Bianchi, N.; Bolognani, S.; Pre, M.D.; Grezzani, G. Design considerations for fractional-slot winding configurations of synchronous machines. *IEEE Trans. Ind. Appl.* **2006**, *42*, 997–1006. [CrossRef]

28. El-Refaie, A.M. Fractional-slot concentrated-windings synchronous permanent magnet machines: Opportunities and challenges. *IEEE Trans. Ind. Electron.* **2010**, *57*, 107–121. [CrossRef]

29. Jian, L.; Chau, K.T.; Jiang, J. A magnetic-geared outer-rotor permanent-magnet brushless machine for wind power generation. *IEEE Trans. Ind. Appl.* **2009**, *45*, 954–962. [CrossRef]

30. Norhisam, M.; Ridzuan, S.; Firdaus, R.; Aravind, C.; Wakiwaka, H.; Nirei, M. Comparative evaluation on power-speed density of portable permanent magnet generators for agricultural application. *Prog. Electromagn. Res.* **2012**, *129*, 345–363. [CrossRef]

Novel Frequency Swapping Technique for Conducted Electromagnetic Interference Suppression in Power Converter Applications

Ming-Tse Kuo [1,*] and Ming-Chang Tsou [2]

[1] Department of Electrical Engineering, National Taiwan University of Science and Technology, No. 43, Section 4, Keelung Road, Da'an District, Taipei 106, Taiwan
[2] Leadtrend Technology Corporation, No. 1, Taiyuan 2nd St., Zhubei City, Hsinchu County 30288, Taiwan; ming@leadtrend.com.tw
* Correspondence: mkuo@mail.ntust.edu.tw

Academic Editor: Ali M. Bazzi

Abstract: Quasi-resonant flyback (QRF) converters have been widely applied as the main circuit topology in power converters because of their low cost and high efficiency. Conventional QRF converters tend to generate higher average conducted electromagnetic interference (EMI) in the low-frequency domain due to the switching noise generated by power switches, resulting in the fact they can exceed the EMI standards of the European Standard 55022 Class-B emission requirements. The presented paper develops a novel frequency swapping control method that spreads spectral energy to reduce the amplitude of sub-harmonics, thereby lowering average conducted EMI in the low-frequency domain. The proposed method is implemented in a control chip, which requires no extra circuit components and adds zero cost. The proposed control method is verified using a 24 W QRF converter. Experimental results reveals that conducted EMI has been reduced by approximately 13.24 dBμV at 498 kHz compared with a control method without the novel frequency swapping technique. Thus, the proposed method can effectively improve the flyback system to easily meet the CISPR 22/EN55022 standards.

Keywords: conduction electromagnetic interference (EMI); frequency swapping; frequency jittering; quasi-resonant flyback (QRF)

1. Introduction

Energy conservation and carbon reduction have received considerable public attention in modern society. Converters have been widely used in renewable energy generations, such as wind generation systems [1], photovoltaic systems and fuel cells [2]. To address such topics, this paper focused on the conversion efficiency of power supplies by investigating a high-efficiency quasi-resonant switching power supply [3–6]. According to the International Special Committee on Radio Interference (CISPR) 22 [7] or European Standard 55022 (EN55022) [7], as well as other relevant regulations, a power factor correction (PFC) mechanism must be installed in power converters with a rated power exceeding 75 W. A PFC mechanism is necessary in devices such as television sets, network adapters, and notebook computer adapters. PFC generally uses a boost architecture to elevate the output voltage to reduce conduction loss. Therefore, a second-stage converter uses quasi-resonant switching power supply to reduce switching loss in switching devices. A quasi-resonant switching power supply operates the power devices at zero voltage switch (ZVS) and zero current switch (ZCS) conditions. Such switching techniques can create highly efficient products and can be widely applied in computer, communication, and consumer electronic products. Until now, no paper has explored the problems related to the electromagnetic interference (EMI) generated from the side-effect while the system efficiency is improved. In this paper, a novel frequency swapping technique is developed to meet CISPR 22/EN55022 standards without incurring additional component cost.

The power-supply switching devices generate very high current and voltage changes during the switching process, yielding high dI/dt and dV/dt and generating strong instantaneous dynamic noise when coupling parasitic inductance and parasitic capacitance [8]. Thus, electromagnetic noise generated in main circuit switching devices and relevant circuits is the main source of EMI in electronic devices. EMI primarily occurs as conduction and near-field interference sources. Certain high-frequency and high-power power supplies, such as high-frequency induction heating power supply and plasma power supply, can generate strong radiated EMI. The generation of EMI in power converters is depicted in Figure 1. Noise can be decomposed as common mode noise and differential mode noise components, which are transmitted through parasitic capacitance.

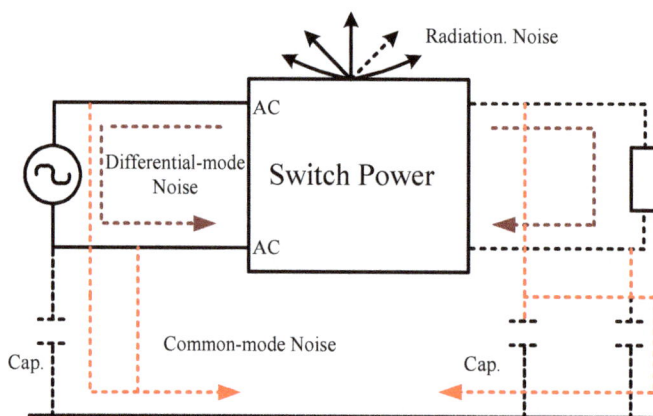

Figure 1. Electromagnetic interference (EMI) induced by a power converter.

High electromagnetic compatibility (EMC) is essential for electronic products. EMC design must be considered at the early stage of product design. Most converters operate with a fixed-frequency pulse-width modulation (PWM) control method. Most radiant energy is transmitted through a fundamental wave and sub-harmonic waves. Harmonic spectra generally have a narrow bandwidth and high amplitude. The amplitude of spectral components is the main item when analyzing EMI. Thus, EMI problems caused by fixed-frequency PWM are prominent. Frequency swapping is a technique that reduces the amplitude of harmonics by spreading spectral energy. In contrast to fixed-frequency PWM, frequency swapping refers to modulating the switching frequency around a fixed frequency during PWM, the principle of which is depicted in Figure 2a. When the signal periods in the time domain differ, spectra in the frequency domain spread with the decline in spectral amplitude, as shown in Figure 2b. The frequency swapping suppresses EMI at the source. It is an active suppression measure that inhibits both conducted EMI and radiant EMI [9].

Based on the spread spectrum principle, pulse frequency modulation (PFM) satisfies the feature of small harmonic amplitudes. However, when the commonly-used PFM control is adopted, the switching frequency is consistent with changes in load or input voltage (current). When the input voltage and load are fixed, PFM becomes another type of fixed-frequency PWM control, in which the frequency is not modulated [10]; thus, the advantage of PFM is lost. By contrast, frequency swapping techniques are unaffected by the output load and voltage. Switching frequency changes are controlled only by modulation signals and can thus easily attain the advantage of low EMI. In addition, when PWM is used for quasi-resonant converters, the voltage-second balance makes it impossible to achieve PWM control with fixed frequency. Instead, the operating frequency in a quasi-resonant system is mainly dominated by the magnetizing inductance. The prior swapping methods are only for fixed-frequency PWM system, and thus this paper provides a novel swapping control for quasi-resonant systems. As the frequency definitely varies, the complexity of the design is not increased by including designs for the magnetic components and the control loop in the system. Besides, how to realize frequency modulation is dominated by controllers, and it is feasible for aforementioned methods to be embedded in a chip. Actually, kinds of control integrated circuits (ICs) inclusive of above functions are available nowadays.

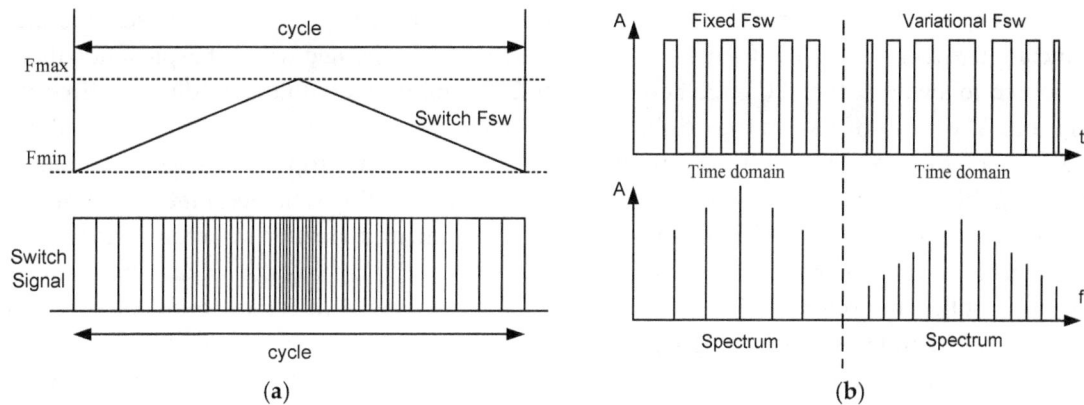

Figure 2. Frequency swapping technique and EMI suppression principle: (**a**) principle of switching frequency swapping; (**b**) fixed-frequency and variable-frequency signal spectra.

In this paper, a literature review is first presented, followed by an introduction of the frequency modulation (FM) principles, and its application in PWM carrier. The spectral characteristics of PWM waves are analyzed in detail. Second, the harmonic spectra generated by periodically-modulated PWM waveforms are analyzed; the control logic with the function of frequency swapping is implemented in an IC controller to verify the feasibilities.

In this paper, a 24 W (12 V/2 A) quasi-resonant flyback (QRF) converter is fabricated. The novel method is adopted to implement frequency swapping. A comparison of switching device voltage and current spectra of fixed-frequency and variable-frequency during operation shows that the amplitude of current and voltage harmonics during frequency swapping yields more favorable performance than that during fixed-frequency control in the low-frequency range. The proposed frequency swapping method reduces harmonic amplitudes by 10.2 dB at 646 kHz, proving the amplitude improvement and EMI suppression effects.

2. Literature Review

Currently, relevant studies can be roughly classified into experimental and commercial product design studies.

2.1. Experimental Study

Because frequency swapping modulates signals, related studies have typically emphasized periodically modulating switching frequencies. Workers at Polytechnic University of Catalonia [11] focused on selecting swapping modulation waveforms and implementing digital techniques. In addition, this research provided modulation waveform algorithms and the corresponding spectra, elaborated theoretical calculation methods, and briefly introduced EMC terminology and measurement instruments. However, the hardware circuit was at a verification stage because an expensive digital function generator was used for the functional circuit. In [12], the application of a frequency swapping technique for selecting rectifier modulation waveforms was discussed; however, no modulation circuit was provided. In [13], a complex theoretical method for determining spectral distribution equations was proposed. In [14], the influence of the resolution bandwidth of measurement instruments on the measurement results of modulated spectra was investigated, and the effect of filters on test results was summarized. The resolution bandwidth affects frequency offset and modulation frequency selection. However, the selection of the frequency swapping range and the inhibition of harmonic amplitude needs a trade-off. Hence, the specific value is difficult to define. In [14], the functional variation problem induced by periodic modulation in the original converter indicated that the output voltage ripple under an open loop condition is worse than that in a fixed-frequency PWM control system. Thus, a boost converter was used as a prototype to verify the argument. By contrast, the output voltage ripple in a close-loop buck converter with periodic modulation switching frequencies indicated suppression (yet incomplete removal) of low-frequency ripples. In [14], it was claimed without explanation that FM contributed to lowering the PFC circuit efficiency by 2.7% compared with that under fixed-frequency operation. The effect of FM on the original converter requires further investigation; the following researches are devoted to improve conduction EMI by choosing appropriate devices materials and optimizing those interconnections on PCB layout. In [15], Grounding schemes, material comparison between ferrite and nanocrystalline cores, and the new integrated filter structure are presented. The integrated structure maximizes the core window area and increases the leakage inductance by integrating both common-mode (CM) and differential-mode (DM) inductances onto one core and to reduce both DM and CM noise using a passive filter in a dc-fed motor drive. In [16], the proposed technique is a rule-based automatic procedure based on suitable databases that consider datasheet information of commercially available passive components (e.g. magnetic cores, capacitors). It allows the minimization of the filter's volume and therefore the improvement of the converter's power density. The size and the performance of EMI filters designed by the proposed procedure have been compared with those of a conventionally designed one. In [17], the proposed technique allows the CM and DM sections of the EMI filter to be properly selected in a more economical way, i.e., without the need of a dedicated hardware or costly radio frequency (RF) instrumentation. The filter has been set up according to a high power-density concept by using a high permeability nanocrystalline magnetic material for the CM choke core that allows a volume and weight reduction. In [18], a new approach for easy and fast modeling of EMI filter in aircraft application is proposed, in order to be used in an optimization process.

The following researches focused on exploring PWM-based random switching modulation techniques and chaotic switching modulation techniques [19]; these techniques have demonstrated satisfactory performance in reducing EMI amplitude and got considerable progress in this regard. However, the harmonic energy spread during modulation became too wide, thus deteriorating the low-frequency characteristics; moreover, it still suffers from uncontrollable energy spread and is inapplicable on QRF converters. Generating random and chaotic signals requires complex circuit structures, which increases the cost of fabrication. Thus, for the frequency swapping modulation method, the presented study applies a simple adjustable periodic modulation method to a QRF converter.

2.2. Commercial Product

In recent years, many corporations have developed power modules or control chips featuring a frequency swapping function. For example, following the single-chip switching power supply IC of the TinySwitch family of Power Integrations (San Jose, CA, USA), the TOP242, a dedicated switching power supply chip, was developed using the frequency swapping technique. The switching frequency varies in the range of 128–136 kHz, with a swapping range of ± 4 kHz [20]. The product parameters are renewed constantly.

The spread oscillators LTC3809 and LTC6902 (Linear Technology, Milpitas, CA, USA) feature the functions of spread FM and pseudo-random noise techniques. The oscillators can generate any frequency between5 kHz and 20 MHz [21]. In addition, the single-chip switching power controller NCP1215A (ON Semiconductor, Phoenix, AZ, USA) [22], ICE3B0365J (Infineon Technologies, Neubiberg, Germany), which has a 67 ± 2.7 kHz oscillation frequency [23], and FSDH0265 (Fairchild

Semiconductor, Sunnyvale, CA, USA) [24] feature similar functions. However, no products applicable to quasi-resonant structures flyback converters are available.

All the aforementioned studies have explored frequency swapping in fixed-frequency controllers. Such an approach cannot be applied in QRF converters because when the system is operated in discontinuous conduction mode or boundary conduction mode, the system operating frequency depends on inductor volt-second balance. This paper proposes a novel frequency swapping control technique to solve these problems in variable FM systems.

3. Frequency Modulation of Pulse-Width Module Signals

This section involves the analysis of single-frequency sinusoidal modulation. The voltage waveform of converter switching devices resembles a rectangular pulse. The line input current is normally a sawtooth waveform. During operation, the current and voltage in a power device become increasingly complex. A simple analysis of sinusoidal modulation in PWM actuating signals is described as follows to predict the voltage characteristics at both ends of power devices.

3.1. Single-Frequency Modulation of Pulse-Width Module Signals

If $K(\omega_c)$ represents the switching function of an angular frequency ω_c, then the unmodulated rectangular wave (PWM wave) signal can be expressed as:

$$u_C(t) = U_{cm} \times K(\omega_c t) \tag{1}$$

Thus, single-frequency sinusoidal modulation of a rectangular wave can be expressed as:

$$U_{fm}(t) = U_{cm} \times K(\omega_c t) \tag{2}$$

The Fourier series of rectangular waves with a duty cycle D is expressed as:

$$u_C(t) = U_{cm}\left\{ D + \sum_{n=0}^{\infty} \frac{2}{n\pi} \sin(Dn\pi)\cos(n\omega_c\tau) \right\} \tag{3}$$

Let $\tau = t + \frac{m_f}{\omega_c}\sin\Omega t$, which can be substituted into (3) To obtain a Fourier series algorithm of FM rectangular waves:

$$u_{fm}(t) = U_{cm}\left\{ D + \sum_{n=0}^{\infty} \frac{2}{n\pi} \sin(Dn\pi)\cos\left[n\left(\omega_c t + m_f \sin\Omega t\right)\right] \right\} \tag{4}$$

Expanded from (4), (5) can be obtained:

$$\begin{aligned}
u_{fm}(t) = U_{cm}D \quad &+ \frac{2}{\pi}U_{cm}\sin(D\pi)\cos\left(\omega_c t + m_f \sin\Omega t\right) \\
&+ \frac{1}{\pi}U_{cm}\sin(2D\pi)\cos\left(2\omega_c t + 2m_f \sin\Omega t\right) \\
&+ \frac{2}{3\pi}U_{cm}\sin(3D\pi)\cos\left(3\omega_c t + 3m_f \sin\Omega t\right) + \cdots
\end{aligned} \tag{5}$$

According to the previous section, FM of square waves is similar to FM of each sub-harmonic wave; however, the modulation index of each sub-harmonic wave in the modulated original carriers increases with the order of sub-harmonic waves. The modulation index of nth sub-harmonic waves is expressed as $m_{f(n)} = n \cdot m_f$, where m_f represents the modulation index of carrier fundamental. Carson's rule is still applicable to the bandwidth of the boundary frequencies of each sub-harmonic wave, which can be expressed as:

$$BW_{(n)} \approx 2\left(n \cdot m_f + 1\right) \cdot f_m \tag{6}$$

where f_m is the frequency of the modulation waves. Generally, FM of non-sinusoidal waves is similar to FM of fundamental waves and each sub-harmonic wave. Because waveforms (or functions) that satisfy convergence conditions theoretically can be expressed as Fourier series expansion, the algorithm of FM current waveforms flowing through power devices can be obtained through similar approaches.

As shown in (6), the bandwidth of each modulated sub-harmonic wave increases with the harmonic order. If the sub-harmonic wave order is sufficiently high, nearby harmonic boundary frequencies will overlap as shown in Figures 3 and 4. Frequency overlap contributes to a uniform spectral distribution of high-order sub-harmonic waves; however, the amplitudes may increase because of component superposition (when the components are in the same phase) or attenuation (because of the reverse phase of nearby frequency components). However, from the perspective of energy diffusion, the amplitudes of spectral components at high frequencies during spectral superposition show an increasing tendency.

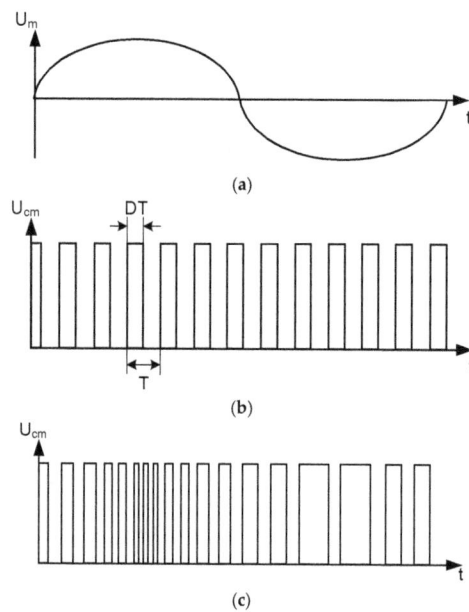

Figure 3. Sinusoidal modulation of FM rectangular waves: (**a**) Modulation Wave; (**b**) Carrier; and (**c**) PFM.

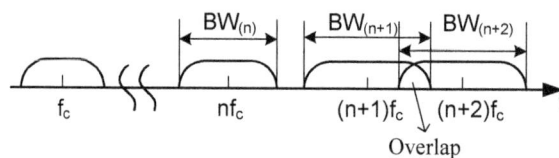

Figure 4. Sinusoidal modulation of FM square waves.

3.2. Calculation of High-Order Sub-Harmonic Superposition

Given that the sub-harmonic order is sufficiently high, the boundary components in the spectra superposition will increase the amplitudes in the affected area. Although the corresponding amplitudes are minimal in accordance with the increased harmonic order, potential interference at certain sensitive frequency bands is still unwanted at all [25].

Increases in amplitudes can be avoided in sensitive frequency bands by deriving the harmonic order during spectral superposition.

Accurate spectral distribution cannot be obtained easily in high-order sub-harmonic waves. Because the determined harmonic order is an integer, to facilitate calculation, the envelope of each sub-harmonic wave is hypothesized to be an ideal square frame [24], as depicted in Figure 5.

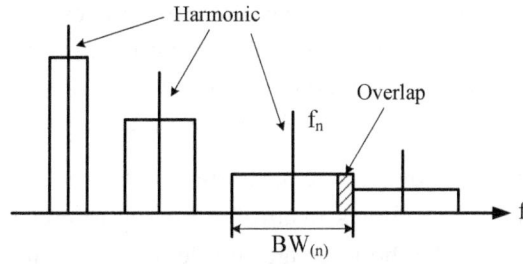

Figure 5. Boundary superposition of high-order sub-harmonic waves in modulated wave spectra.

The central frequency of each sub-harmonic wave of the original carrier is f_n, and the carrier fundamental frequency is f_c; thus, $f_n = n \cdot f_c$. Let the bandwidth of each sub-harmonic boundary of carriers be denoted as $BW_{(n)}$ and bandwidth of the fundamental wave be denoted as BW. Combining (6) and $m_f = \Delta f_{cmax}/f_m$ yields:

$$BW_{(n)} = 2\left(n \cdot m_f \cdot f_m + f_m\right) = 2(n \cdot \Delta f_{cmax} + f_m) \tag{7}$$

When the first superposition occurs, (8) is established for the superposition area:

$$f_n + \frac{BW_{(n)}}{2} = f_{n+1} - \frac{BW_{(n+1)}}{2} \tag{8}$$

where n is the harmonic order.

Substituting (7) into (8) yields:

$$n \cdot f_c + n \cdot \delta \cdot f_c + f_m = (n+1)f_c - (n+1) \cdot \delta \cdot f_c - f_m \tag{9}$$

From (9), (10) is derived as follows:

$$f_c = \frac{2}{1 - \delta \cdot \left(1 + 2 \cdot n_{overlap}\right)} \cdot f_m \tag{10}$$

where f_c is the fundamental frequency of the carrier, f_m is the modulated wave frequency, and $n_{overlap}$ is the carrier harmonic order during superposition. The $n_{overlap}$ is derived in (10) to obtain (11):

$$n_{overlap} = \frac{1}{\delta} \cdot \left(\frac{1}{2} - \frac{f_m}{f_c}\right) - \frac{1}{2} \tag{11}$$

According to (11), when the carrier fundamental wave frequency decreases, or when the modulation frequency f_m and frequency offset δ increase, the harmonic order during superposition decreases. Generally, carrier fundamental wave frequencies are considerably higher than modulated wave frequencies (i.e., $f_c > f_m$). Thus, (11) can be approximately expressed as:

$$n_{overlap} = \frac{1}{\delta} \cdot \left(\frac{1}{2}\right) - \frac{1}{2} \tag{12}$$

In (12), considerable frequency offsets tend to generate spectral superposition in low harmonic waves. The trivial effect of modulated wave frequency f_m on the superposition can be ignored.

Calculations after spectral superposition can be employed to predict the general distribution of high-order sub-harmonic boundaries. As stated previously, to prevent the potential hazard of high amplitudes in certain high-frequency bands, modulation frequency f_m and frequency offset δ can be reasonably selected according to (12). Postponing spectral superposition until after the sensitive frequencies have been identified can avoid potential interference in the specific frequency bands.

In reality, modulated waveforms cannot be single-frequency sinusoidal waves. When the modulated waveform contains only two modulated frequencies $\Omega 1$ and $\Omega 2$, which is already the simplest situation, the corresponding modulation indices will be mf_1 and mf_2, respectively. The analytic expression of a modulated wave is complex and the following components are included in the spectra:

(1)   Carrier frequency ω_c shows an amplitude proportional to $J_0(mf_1)J_0(mf_2)$.
(2)   Boundary frequency $(\omega_c \pm n_{\Omega 1})$ shows an amplitude proportional to $J_n(mf_1)J_0(mf_2)$.
(3)   Boundary frequency $(\omega_c \pm n_{\Omega 2})$ shows an amplitude proportional to $J_0(mf_1)J_n(mf_2)$.
(4)   Additional boundary frequency (combined frequency; $\omega_c \pm n_{\Omega 1} \pm n_{\Omega 2}$) shows an amplitude proportional to $J_n(mf_1)J_0(mf_2)$, where n and p are random integers.

Clearly, if the modulated waveform frequency is not a single frequency, then the generated spectral structure can be very complex. Thus, the present paper explores only the single-FM waveforms and corresponding spectra. Given that the modulated waveform is a single-frequency non-sinusoidal wave, it can be decomposed into a Fourier series. Because the frequency of each harmonic and the fundamental wave have an integer multiple relationship, mf_1 and mf_2 are proportional reasonably. Thus, these four types of boundary frequency components can be simplified. Nonetheless, spectrum analytic expression is intricate. In particular, when the modulated signal is discontinuous or presents irregular patterns, expressing it in analytic form is difficult. To address this, discrete methods can be employed to formulate a suitable equation.

In summary, this section explains the basic theory of frequency swapping and the principles of FM, and modulated PWM wave spectral characteristics are analyzed as well. Critical parameters related to FM theory are provided. In addition, the FM PWM wave spectral superposition at high-order sub-harmonic waves is explained and calculated carefully.

4. Design of Frequency Swapping Circuits

Figure 6 shows the close-loop control block diagram for a QRF converter. This is the generalized system shown above. The top part, $G_o(s)$ represents all the systems and all the controllers on the forward path. The bottom part, $G_f(s)$ represents all the feedback processing elements of the system. The letter "$G_i(s)$" in the beginning of the system is called the Gain. We can define the Closed-Loop Transfer Function as follows:

$$\frac{V_{OUT}(s)}{V_{IN}(s)} = \frac{G_i(s)G_o(s)}{1 + G_o(s)G_f(s)} \tag{13}$$

Perturbation feedback signals are added at the feedback end to modulate on-time of power switches. This new method doesn't affect the close-loop control system. The expected frequency swapping effect is obtained because the QRF converter is operated and based on ampere-second balance to reach ZCS.

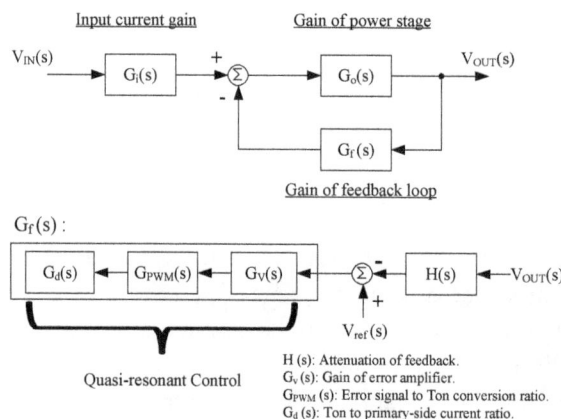

Figure 6. Control block diagram of the modulation waves.

4.1. Selection and Analysis of Frequency Swapping Techniques

As indicated in Section 3, the amplitude, boundary bandwidth, and modulation indices of spectral components are closely related. Thus, modulation of the switching frequency proportionate to modulation of each sub-harmonic frequency can facilitate favorable EMI. The following sections explore the application methods. The methods and logic concepts share a common goal of adjusting the system operating frequencies to facilitate favorable EMI conduction. A simple analysis of the three methods are provided as follows.

4.1.1. Change in Quasi-resonant Detection Delay Time

As depicted in Figure 7, the quasi-resonant detection (QRD) mechanism detects whether the metal-oxide-semiconductor (MOS) drain to source (Vds) reaches the lowest point (valley). When the control IC detects a valley, the power switch is turned on to activate the valley switch and reduce the switching loss. Method 1 involves changing the interval between the QRD delay times (Points 1, 2, 3, 4... repeat changes); subsequently, the system period and frequency can be adjusted. However, this method has two major disadvantages. First, the changeable period difference is only ±300 ns, yielding only small changes in the modulation index. Second, the QRF features enabling the activation of the valley switch while the power switch is on. When the QRD delay time is adjusted, switches in certain cycles cannot be activated at a valley time point, thereby lowering system efficiency, which can be worse than that of the original system. Thus, adjusting the QRD delay time is not applicable for the QRF structure.

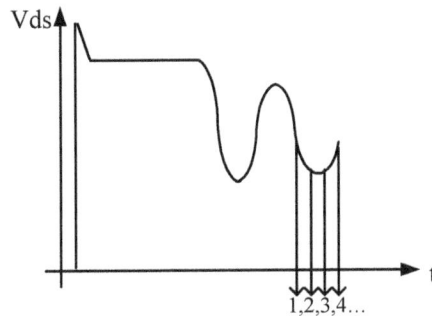

Figure 7. Change in quasi-resonant detection (QRD) delay signals.

4.1.2. Change in Quasi-Resonant Flyback Valley Order

Figure 8 shows the Vds waveform in the inductance and capacitance resonant state when the first-order power MOS and second-order diode in a flyback converter are switched off. Method 2 involves applying the control valley order to determine whether the power switch is to be switched on. In other words, in State 1, the third valley indicates that the power switch is to be switched on; in State 2, the second valley indicates that the power switch is to be switched on; and in State 3, the first valley indicates that the power switch is to be switched on. Subsequently, the state returns to State 1 and repeats continuously. Thus, the system period and frequency can be adjusted. However, this method has three major disadvantages. First, the changeable period difference is considerable, which easily causes a side effect of a large output ripple. Second, the resonant frequency relies on system inductance and MOS parasitic capacitance; modulation index variation between different systems cannot be controlled. Third, when the system is operated during the occurrence of the first valley, the control IC cannot change the valley switching order because of the volt-second balance principle. Thus, the method of adjusting the QRF valley order is not applicable to the QRF structure.

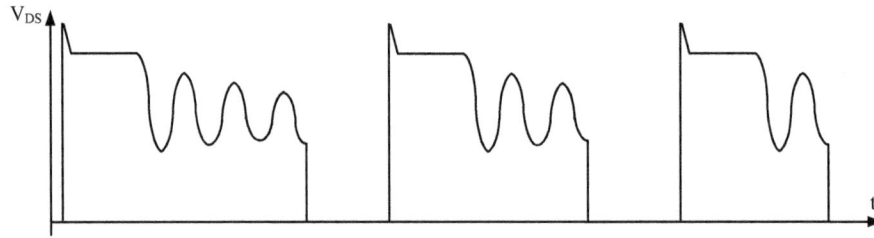

Figure 8. Valley order change.

4.1.3. Disturbance on Voltage Feedback Signal

As depicted in Figure 9, Method 3 causes direct disturbance on the voltage feedback signals. When the disturbance voltage is positive, the gate turn on (T_{on}) increases and the gate turn off (T_{off}) time increases, thereby enlarging the periods. By contrast, when the disturbance voltage is negative, the gate turn on (T_{on}) decreases and the gate turn off (T_{off}) time decreases, thereby changing the system periods and frequencies. This method has a disadvantage. Figure 10 defines the relative parameters of triangular modulation signals.

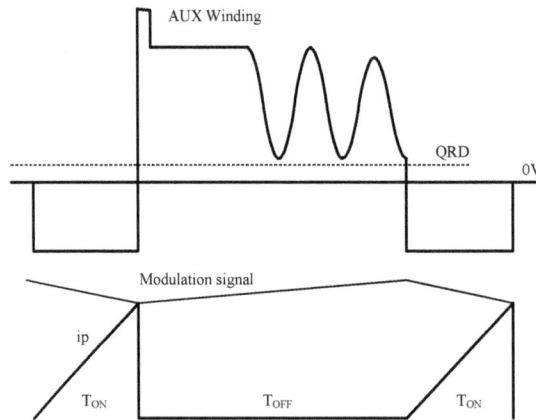

Figure 9. Disturbance voltage feedback signal.

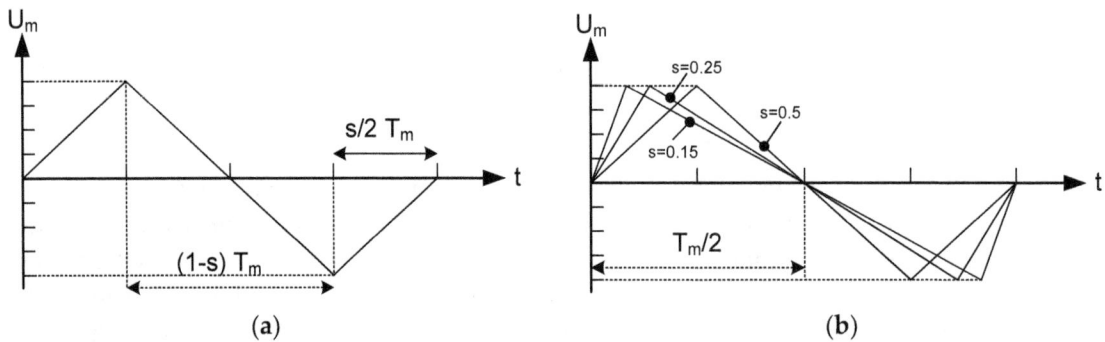

Figure 10. Triangular modulation signals: (**a**) signals; (**b**) different s ratio values.

The triangular wave expression is:

$$u(t) \begin{cases} U_m \cdot f_m \cdot \frac{2}{s} t & 0 \leq t \leq s \cdot \frac{T_m}{2} \\ \frac{U_m}{1-s} \cdot (1 - 2 \cdot f_m \cdot t) & s \cdot \frac{T_m}{2} \leq t < (1 - s/2) \cdot T_m \\ \frac{2}{s} \cdot U_m \cdot (f_m \cdot t - 1) & (1 - s/2) \cdot T_m \leq t < T_m \end{cases} \tag{14}$$

In Equation (14), U_m, T_m, and f_m denote the amplitude, period, and frequency of triangular waveforms respectively. The s parameter controls the location of the triangular peak between the original of the axis and the position of $T_m/2$, named as peak parameter, and the range is restrained from 0 to 1. The typical peak parameter is 0.5 commonly, which makes the triangular waveforms central-symmetrical and axis-symmetrical. The spectral composition of triangular modulation still meets the rules that the attenuation of bandwidth and amplitude increases when the modulation parameter increases. Compared with sinusoidal modulation, the spectral of triangular modulation is smoother. Next, the frequency of modulation waveforms needs to be defined. Because the frequency modulation spreads the system intrinsic frequency, parts of spread frequency enter the region lower than operating frequency of the system. Particularly, the spread frequency is forbidden to be lower than acoustic frequency. The acoustic noise often occurs at fundamental waves which own the lowest frequency among all harmonic waves. Hence, Equation (15) must be satisfied as:

$$f_c - \frac{BW}{2} = f_c - (\Delta f_{cmax} + f_m) > f_{amax} \qquad (15)$$

In Equation (15), f_c is the frequency of the carrier waveforms; BW is the bandwidth; f_{amax} is the maximum of acoustic frequency about 20 kHz. The switching frequencies of low-power converters are commonly located between hundreds of kHz and numbers of MHz, which makes the acoustic frequency easier to be avoided. However, for converters with tens of kHz frequency, the condition of Equation (15) must be cared and observed. Therefore, the thesis uses the triangular modulation with s parameter as 0.5 and modulation frequency as 1 kHz.

However, this method does not have the flaws of Method 1 or 2, nor is it confined by the system components. To achieve the optimal frequency swapping method and minimize induced problems on systems, the disturbance voltage feedback-signal method is the most suitable solution for application on the QRF structure. In the following exploration and analysis section, the disturbance voltage feedback signal structure is employed to design a converter.

4.2. Embedded Logic Circuit of Adjustable Frequency Swapping

Before the control IC is activated, the auxiliary winding of a transformer is considered as a route to ground. The resistance of copper coils of the auxiliary winding is measured as 106 mΩ (50 Hz) which is far smaller than RA resistor in Figure 11. Thus, parallel resistors (R_{total} = RA//RB in Figure 11) can be applied to calculate the total external impedance from the FB pin in Figure 11. When VCC in Figure 12 encounters under-voltage-lockout (UVLO ON), the IC generates a current source of approximately 50 μA. As shown in Figures 11 and 12, this state enables the FB pin to determine the adjustable frequency swapping range. The adjustable range and setting values are listed in Table 1. This adjustable frequency swapping range enables users to control the disturbance and optimize the output ripple standards.

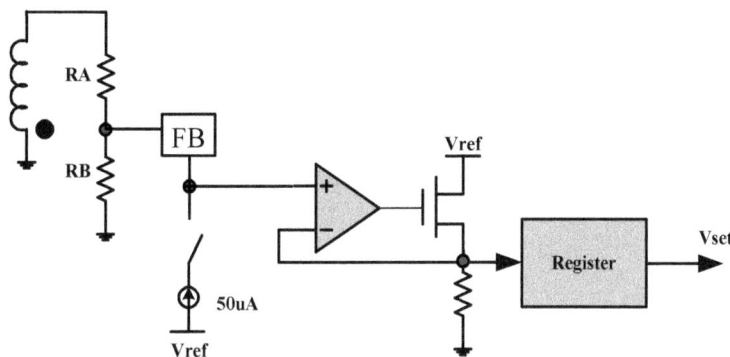

Figure 11. Circuit for adjusting the frequency swapping range.

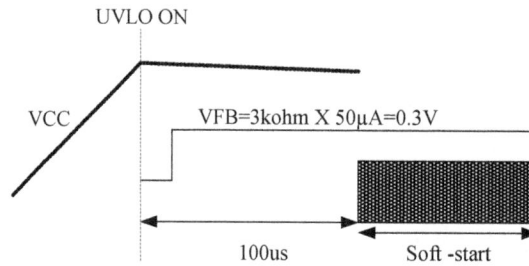

Figure 12. Embedded logic circuit for adjusting the frequency swapping range.

Table 1. Comparison of swapping range settings.

Ideal R_{total}	V_{FB} Range	Suggested R_{total} Value	Swapping Range (%)
40 k < R_{total}	1.7 < V_{FB}	42 k	±9
22 k < R_{total} < 40 k	1.1 < V_{FB} < 1.7	30 k	±6
14 k < R_{total} < 22 k	0.7 < V_{FB} < 1.1	20 k	±3
R_{total} < 14 k	V_{FB} < 0.7	10 k	±0

4.3. Embedded Logic Circuit of Feedback Disturbance

Based on Figure 13 and Equation (16), after choosing adequate R_C in Figure 13 and determining frequency swapping, this logic circuit will adjust R_D to alter frequency swapping range, and the resistance of R1 is decided by an 8-bit binary code.

This switch array switches in 16 steps. The clock signal in Figure 13 is the FM disturbance period (in cycles) for adjusting the R_C resistance.

$$C_{comp\_J} = V_{comp} \times \left(1 \pm \frac{R_C}{R_D} \right) \tag{16}$$

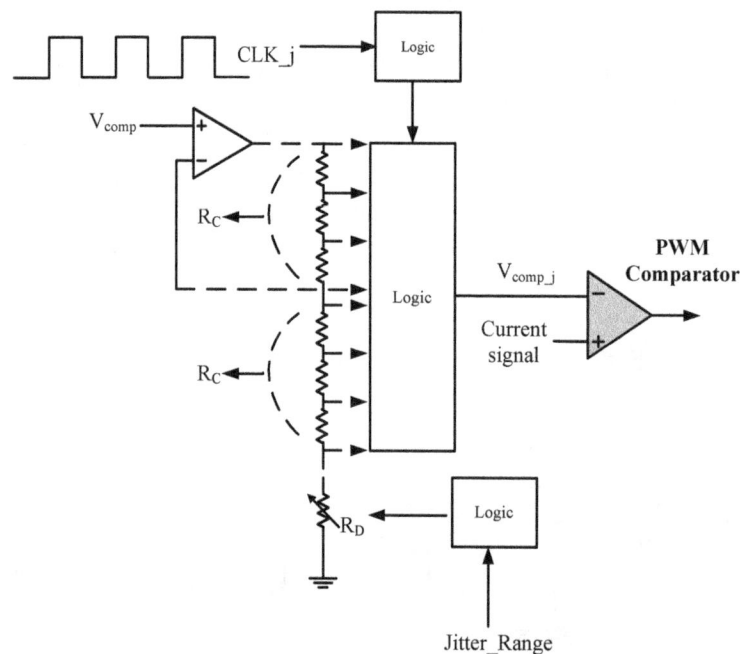

Figure 13. Frequency swapping logic circuit.

5. Results and Analysis

The theoretical analysis validates the feasibility of the proposed technique and the results are compared in two parts. First, SIMPLIS simulation software (SIMPLIS Technologies, Inc., Portland, OR, USA) was used to verify and determine the optimal parameters. Second, a novel frequency swapping technique was incorporated into the control chip (Figure 14) to fabricate a practical prototypical converter for comparison (Figure 15).

Figure 14. Integrated circuit (IC) layout of the control chip (Taiwan Semiconductor Manufacturing Company; 0.6 μm 7 V/12 V) featuring the novel frequency swapping technique.

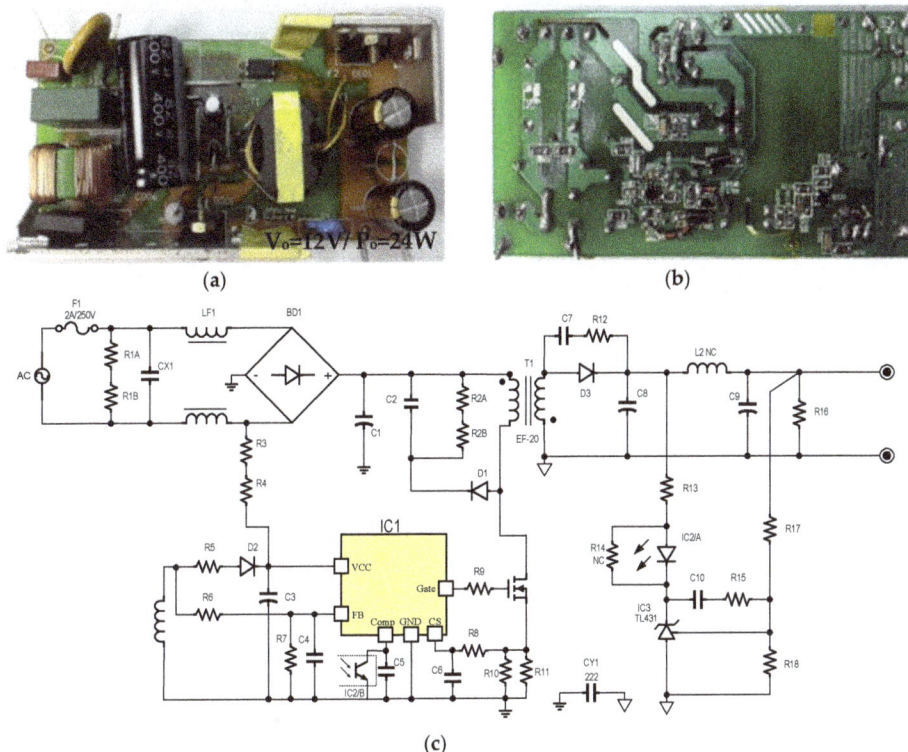

Figure 15. A photograph of the prototypical converter: (**a**) top layer; (**b**) bottom layer; and (**c**) detailed schematic of QRF.

The typical circuit of QRF converter is shown in Figure 15c. It mainly depends on a controller that provides necessary QRD (FB PIN) signals and voltage mode functions: under current mode control, this controller detects the inductance current through a current-sensing resistor (CS PIN) and catches output voltage signal through optical coupler to determine on time duty cycle. When $R6$ and $R7$

will divide the voltage signal into the FB pin by the auxiliary winding. Chip will be able to set the swapping range (R_{total} = R6//R7) and determine the QRD signal. By the way, the circuit involves the CM choke (LF1) and filtering X-capacitor (CX1) to inhibit conduction EMI. In the following, our discussion will focus on these devices and analyze their pros and cons. Table 2 lists the external system component parameters.

Table 2. Hardware and specifications.

Hardware/Simulation Parameters	Specifications
Input Voltage	V_{in} = 90 V–264 V
Output Voltage	V_o = 12 V
Output Current	I_o = 0–2 A
Switch Frequency	f_s = 65 kHz
Max. Duty Cycle	D_{max} = 0.48
Transformer	Lp = 1.6 mH
Ratio of Transformer	Nm = 0.8
Output Capacitance	C_o = 680 uF × 2; ESR = 0.01 ohm

5.1. Simulation Results

At first, the functional circuit for frequency swapping block is built by SIMPLIS as shown in Figure 16a shows that the binary code for frequency swapping was used to control the R_1 resistance and an eight-step variable-frequency periodic cycle was selected to make the operating frequency changing from 130 to 0.7 kHz within a single step. By contrast, Figure 16b shows how to control the R1 resistance through the switches. Through the binary signals (Qib and Qi), changes in the sink and source current are switched to generate positive and negative voltage differences from the disturbed voltage feedback, thereby achieving voltage compensation disturbance. Finally, Figure 16c shows how FM signals are generated by using a low-frequency triangular wave generator, and subsequently the cyclic frequency swapping repeats.

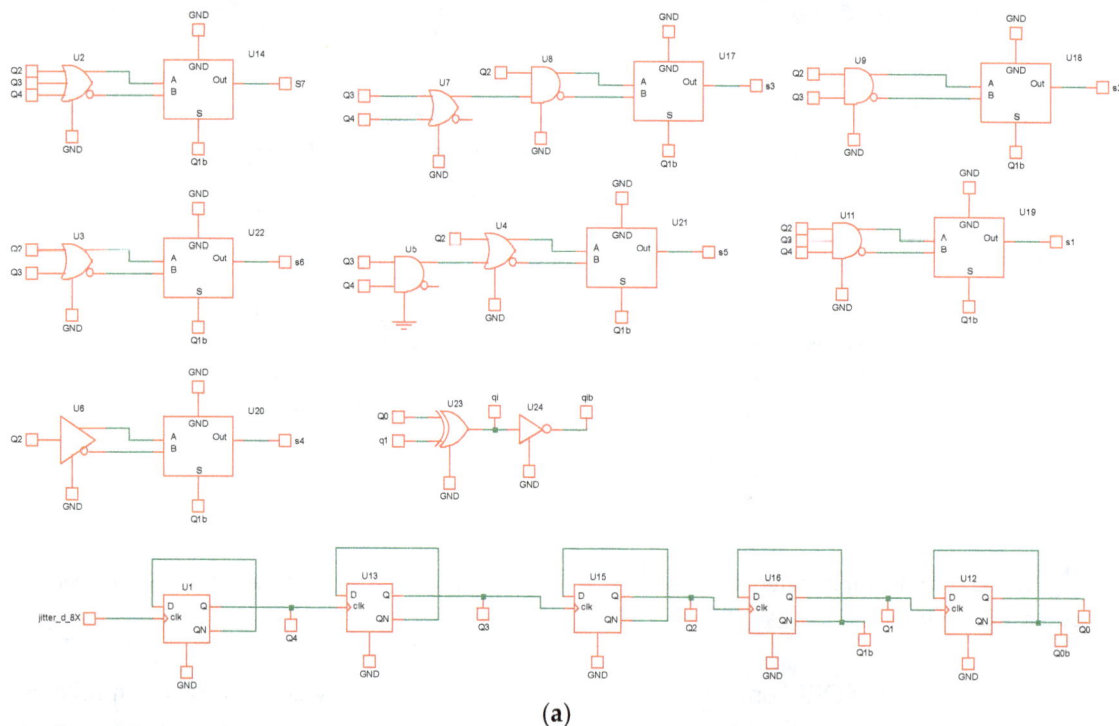

(a)

Figure 16. Cont.

(b)

(c)

Figure 16. The novel frequency swapping circuit: (**a**) binary code for frequency swapping; (**b**) ratio for frequency swapping; and (**c**) frequency swapping modulation.

Figure 17 and Table 3 further indicate that larger frequency offsets (ΔF_{sw}) result in greater amplitude attenuation. However, the frequency offset is limited by the circuit design in actual converters.

(1) When the frequency offset is too high, although the duty cycle remains unchanged, the corresponding component conduction time decreases. The T_{on} and T_{off} time occupy a major portion of duty cycle, thus hindering energy transmission. When the switching frequency is too low, the corresponding conduction time increases, which probably results in magnetic component saturation.

(2) When the frequency offset is too high, energy transmission becomes uneven, which possibly induces large output ripples if the system cannot response this voltage feedback disturbance rapidly.

Although the simulation results cannot verify the statement in Passage (1), when $R_{total} = 52$ kΩ the ΔF_{sw} disturbance is within $\pm 12\%$ (Table 3), the output ripple exceeds the regulated tolerance (100 mV). Thus, in the following section, the maximum frequency swapping adjustment range is set as $\pm 9\%$ in the IC design ($R_{total} = 40$ kΩ). In response to the modulation index, this paper sets the modulation index to 1 ms, because the system voltage feedback mechanism can track and compensate the index, consequently mitigating the disturbance effect if the modulation index is less than 1 ms. By contrast, when the modulation index is higher than 1 ms, it easily enters the hearing range (1–20 kHz), which forbids its application in computer, communication, and consumer electronic products. Thus, in the IC FM parameters, the modulation index is set at 1 ms.

Table 3. Comparison of modified swapping ratios.

X-Capacitor Value	Comparison of Swapping Range Settings	Frequency Swapping Ratio	Frequency Swapping Period	Output Ripple	
				90 V/60 Hz	264 V/50 Hz
	8.61 kΩ	±0%	1 ms	Δ3 mV	Δ5.2 mV
	18.18 kΩ	±3%	1 ms	Δ21 mV	Δ41 mV
0.22 μF	29 kΩ	±6%	1 ms	Δ32 mV	Δ71 mV
	40 kΩ	±9%	1 ms	Δ48 mV	Δ98 mV
	52 kΩ	±12%	1 ms	Δ102 mV	Δ143 mV

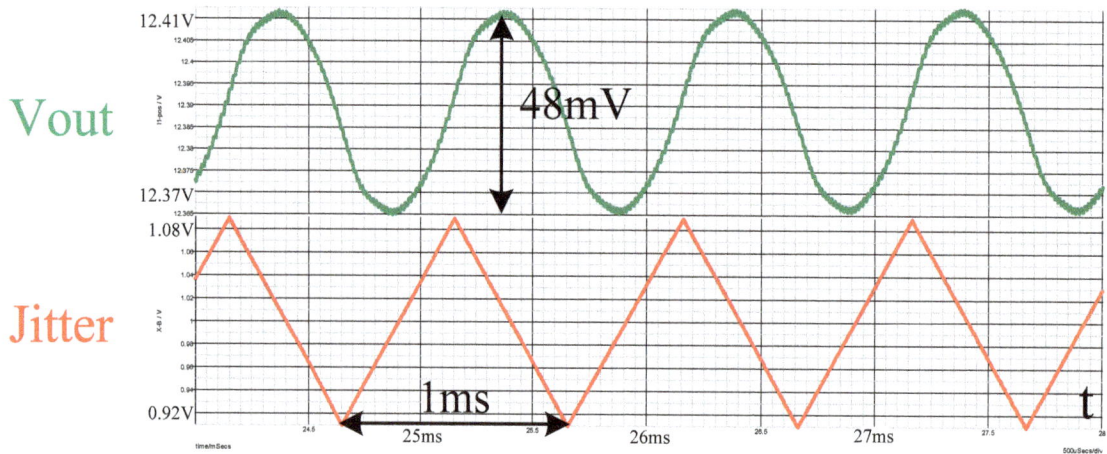

(a)

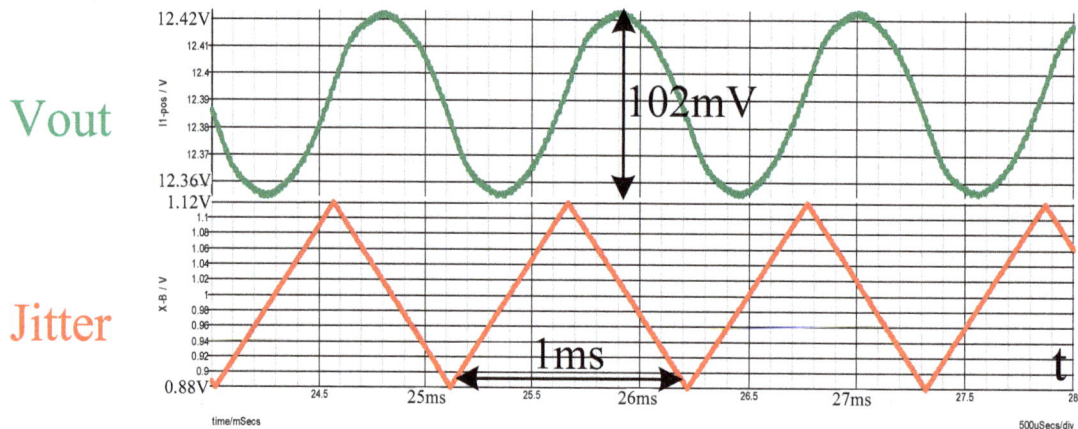

(b)

Figure 17. Modified frequency swapping ratio comparison: (**a**) output ripple for ±9% frequency swapping range (90 V/60 Hz); (**b**) output ripple for ±12% frequency swapping range (90 V/60 Hz).

5.2. Waveform Measurement

As shown in Figure 18, the system sets impedance from the FB pin to adjust the frequency swapping range as R_{total} = 40 kΩ, 29 kΩ, 18.18 kΩ , and 8.61 kΩ to achieve the expected frequency swapping range.

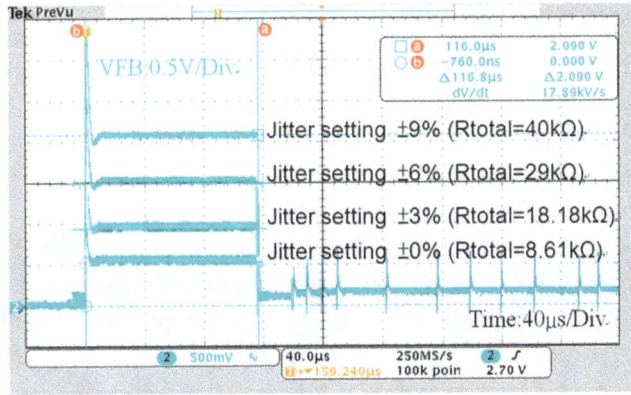

Figure 18. Using the feedback (FB) pin to set frequency swapping range.

The spectral analysis function of the oscilloscope is used to intercept the time domain waveform of the PWM wave at a fixed frequency swapping modulation and the spectrum within the frequency swapping range (Figure 19). Figure 19a shows there is no frequency amplitude for a system working frequency of 45.6 kHz; thus, this setting is used for the control group. Figure 19b shows that for a system working frequency of 45.6±1.45 kHz, the frequency swapping amplitude is ±3%. Figure 19c shows that for a system working frequency of 45.6 ± 2.9 kHz, the frequency swapping amplitude is ±6%. Figure 19d shows that for a system operating frequency of 45.6 ± 4.4 kHz means the maximum amplitude of the frequency swapping of ±9%; therefore, this frequency swapping range (±4.4 kHz) is adopted for the comparison group in this paper. According to the aforementioned measurement data, the FM parameters and frequency swapping range are identical with simulated conditions.

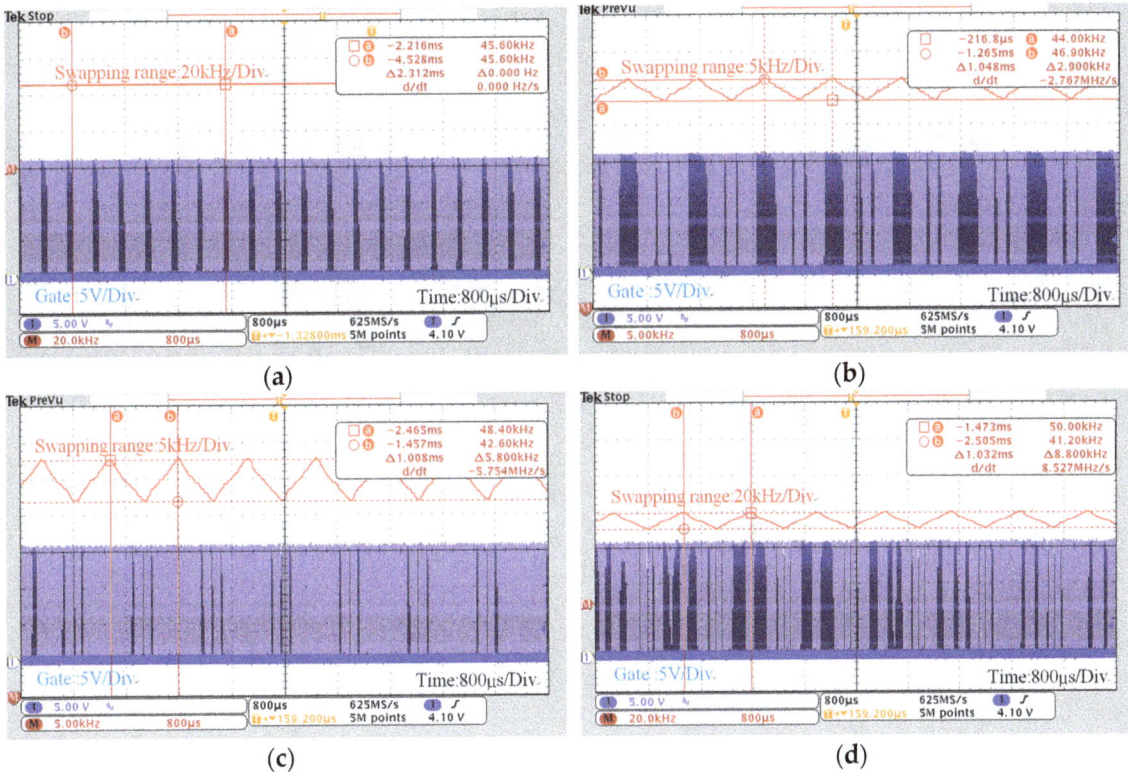

Figure 19. Measurement of operating frequencies. (**a**) no frequency swapping; (**b**) frequency swapping range set at ±3%; (**c**) frequency swapping range set at ±6%; and (**d**) frequency swapping range set at ±9%.

5.3. Testing Conducted Electromagnetic Interference

The 24 W prototypical converter used in this thesis is applied to high resistive load, which commonly uses π filter to filter high-frequency noise. Based on the frequency of the unwanted noise to be inhibited, the crossover frequency and the components (common mode choke and X-capacitor) of the filter can be designed to improve the noise immunity. To test the conducted EMI is divided into two parts at high input voltage (230 V/50 Hz).

First, the novel frequency swapping technique was incorporated into the quasi-peak (QP) amplitude and average (AVG) amplitude parameters with ±9% frequency swapping range, and the performance of this system was compared with another system without the technique. Because the two windings of a common mode choke are connected in series to two output lines, when the output currents flow through each winding, the magnetic flux direction formed in the iron core is inverse and can be neutralized. The magnetic flux in the iron core at balance is zero. Thus, the change in common mode inductance must consider the saturation from the increasing input current. Accordingly, the common mode choke (choke = 20 mH) was set as a fixed value to avoid other factors from causing spurious comparisons of the conducted EMI. In the following comparison, the X-capacitor capacitance is the only value that is adjusted in order to observe whether the X-capacitor capacitance amplitudes can be further decreased with the frequency swapping range of ±9%, and the EN55022 standards is still complied with in the meantime.

For the EMI measurement, the spectral range of 150 kHz–30 MHz was observed. In Figure 20a, no frequency swapping signals were added to the neutral end measurement. The QP amplitude testing value corresponding to the EN55022 standard satisfied the QP margin of more than 6 dBμV. In comparison, the AVG amplitude testing value corresponding to the EN55022 standard yielded an average margin of less than 6 dBμV at 428, 498, and 1499 kHz. The lowest readings were 4.28, −2.79, and 3.59 dBμV, respectively. Thus, EMI shielding components, including an augmented common mode choke, X-capacitor, and Y1-capacitor, must be added to the system to improve the AVG amplitude.

Figure 20b shows that the measurement at the line end without the frequency swapping signals attains a QP amplitude that corresponds to the EN55022 standard and satisfies a QP margin of more than 6 dBμV. By contrast, the AVG amplitude corresponding to the EN55022 standard does not exceed the AVG margin of more than 6 dBμV at 498 and 3867 kHz, yielding the lowest readings at −0.25 and 0.23 dBμV, respectively. Thus, systems must be equipped with EMI shielding components to improve the AVG amplitude.

How to set the frequency swapping range is described in this passage. The FB pin impedance (R_{total}) is changed from 8.61 kΩ to 40 kΩ. Figure 21a shows that in the neutral end measurement with the ±9% frequency swapping signals, the QP amplitude corresponding to the EN55022 standard satisfies the QP margin of more than 6 dBμV. By comparison, the AVG amplitude corresponding to the EN55022 standard satisfies the QP margin of more than 6 dBμV in the range of 150 kHz–30 MHz. When all other system components were identical, adjusting the frequency swapping signals substantially improved QP and AVG amplitude. Moreover, the EMI shielding components did not require reinforcement. Compared with Figure 20a, the AVG amplitude at approximately 498 kHz was improved by 13.24 dBμV, which was the largest improvement. In addition, the AVG amplitude at approximately 14.9 MHz was improved by 13.18 dBμV. Figure 20b shows that in line end measurement added with ±9% frequency swapping signals, QP amplitude testing value corresponding to the EN55022 standard can satisfy the QP margin of higher than 6 dBμV. Compared with Figure 20b, this result shows that the AVG amplitude was improved by 12.65 and 11 dBμV at approximately 498 and 3867 kHz, respectively.

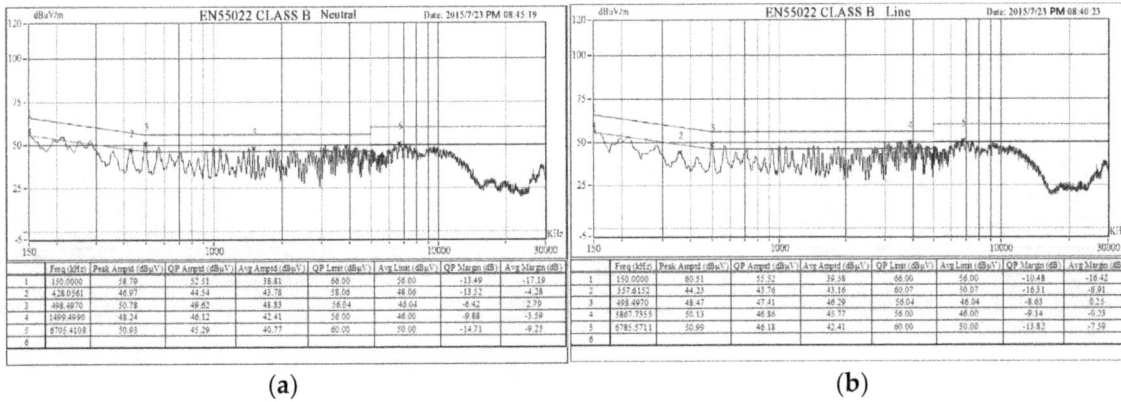

Figure 20. Conducted EMI measurements in the X-capacitor of 0.33μF (fixed frequency): (**a**) neutral; (**b**) line.

Comparison of the spectral amplitudes with and without frequency swapping shows that the harmonic amplitudes of switch voltage Vds were clearly suppressed with frequency swapping technique. The amplitudes from the fundamental wave to the 8th-order sub-harmonic wave decreased. For certain orders of harmonic waves (e.g., 2nd, 3rd, 5th and 7th orders), a maximum difference of 13.24 dB in the average amplitude was generated, showing that the novel frequency swapping technique is effective in suppressing conducted EMI.

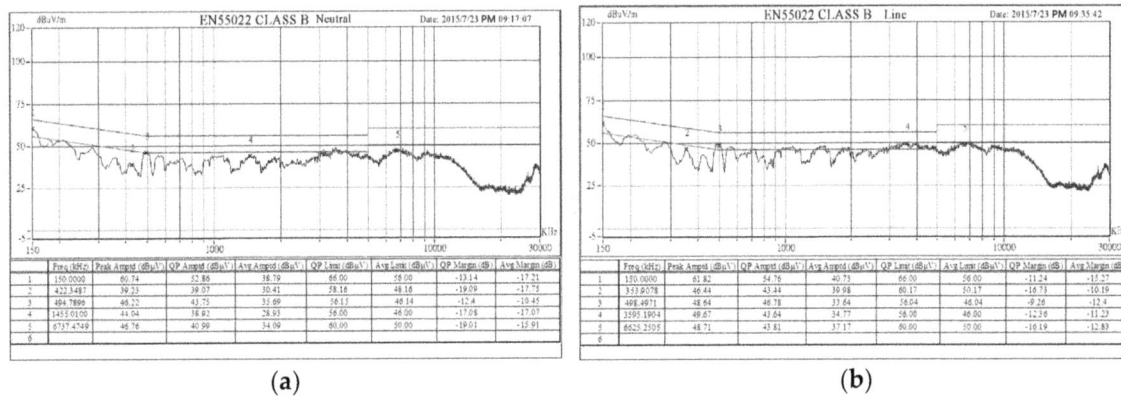

Figure 21. Conducted EMI measurements in the X-capacitor of 0.33 μF (±9% frequency swapping): (a) neutral; (b) line.

Comparison of Conducted Electromagnetic Interference with Distinct X-Capacitor Values

X-capacitors are a common capacitor in EMI filtration. Because X-capacitors are cross-connected to the line and neutral ends, X-capacitors do not have current leakage problems. The primary goal of X-capacitors is to suppress differential mode noise. Because the interference frequency is low, X-capacitors must be augmented to filter interference noise. Subsequent analysis involves identifying whether adding the novel frequency swapping technique on an original system can decrease the X-capacitor capacitance to conserve components and reduce layout area.

Figure 22a,b shows that in the measurements derived at the conduction EMI, the X-capacitor was reduced from 0.33 to 0.22 μF (fixed frequency), which tells when the system is equipped with 0.22 μF X-capacitor, the EMI-inhibition performance of π filter becomes worse. Compared with systems with 0.33 μF X-capacitor, the conduction EMI measurement raises up wholly in the range of 150 kHz to 30 MHz. There is less margins for EN55022 standards. Hence, the proposed frequency swapping is involved in the system with 0.22 μF X-capacitor to prove the improvements.

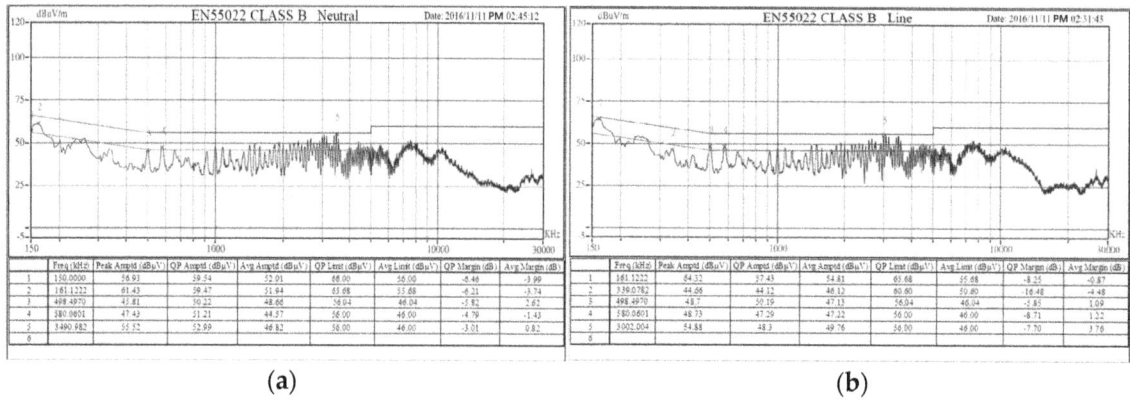

	Freq (kHz)	Peak Ampltd (dBµV)	QP Ampltd (dBµV)	Avg Ampltd (dBµV)	QP Limit (dBµV)	Avg Limit (dBµV)	QP Margin (dB)	Avg Margin (dB)
1	150.0000	56.93	39.54	52.91	66.00	56.00	-6.46	-3.99
2	161.1222	61.43	59.47	51.94	65.68	55.88	-6.21	-3.74
3	498.4970	45.81	30.22	48.66	56.04	46.04	-5.82	2.62
4	580.0601	47.42	31.21	44.57	56.00	46.00	-4.79	-1.43
5	3490.982	55.52	52.99	46.82	56.00	46.00	-3.01	0.82

(a)

	Freq (kHz)	Peak Ampltd (dBµV)	QP Ampltd (dBµV)	Avg Ampltd (dBµV)	QP Limit (dBµV)	Avg Limit (dBµV)	QP Margin (dB)	Avg Margin (dB)
1	161.1222	64.32	37.43	54.81	61.68	55.68	-8.25	-0.87
2	339.0782	44.66	44.12	46.12	60.80	50.80	-16.48	-4.48
3	498.4970	43.7	30.29	47.13	56.04	46.04	-2.85	1.09
4	580.0601	48.73	47.29	47.22	56.00	46.00	-8.71	1.22
5	3002.004	54.88	48.3	49.76	56.00	46.00	-7.70	3.76

(b)

Figure 22. Conducted EMI measurements in the X-capacitor of 0.22 µF (fixed frequency): (**a**) neutral; (**b**) line.

Figure 23a shows that in the measurements derived at the neutral end, the X-capacitor was reduced from 0.33 to 0.22 µF (±9% frequency swapping), and the QP amplitude corresponding to the EN55022 standard can satisfy the QP margin of more than 6 dBµV. However, the filtration frequency declined 0.5-fold. At 150 kHz, the QP amplitude was decreased by 1.93 dBµV, indicating that decreasing the X-capacitor affects the QP amplitude. In addition, the AVG amplitude corresponding to the EN55022 standard can satisfy the QP margin of more than 6 dBµV in the range of 150 kHz–30 MHz.

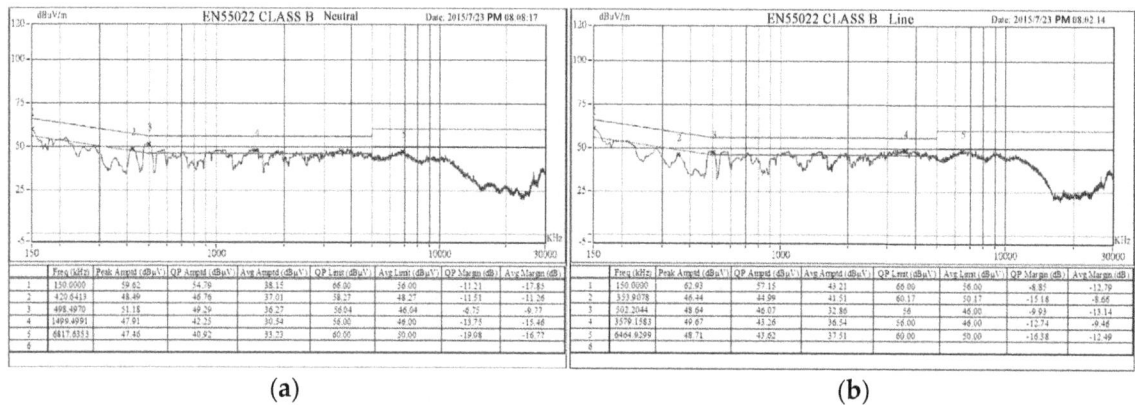

	Freq (kHz)	Peak Ampltd (dBµV)	QP Ampltd (dBµV)	Avg Ampltd (dBµV)	QP Limit (dBµV)	Avg Limit (dBµV)	QP Margin (dB)	Avg Margin (dB)
1	150.0000	59.62	54.79	38.15	66.00	56.00	-11.21	-17.85
2	420.6413	48.49	46.76	37.01	58.27	48.27	-11.51	-11.26
3	498.4970	51.18	49.29	36.27	56.04	46.04	-6.75	-9.77
4	1498.4991	47.91	42.25	30.54	56.00	46.00	-13.75	-15.46
5	6817.8353	47.46	40.92	33.23	60.00	50.00	-19.08	-16.77

(a)

	Freq (kHz)	Peak Ampltd (dBµV)	QP Ampltd (dBµV)	Avg Ampltd (dBµV)	QP Limit (dBµV)	Avg Limit (dBµV)	QP Margin (dB)	Avg Margin (dB)
1	150.0000	62.93	57.15	43.21	66.00	56.00	-8.85	-12.79
2	333.9078	46.44	44.99	41.51	60.17	50.17	-15.18	-8.66
3	502.2044	43.64	46.07	32.86	56	46.00	-9.93	-13.14
4	3579.1563	49.67	43.26	36.54	56.00	46.00	-12.74	-9.46
5	6464.9299	48.71	43.62	37.51	60.00	50.00	-16.38	-12.49

(b)

Figure 23. Conducted EMI measurements in the X-capacitor of 0.22µF (±9% frequency swapping): (**a**) neutral; (**b**) line.

The lowest margin is 9.77 dBµV at 498 kHz, which still conforms to EN55022 standards. Figure 23b reveals that in the line end measurement, the X-capacitor is decreased from 0.33 to 0.22 µF, and the QP amplitude corresponding to the EN55022 standard can satisfy the QP margin of more than 6 dBµV. Similarly, the filtration frequency is decreased by 0.5-fold. At 150 kHz, the QP amplitude is decreased by 2.39 dBµV. The testing results show that decreasing the X-capacitor affects the QP amplitude. Moreover, the average amplitude corresponding to the EN55022 standard can satisfy the QP margin of more than 6 dBµV in the range of 150 kHz to 30 MHz. The lowest margin is 8.66 dBµV at 353 kHz, which still conforms to EN55022 standards. Thus, the X-capacitor of the system π filter can be decreased to 0.22 µF.

In response to the LC filter order, the first-order LC filter owns the capability of 40 dB decay. EMI refers to the frequency to attenuate a maximum conducted common mode noise measurement. In 12 V/24 W applications, adding the novel frequency swapping technique can reduce the X-capacitor to 0.22 µF. Experimental results show Table 4 that the conducted EMI at both the neutral and line ends can pass the EN55022 standard; in addition, both the QP and AVG amplitude margins are more than 6 dBμV.

Table 4. Comparison of conducted EMI measurements. QP: Quasi-peak.

Items		Frequency	QP Limit	Average Limit	QP Margin	Average Margin
		kHz	dBμV	dBμV	dB	dB
X-capacitor of 0.33 μF (fixed frequency)	Neutral	428.0561	58.06	48.06	−13.52	−4.28
		498.4970	56.04	46.04	−6.42	2.79
	Line	357.6152	60.07	50.07	−16.31	−6.91
		498.497	56.04	46.04	−8.63	0.25
X-capacitor of 0.33 μF (±9% frequency swapping)	Neutral	422.3487	58.16	48.16	−19.09	−17.75
		494.7896	56.15	46.14	−12.4	−10.45
	Line	353.9078	60.17	50.17	−16.73	−10.19
		498.4971	56.04	46.04	−9.26	−12.4
X-capacitor of 0.22 μF (fixed frequency)	Neutral	498.4970	45.81	50.22	−5.82	2.62
		580.0601	47.43	51.21	−4.79	−1.43
	Line	339.0782	44.66	44.12	−16.48	−4.48
		498.4970	48.7	50.19	−5.85	1.09
X-capacitor of 0.22 μF (±9% frequency swapping)	Neutral	420.6413	58.27	48.27	−11.51	−11.26
		498.497	56.04	46.04	−6.75	−9.77
	Line	353.9078	60.17	50.17	−15.18	−8.66
		502.2044	56	46	−9.93	−13.14

6. Conclusions

This paper explored the effect of frequency swapping techniques on suppressing EMI in a converter. FM PWM wave spectral characteristics were analyzed in detail. The spectra of the modulated waveform were calculated. A novel adjustable QRF converter frequency swapping method was proposed. Through analog modulation circuits and ICs, a frequency swapping function was implemented. In addition, testing of the QRF converter was conducted to verify the proposed technique. The proposed novel frequency swapping technique was verified experimentally. EMI testing revealed that the average (150 kHz–10 MHz) margin was attenuated by approximately 13.24 dBμV, effectively reducing the EMI caused by the converter. The conclusions of this paper are summarized as follows:

(1) The novel frequency swapping technique can effectively reduce the EMI caused by the converters. EMI testing showed that the average (150 kHz to 10 MHz) margin was attenuated by approximately 13.24 dBμV.

(2) The spectral distribution of the modulation wave is related to the slope of the modulation ratio. A well-performed spectral distribution is obtained by setting the swapping range to ±9%. When the spectrum of the modulation wave is non-modulated, there is no swapping frequency mechanism and vice versa. This pattern could be used to predict the spectral structures for the modulation waves of other shapes.

(3) When close-loop oscillation and oscillation noise factors are considered, the switching frequency offset and modulation frequency should not be excessive. In the selection of the modulation parameters, the frequency offset must be determined first. The offset adopted in this experiment was approximately 9%, which caused zero abnormality in the original circuit.

(4) Applying the novel frequency swapping technique can reduce the X-capacitor in the system π filter to 0.22 μF. The EN55022 standard was met regardless of testing the conducted EMI at the neutral or line ends. Both the QP and AVG amplitude margins were more than 6 dBμV. Thus, the proposed technique has the advantages of conserving the system components and reducing the layout area.

Acknowledgments: This work was supported by the National Science Council of Taiwan under Contract Ministry of Science and Technology (MOST) 104-2221-E-011-077. The authors would like to thank Leadtrend Technology Corp. for the support on chip fabrication, measurement, and verifications.

Author Contributions: Both of the authors played important roles during the design of the novel frequency swapping technique and partially wrote the paper. Ming-Chang Tsou ran the simulation and did the experiment and Ming-Tse Kuo reviewed and revised the manuscript.

Conflicts of Interest: The authors declare no conflict of interest.

References

1. Wang, Y.-F.; Yang, L.; Wang, C.-S.; Li, W.; Qie, W.; Tu, S.-J. High step-up 3-phase rectifier with fly-back cells and switched capacitors for small-scaled wind generation systems. *Energies* **2015**, *8*, 2742–2768. [CrossRef]

2. Arango, E.; Ramos-Paja, C.A.; Calvente, J.; Giral, R.; Serna, S. Asymmetrical interleaved DC/DC switching converters for photovoltaic and fuel cell applications—Part 1: Circuit generation, analysis and design. *Energies* **2012**, *5*, 4590–4623. [CrossRef]

3. Karveis, G.A.; Manias, S.N. Analysis and design of a flyback zero-current switched (ZCS) quasi-resonant (QR) AC/DC converter. In Proceedings of the 27th Annual IEEE Power Electronics Specialists Conference, Baveno, Italy, 23–27 June 1996.

4. Hsieh, P.C.; Chang, C.J.; Chen, C.L. A primary-side-control quasi-resonant flyback converter with tight output voltage regulation and self-calibrated valley switching. In Proceedings of the 2013 IEEE Energy Conversion Congress and Exposition, Denver, CO, USA, 15–19 September 2013.

5. Wang, Z.; Wu, X.; Chen, M.; Zhang, J. Optimal design methodology for the current-sharing transformer in a quasi-resonant (QR) flyback LED driver. In Proceedings of the Twenty-Seventh Annual IEEE Applied Power Electronics Conference and Exposition, Orlando, FL, USA, 5–9 February 2012.

6. Kuo, M.-T.; Tsou, M.-C. Simulation of standby efficiency improvement for a line level control resonant converter based on solar power systems. *Energies* **2015**, *8*, 338–355. [CrossRef]

7. EN 55022, CISPR 22. Available online: http://www.rfemcdevelopment.eu/en/en-55022-2010 (accessed on 22 December 2016).

8. González, D.; Gago, J.; Balcells, J. New simplified method for the simulation of conducted EMI generated by switched power converters. *IEEE Trans. Ind. Electron.* **2003**, *50*, 1078–1084. [CrossRef]

9. Santolaria, A.; Balcells, J.; González, D. Theoretical and experimental results of power converter frequency modulation. In Proceedings of the IECON 02 IEEE 2002 28th Annual Conference of the Industrial Electronics Society, Seville, Spain, 5–8 November 2002.

10. Lin, F.; Chen, D.Y. Reduction of power supply EMI emission by switching frequency modulation. *IEEE Trans. Power Electron.* **1994**, *9*, 132–137.

11. Paul, C.R. *Transmission Lines in Digital Systems for EMC Practitioners*; Wiley: Hoboken, NJ, USA, 2012.

12. Zhang, D.; Chen, D.; Sable, D. Non-intrinsic differential mode noise caused by ground current in an off-line power supply. In Proceedings of the 29th Annual IEEE Power Electronics Specialists Conference, PESC 98 Record, Fukuoka, Japan, 22–22 May 1998; pp. 1131–1133.

13. Kchikach, M.; Yuan, Y.S.; Qian, Z.M.; Pong, M.H. Simple modeling for conducted common-mode current in switching circuits. In Proceedings of the PESC Record—IEEE 32nd Annual Power Electronics Specialists Conference, Vancouver, BC, Canada, 17–22 June 2001.

14. Kuisma, M. Variable frequency switching in power supply EMI-control: An overview. *IEEE Aerosp. Electron. Syst. Mag.* **2003**, *18*, 18–22. [CrossRef]

15. Maillet, Y.; Lai, R.; Wang, S.; Wang, F.; Burgos, R.; Boroyevich, D. High-density EMI filter design for dc-fed motor drives. *IEEE Trans. Power Electron.* **2010**, *25*, 1163–1172. [CrossRef]

16. Ala, G.; Giaconia, G.C.; Giglia, G.; Di Piazza, M.C.; Luna, M.; Vitale, G.; Zanchetta, P. Computer aided optimal design of high power density EMI filters. In Proceedings of the 2016 IEEE 16th International Conference on Environment and Electrical Engineering, Florence, Italy, 6–8 June 2016.

17. Ala, G.; Di Piazza, M.C.; Giaconia, G.C.; Giglia, G.; Vitale, G. Design and performance evaluation of a high power-density emi filter for PWM inverter-fed induction-motor drives. *IEEE Trans. Ind. Appl.* **2016**, *52*, 2397–2404. [CrossRef]

18. Touré, B.; Schanen, J.L.; Gerbaud, L.; Meynard, T.; Roudet, J.; Ruelland, R. EMC modeling of drives for aircraft applications: Modeling process, EMI filter optimization, and technological choice. *IEEE Trans. Power Electron.* **2013**, *28*, 1145–1156. [CrossRef]

19. Hardin, K.B.; Fessler, J.T.; Bush, D.R. A study of the interference potential of spread spectrum clock generation techniques. In Proceedings of the IEEE International Symposium on Electromagnetic Compatibility, Atlanta, GA, USA, 14–18 August 1995.

20. TOP242-TOP250 Reference Design. Available online: http://chinese.powerint.com/dak.zh-cn.html (accessed on 8 March 2016).

21. LTC6902 Datasheet. Available online: http://datasheet.eeworld.com.cn/pdf/28512_LINER_LTC6902.html (accessed on 9 August 2016).

22. NCP1215A Datasheet. Available online: http://datasheet.eeworld.com.cn/pdf/12824_ONSEMI_NCP1215.html (accessed on 9 August 2016).

23. ICE3B0365J Datasheet. Available online: http://datasheet.eeworld.com.cn/pdf/151232_INFINEON_ICE3B0365J.html (accessed on 9 August 2016).

24. FSDH0265 Series Datasheet. Available online: http://datasheet.eeworld.com.cn/pdf/37029_FAIRCHILD_FSDH0265RL.html (accessed on 9 August 2016).

25. Sendra, J.B.I. SSCG Methods of EMI Emission Reduction applied to Switching Power Converters. Ph.D. Thesis, BarcelonaTech, Barcelona, Spain, 2004.

3

Hybrid Off-Grid SPV/WTG Power System for Remote Cellular Base Stations Towards Green and Sustainable Cellular Networks in South Korea

Mohammed H. Alsharif * and Jeong Kim

Department of Electrical Engineering, College of Electronics and Information Engineering, Sejong University, 209 Neungdong-ro, Gwangjin-gu, Seoul 05006, Korea; kimjeong@sejong.ac.kr
* Correspondence: malsharif@sejong.ac.kr

Academic Editor: Adrian Ilinca

Abstract: This paper aims to address the sustainability of power resources and environmental conditions for telecommunication base stations (BSs) at off-grid sites. Accordingly, this study examined the feasibility of using a hybrid solar photovoltaic (SPV)/wind turbine generator (WTG) system to feed the remote Long Term Evolution-macro base stations at off-grid sites of South Korea the energy necessary to minimise both the operational expenditure and greenhouse gas emissions. Three key aspects have been discussed: (i) optimal system architecture; (ii) energy yield analysis; and (iii) economic analysis. In addition, this study compares the feasibility of using a hybrid SPV/WTG system vs. a diesel generator. The simulation results show that by applying the proposed SPV/WTG system scheme to the cellular system, the total operational expenditure can be up to 48.52% more efficient and sustainability can be ensured with better planning by providing cleaner energy.

Keywords: hybrid energy system; remote sites; cellular networks; operational expenditure; South Korea

---

1. Introduction

Mobile communication is one of the most successful technological innovations in modern history. However, the Long Term Evolution (LTE, also known as pre-4G, i.e., 3.9G) of a cellular network is considered to be the fastest developing technology in mobile communication [1]. In addition, South Korea has witnessed tremendous development in the LTE cellular network in the last five years [2]. Today, South Korea has an LTE cellular network unrivalled in both technology and reliability, with the best global coverage (97% of the time on LTE) [3]. The LTE network offers data-oriented services that include, but are not limited to, multimedia communications, online gaming, and high-quality video streaming. As a result, the number of mobile subscribers has increased exponentially, prompting the mobile operators to significantly increase the number of LTE BSs to meet the needs of mobile subscribers. In 2013, the number of LTE BSs was 35,255 units; that number increased by 4.7-fold in 2015 to 165,193 LTE BSs [2]. This increase of LTE BSs led to increases in both the energy consumption and operational expenditure (OPEX) as a result of the LTE BS being the primary source of energy consumption in cellular networks, accounting for 57% of the total energy used [4,5]. Moreover, cellular network operators are continually seeking to increase their network coverage areas, provide services to potential customers in rural areas, and open new markets to increase profitability. Diesel generators (DGs) are typically used for power at off-grid BS sites because extending the grid connection to power off-grid BS sites is not economically attractive or existing grid electricity is not available because of geographical limitations [6]. Today, the idea of using diesel generators as a primary power supply has become less favourable because of the challenges linked to their reliability, availability, high operational and maintenance (O&M) costs, and significant environmental impacts [7].

It is now acknowledged that the LTE cellular communication network in South Korea will have greater economic and ecological impact in the coming years. The key features for power sources, such as economic, environmental, and social sustainability, of BS sites are a critically important issue. Power shortages are not allowed in the cellular mobile sector or during service outages. The specific power supply requirements for off-grid BSs, such as cost-effectiveness, efficiency, sustainability and reliability, can be met by utilising the technological advances in renewable energy [6]. The Korean government continually seeks to increase energy independence by expanding renewable energy to enable more efficient development and ensure sustainability with better planning to provide cleaner energy [8].

This study aims to give a holistic view of the sustainable power supply solution for an off-grid LTE-macro BS based on the characteristics of South Korea average solar radiation exposure and wind speed. The key contributions in this paper are summarised as follows:

1. To determine the optimal size and technical criteria of the hybrid SPV/WTG system to feed LTE-macro BS deployment at off-grid sites of South Korea. The optimum criteria, including economic, technical, and environmental feasibility parameters, were analysed by the *Hybrid Optimisation Model for Electric Renewable* (HOMER).

2. To analyse and evaluate the feasibility of using a hybrid SPV/WTG system in terms of the energy yield and economic feasibility over the project lifetime.

3. To analyse and compare the implications of choosing a hybrid SPV/WTG system with respect to a classical DG powered solution in terms of the (i) OPEX savings to maintain profitability for cellular network operators and (ii) GHG emissions that have a bad effect on the environment.

The rest of this paper is organised as follows: Section 2 presents the related work. In Section 3, the system architecture for the hybrid SPV/WTG system to supply LTE-macro BS is described, and Section 4 presents the mathematical model. Section 5 includes the system implementation and configuration, and Section 6 presents the optimization results and discussion. Section 7 presents the comparison and estimation of the feasibility of using the SPV/WTG system approach with the classical solution DG, and Section 8 concludes this paper.

2. Power Supply and Energy Storage Solutions for Off-Grid Base Stations

Following the emerging concept of green telecommunication networks, the realization of powering BS sites using sustainable solutions has started to receive significant attention. Therefore, various studies and developments have been done to help telecom operators shift away from using diesel generators as their primary power supply solution for BSs. It is being realized that by moving away from diesel generators, the unreliability factors and the high O&M costs usually associated with this solution can be avoided. This section summarises the renewable energy solutions in the telecommunication sector and highlights the various power supply and energy storage solutions for off-grid BSs that have been proposed. Figure 1 provides a summary of related works that have investigated green wireless network optimisation strategies within smart grid environments. In addition, Derrick et al. [9] have studied resource allocation algorithm design for energy-efficient communication in an orthogonal frequency-division multiple access (OFDMA) downlink network with hybrid energy harvesting BS. Reference [10] has discussed a point-to-point communication link where the transmitter has a hybrid supply of energy. Specifically, the hybrid energy is supplied by a constant energy source and an energy harvester, which harvests energy from its surrounding environment and stores it in a battery which suffers from energy leakage.

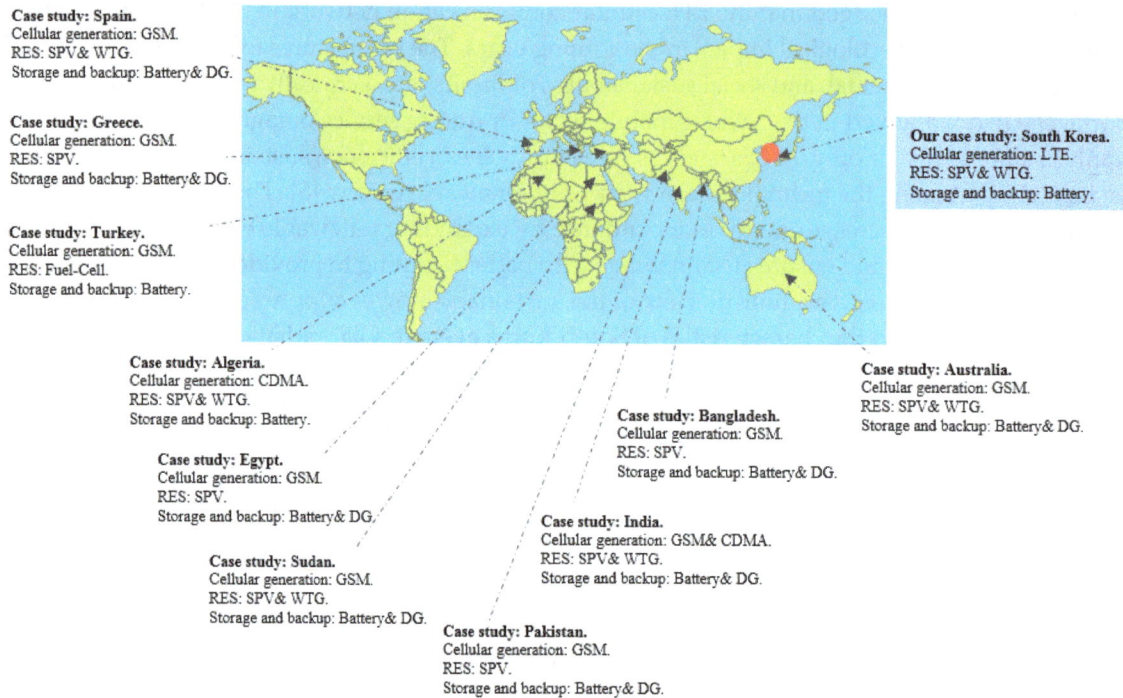

Figure 1. Summary of related works of green wireless network optimisation strategies within smart grid environments [11–19].

In addition, in a study conducted by the GSMA, which is a mobile trade organization, 320,100 renewable-based off-grid BS sites have already been rolled out in 2014. This number is expected to increase further in 2020 with about 389,800 sites [20].

3. System Architecture

Figure 2 is a schematic showing two subsystems: the LTE-macro BS and the hybrid SPV/WTG power subsystem.

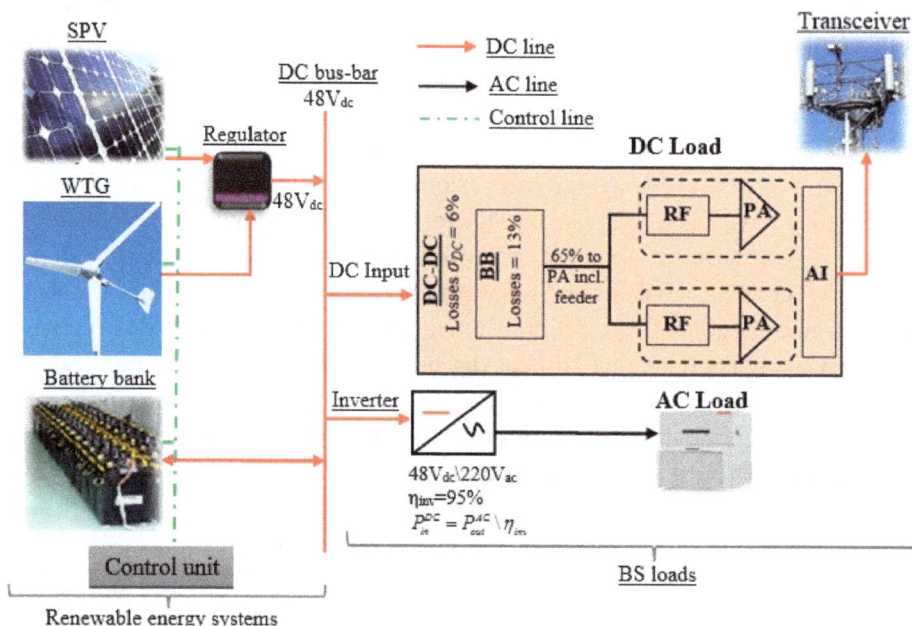

Figure 2. System model of the hybrid SPV/WTG power scheme for an LTE-macro BS.

3.1. LTE Macro-BS Subsystem and Power Consumption Modeling

The BS, a centrally located set of equipment used to communicate with mobile units and the backhaul network, consists of multiple transceivers that in turn consist of a power amplifier that amplifies the input power, a radio-frequency small-signal transceiver section, a baseband for system processing and coding, a DC-DC power supply, and a cooling system. More details on the BS components can be found in [21].

An LTE-macro BS type is described in [22] with a focus on the component level for three sectors; each BS contains two antennas. BS operating power is expressed as $P_{op} = N_{TRX} \times (P_{PA}^{DC} + P_{RF}^{DC} + P_{BB}^{DC})/(1 - \sigma_{DC})(1 - \sigma_{cool})$, where N_{TRX} is the number of transceivers (i.e., transmit/receive antennas per site); P_{PA}^{DC}, P_{RF}^{DC}, and P_{BB}^{DC} are the power amplifier (PA), radio frequency (RF), and baseband power (BB), respectively. Losses incurred by the DC-DC regulator and active cooling (air conditioner) scale linearly with the power consumption of the other components and may be approximated by loss factors $\sigma_{DC} = 6\%$, and $\sigma_{cool} = 10\%$. The power consumption of the air-conditioning unit depends on the internal and ambient temperature of the BS cabinet. Typically, an internal and ambient temperature of 25 °C is assumed, resulting in constant power consumption for the air conditioning. P_{PA}^{DC} is a linear function of the BS transmission power $P_{tx}^{\max}$ and is expressed as $P_{tx}^{\max}/\eta_{PA}$, where η_{PA} is the PA efficiency. In general, the BS transmission power depends on the radius of coverage and the signal propagation fading. To simplify the model derivation, the macro BS transmission power is normalized as $P_o = 40$ W with a coverage radius of $R_o = 1$ km. Similarly, the BS transmission power with coverage radius R is denoted by $P_{tx}^{\max} = P_o \times (R/R_o)^{\alpha}$, where α is the path loss coefficient. Therefore, the BS operating power with coverage radius R is expressed as $P_{op} = N_{TRX}[P_o \cdot (R/R_o)^{\alpha}/\eta_{PA} + P_{RF}^{DC} + P_{BB}^{DC}]/(1 - \sigma_{DC})(1 - \sigma_{cool})$. In addition, most sites use a microwave backhaul (P_{mc}), and auxiliary equipment located at the BS site include lighting (P_{lm}). Then, a mathematical expression for the power consumption of a BS site is given by Equation (1):

$$P_{BS} = \frac{N_{TRX}\left(\frac{P_o\left(\frac{R}{R_0}\right)^{\alpha}}{\eta_{PA}} + P_{RF}^{DC} + P_{BB}^{DC}\right)}{(1 - \sigma_{DC})(1 - \sigma_{cool})} + P_{mc} + P_{lm} \qquad (1)$$

Table 1 summarises the power consumption of the different pieces of LTE-macro BS equipment for a 2×2 multiple-input and multiple-output (MIMO) antenna configuration with three sectors.

Table 1. Power consumption of the LTE-macro BS hardware elements [21].

Item	Notation	Unit	LTE Macro-BS
PA	Max transmit (*rms*) power, $P_{\max}$	W	39.8
	Max transmit (*rms*) power	dBm	46.0
	Peak average power ratio (PAPR)	dB	8.0
	Peak output power	dBm	54.0
	PA efficiency, μ	%	38.8
	Total PA $P_{PA}^{DC} = \frac{P_{\max}}{\mu}$	W	102.6
TRX	P_{TX}	W	5.7
	P_{RX}	W	5.2
	Total RF P_{RF}^{DC}	W	10.9
BB	Radio (inner *Rx/Tx*)	W	5.4
	Turbo code (outer *Rx/Tx*)	W	4.4
	Processor	W	5.0
	Total BB P_{RF}^{DC}	W	14.8
	DC-DC loss, σ_{DC}	%	6.0
	Cooling loss, σ_{cool}	%	10.0
Total per TRX $= \frac{P_{PA}^{DC} + P_{RF}^{DC} + P_{BB}^{DC}}{(1 - \sigma_{DC})(1 - \sigma_{cool})}$		W	151.65

Table 1. *Cont.*

Item	Notation	Unit	LTE Macro-BS
Number of sectors (N_{Sect})		#	3
Number of antennas (N_{Ant})		#	2
Number of carriers (N_{Carr})		#	1
Number of transceivers ($N_{TRX} = N_{Sect} \times N_{Ant} \times N_{Carr}$)		#	6
Total number of N_{TRX} chains, $P_{op} = N_{TRX} \times$ Total per TRX		W	909.93
Microwave link (P_{mc})		W	80
Lamps P_{lm}^{AC}		W	40

In addition, Figure 3 shows the hourly load profiles. The alternating current (AC) load includes a 91 W air conditioner that represents 10% of P_{op} and 40 W lighting that operates from 7 PM to 7 AM. The direct current (DC) load includes a BS (P_{op} minus air conditioner equals 819 W and microwave backhaul equals 80 W).

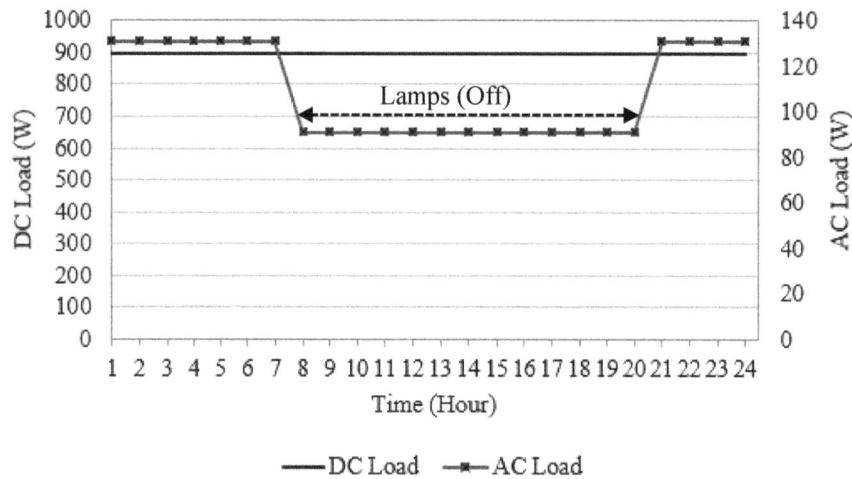

Figure 3. Hourly load profile for the LTE macro-BS.

3.2. SPV/WTG Power Subsystem

The solar power supply system consists of various elements (as listed below) that all contribute to energy savings, and have to be designed in a way that allows easy disassembling and component separation for recycling.

a.   Solar panels: responsible for absorbing shortwave irradiance and converting light into direct current (DC) electricity [23].

b.   WTG: responsible for converting wind energy to a regulated power which can be connected to the DC-power bus. Generally, for a small power load, vertical style windmills show some special benefit [23].

c.   Regulator charger: the highest power demand in a typical BS is based on 48 V_{dc} voltage. Therefore, it is beneficial to use DC/DC solar regulator converters that can directly convert the unregulated DC output voltage and current from a solar panel to a regulated output voltage for the BS equipment to protect the battery bank.

d.   Battery bank: stores excess electricity for future consumption by the BS at night, during load-shedding hours or if the available solar energy is not sufficient to feed the BS load completely. To protect the battery, inclusion of a charge controller is recommended. A charge controller or battery regulator limits the rate at which the electric current is added to or drawn from electric batteries, preventing overcharging and potentially protecting against overvoltage that can reduce battery performance or lifespan and may pose a safety risk. A charge controller may also prevent complete battery draining ("deep discharging") or perform controlled discharges, depending on the battery technology, to protect battery life [23].

e.　Inverter: An inverter is a device that changes a low DC-voltage into usable 220 V AC voltage. It is one of the system's main elements. Inverters differ by the output wave format, output power and installation type. It is also called a power conditioner because it changes the form of the electric power. There are two types of output wave format: modified sine-wave (MSW) and pure sine-wave. The MSW inverters are economical and efficient; the sine wave inverters are usually more sophisticated with high-end performance that can operate virtually any type of load [23].

f.　Control system: serves as the brains of a complex control, regulation, and communication system. The most common communication units in the remote interface are wireless modems or network solutions. In addition to the control functions, the data logger and alarm memory capabilities are of high importance. All power sources working in parallel are managed by a sophisticated control system and share the load with their capabilities to accommodate the fact that power shortages are not admissible in the cellular telephony sector.

4. Mathematical Model

This section addresses the details of the mathematical model of a hybrid SPV/WTG system proposed to feed the LTE macro-BS.

4.1. Photovoltaic System

The SPV generator contains modules that are composed of many solar cells interconnected in series/parallel to form a solar array. HOMER calculates the energy output of the SPV array (E_{PV}) by using the following equation [24],

$$E_{PV} = Y_{PV} \times PSH \times f_{PV} \tag{2}$$

where Y_{PV} is the rated capacity of the SPV array (kW), and PSH is a peak solar hour which is used to express solar irradiation in a particular location when the sun is shining at its maximum value for a certain number of hours. Because the peak solar radiation is 1 kW/m^2, the number of peak sun hours is numerically equal to the daily solar radiation in kWh/m^2 [25] and f_{PV} is the SPV derating factor (sometimes called the performance ratio), a scaling factor meant to account for effects of dust on the panel, wire losses, elevated temperature, or anything else that would cause the output of the SPV array to deviate from the expected output under ideal conditions. In other words, the derating factor refers to the relationship between actual yield and target yield, which is called the efficiency of the SPV. Today, due to improved manufacturing techniques, the performance ratio of solar cells increased to 85%–95%.

4.2. Wind Conversion System

This system produces energy by converting the flowing wind speed into mechanical energy and then into electricity. The power contained in the wind kinetic energy is expressed by [24]:

$$P = \frac{1}{2}\rho V^3 C_p \tag{3}$$

where V is the monthly wind speed (m/s), C_p is the coefficient of the Betz limit, which can achieve a maximum value of 59% for all types of wind turbines, and ρ is the corrected monthly air density (kg/m^3). HOMER assumes that the power curve applies a standard air density of 1.225 kg/m^3, which corresponds to standard temperature and pressure conditions.

4.3. Battery Model

The battery characteristics that play a significant role in designing a hybrid renewable energy system are battery capacity, battery voltage, battery state of charge, depth of discharge, days of autonomy, efficiency, and lifetime of battery.

The nominal capacity of the battery bank is the maximum state of charge SOC_{max} of the battery. The minimum state of charge of the battery, SOC_{min}, is the lower limit that does not discharge below the minimum state of charge. The DOD is used to describe how deeply the battery is discharged and is expressed in Equation (4) [24]:

$$DOD = 1 - SOC_{min} \tag{4}$$

Based on Equation (4), the DOD for the *"Trojan L16P"* battery is 70%, which means that the battery has delivered 70% of its energy and has 30% of its energy reserved. DOD can always be treated as how much energy the battery delivered. A battery bank is used as a backup system and is sized to meet the load demand when the renewable energy resources failed to satisfy the load. The number of days a fully charged battery can feed the load without any contribution of auxiliary power sources is represented by days of autonomy. The battery bank autonomy is the ratio of the battery bank size to the electric load (LTE-macro BS). HOMER calculates the battery bank autonomy (A_{batt}) by using the following equation [24]:

$$A_{batt} = \frac{N_{batt} \times V_{nom} \times Q_{nom}\left(1 - \frac{SOC_{min}}{100}\right)(24\,\text{h/d})}{L_{prim-avg}(1000\,\text{Wh/kWh})} \tag{5}$$

where N_{batt} is the number of batteries in the battery bank, V_{nom} is the nominal voltage of a single battery (V), Q_{nom} is the nominal capacity of a single battery (Ah), and $L_{prim,ave}$ is the average daily LTE-macro BS load (kWh). Battery life is an important factor that has a direct impact on replacement costs. Two independent factors may limit the lifetime of the battery bank: the lifetime throughput and the battery float life. HOMER calculates the battery bank life (R_{batt}) based on these two factors as given in the following equation [24]:

$$R_{batt} = \min\left(\frac{N_{batt} \times Q_{lifetime}}{Q_{thrpt}}, R_{batt,f}\right) \tag{6}$$

where $Q_{lifetime}$ is the lifetime throughput of a single battery (kWh), Q_{thrpt} is the annual battery throughput (kWh/year), and $R_{batt,f}$ is battery float life (year).

4.4. Economic Mathematical Model

HOMER calculates the net present cost (NPC) according to the following equation [24]:

$$NPC = \frac{TAC}{CRF} \tag{7}$$

where TAC is the total annualised cost ($). The capital recovery factor (CRF) is given by [24],

$$CRF = \frac{i(1+i)^n}{(1+i)^n - 1} \tag{8}$$

where n is the project life time, and i is the annual real interest rate. HOMER assumes that all prices escalate at the same rate and applies an annual real interest rate rather than a nominal interest rate.

The discount factor (f_d) is a ratio used to calculate the present value of a cash flow that occurs in any year of the project lifetime. HOMER calculates the discount factor by using the following equation [24]:

$$f_d = \frac{1}{(1+i)^n} \tag{9}$$

NPC estimation in HOMER also considers the salvage cost, which is the residual value of the power system components at the end of the project lifetime. The equation used to calculate the salvage value (*S*) [24] is:

$$S = rep\left(\frac{rem}{comp}\right) \tag{10}$$

where *rep* is the replacement cost of the component, *rem* is the remaining lifetime of the component, and *comp* is the lifetime of the component.

5. System Implementation and Configuration

The implementation of a hybrid SPV/WTG system within the HOMER software and the configurations for the various elements in the system are given in Figure 4.

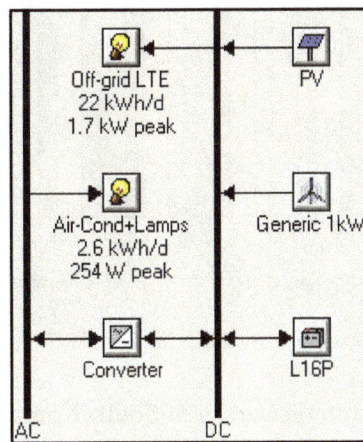

Figure 4. Micro-power system modelling in HOMER.

HOMER model the operation of a system by generating energy balance calculations for each of the 8760 h in a year. For every hour, HOMER compares the electric demand of BS in one hour to the energy that the system can supply in that hour, calculates the energy flows to and from each component of the system, and determines whether to charge or discharge the batteries. The model then decides whether or not a configuration is feasible, i.e., whether it can meet the electric demand under the specified conditions, and estimates the cost of installing and operating the system throughout the lifetime of the project. Figure 5 summarizes the main parts of HOMER simulation, inputs, optimization, and outputs; to finding the best optimized system with low NPC. To perform simulation and optimisation of a power system using the HOMER tool, information and data on natural resouces (wind speed and solar irradiance data), LTE BS load, economic constraints (real interest rate), equipment costs and lifetime, project lifecycle are required, which will provide in the following subsections.

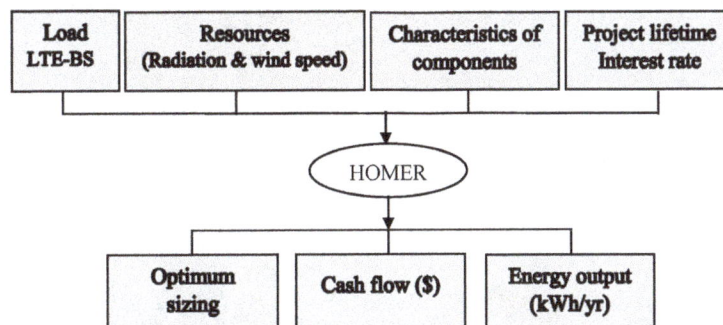

Figure 5. Architecture of HOMER software.

5.1. Solar Radiation

The monthly average solar radiation values used in this study are obtained based on both the Korea Meteorological Administration (KMA) and the NASA Surface Meteorology and Solar Energy using the longitude and latitude of South Korea [26,27].

The average daily solar radiation in South Korea, which is located at a latitude between 34° and 38° north, is estimated to be 4.01 kWh/m^2 and varies from 2.474 kWh/m^2 in December to 5.622 kWh/m^2 in May [26,27]. The monthly variation, as shown in Figure 6, is largely due to the shift in the elevation angle of the sun.

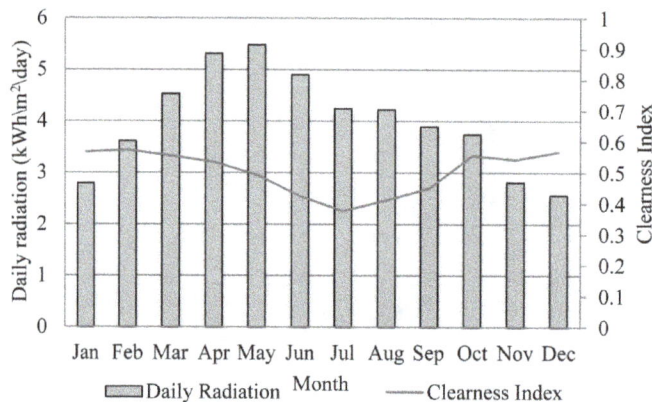

Figure 6. Monthly average solar global horizontal irradiance at 35.5° latitude and 127.5° longitude [26,27].

Moreover, solar radiation in different areas of South Korea is shown in Figure 7. Relatively higher solar radiation of over 5 kWh/m^2/day can be obtained in the southwestern coastal area including Jeju island. In contrast, in the northwestern region around Seoul, solar radiation is lowered to approximately 4.7 kWh/m^2/day, and Gochang, located at the western coast of South Korea, shows the lowest solar radiation of 4.48 kWh/m^2/day. Accordingly, this study will investigate different values of average daily solar radiation values to cover all of the areas of South Korea: 4.0, 4.5, 5.0, and 5.5 kWh/m^2.

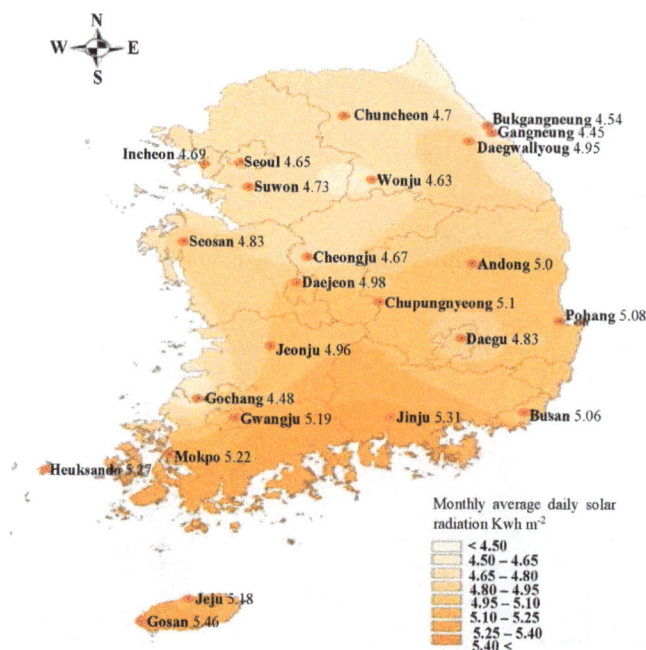

Figure 7. Monthly average daily solar radiation [28].

5.2. Wind Speed

The wind energy resource data used in this paper was mainly obtained from the Weather Resource Maps of the National Institute of Meteorological Science (NIMS). Figure 8 shows the average wind speed at 50 m above the surface of the earth in South Korea [29]. The average wind speed in the most of the interior of South Korea does not exceed 4 m/s. However, the wind speed above 7.5 m/s can be observed in the mountainous regions nearby east coast, the southeastern coast, and Jeju Island which is located at the below of the peninsula. In particular, almost every region in Jeju Island shows the wind speed above 5.5 m/s, which is the main reason why a lot of wind turbines have been installed in Jeju Island.

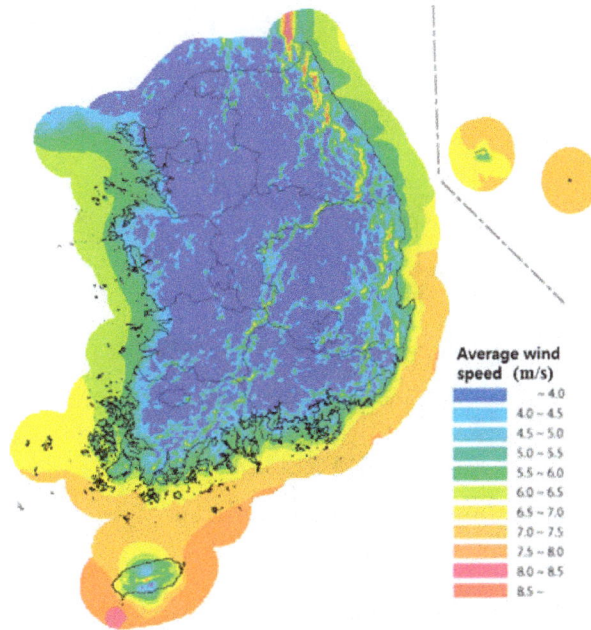

Figure 8. Average wind speed in South Korea [29].

The monthly average wind speed for South Korea is shown in Figure 9 [26,27]. In summer the wind speed considerably decreases compared with the wind speed in winter. This seasonal variation of the wind speed is deeply related with the atmospheric pressure around South Korea and finally the direction of the wind. In summer the North Pacific atmospheric pressure is to the south of South Korea and affects to every region in South Korea. This causes the direction of the wind in summer is largely southeast or southwest. On the contrary, the northwestern wind is mainly occurred in winter, which is originated from the high Siberia atmospheric pressure located on the north of South Korea. The change of the wind direction with the season is one of the weak points for using the wind energy in South Korea.

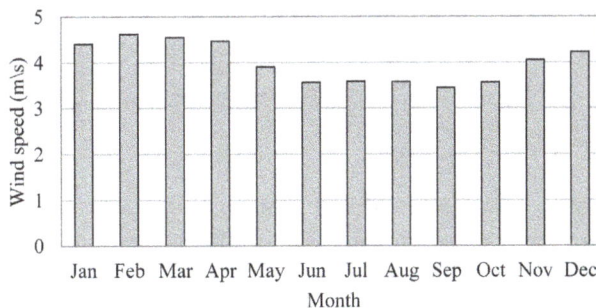

Figure 9. Monthly average wind speed for South Korea [26,27].

5.3. Load Profile

The LTE-macro BS load is critical for designing a reliable and efficient system. Sizing and modelling of the solar power system depend on the load profile. We used 24-hour load values for 365 days for an accurate analysis. The seasonal load profiles for both the DC and AC loads are shown in Figure 10.

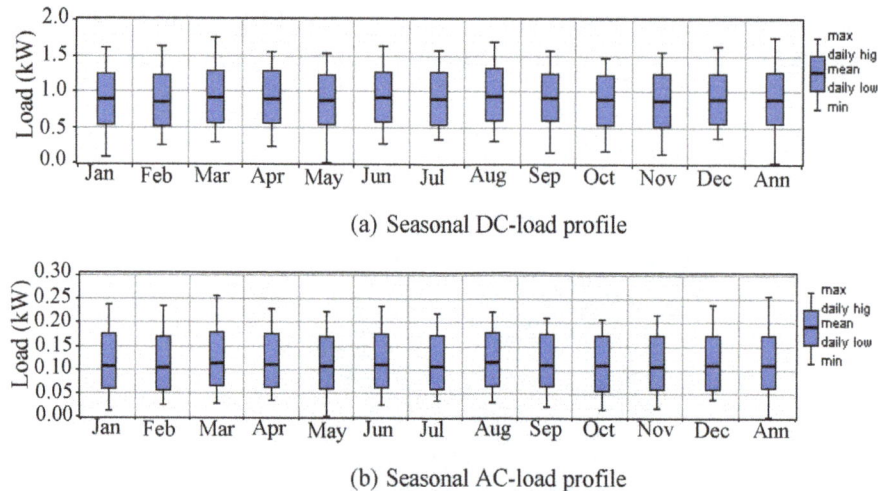

(a) Seasonal DC-load profile

(b) Seasonal AC-load profile

Figure 10. Seasonal load profiles: (**a**) DC-load profile; (**b**) AC-load profile.

5.4. Technical and Economic Criteria of the SPV/WTG System

The control parameters and the technical and economic criteria of the SPV/WTG system's components are provided in the following subsections.

5.4.1. Annual Real Interest Rate and Project Lifetime

The South Korea annual real interest rate was 1.25% in June 2016 [30]. However, the lifetime of the project is 10 years, representing the lifetime of the BS equipment [31].

5.4.2. SPV

The "*Sharp*" solar model is considered in this study; this model has highly efficient, affordable systems. Moreover, the Sharp solar module incorporates an advanced surface to increase light absorption and improve efficiency [32]. The technical specifications and costs used in the HOMER simulation based on the proposed solar model are listed below:

a.   Economic issues: the initial installation costs of SPV, the replacement costs, and the annual O&M costs per 1 kW are $1000, $1000, and $10, respectively.

b.   Technical issues: the lifetime of an SPV array is 25 years; a derating factor of 0.9, a reflectance of 20%, and a dual-axis tracking system for the SPV array are used in this paper.

c.   SPV size: the sizes of the simulation values of the applied SPV panels are 4, 4.5, 5, 5.5, 6, 6.5, and 7 kW.

5.4.3. WTG

The FT-1000L WTG model is considered in this study; this model is affordable [33]. In addition, Figure 11 summarises the output power at different wind speed values. The technical specifications and costs used in the HOMER simulation are based on the proposed WTG model and are listed below:

a.   Economic issues: the initial installation costs of a WTG, the replacement costs, and the annual O&M costs per 1 kW are $600, $600, and $50, respectively.

b.   Technical issues: the lifetime of a WTG and the height of the hub are 15 years and 50 m, respectively.

c.   WTG size: the sizes of the simulation values of the applied WTG are 1, 2, and 3 kW.

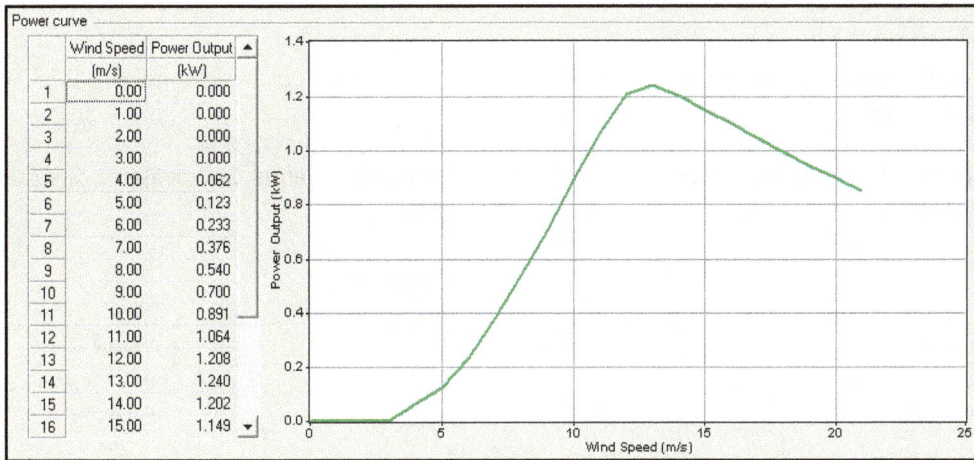

Figure 11. The output power over a different wind speed values.

5.4.4. Battery

The *"Trojan L16P"* battery model is considered in this study, which provides good characteristics combined with low cost. Figure 12 summarises the technical specifications to *"Trojan L16P"* battery model. More details can be found in [34]. The technical specifications and costs used in the HOMER simulation based on the proposed battery model are listed below:

a.   Economic issues: the initial installation costs of a battery, the replacement costs, and the annual O&M costs per unit are $300, $300, and $10, respectively.

b.   Technical issues: the lifetime of the *"Trojan L16P"* battery and the efficiency are set as 5 years and 85%, respectively.

c.   Battery size: the sizes of the simulation-values of the applied inverter are 32, 40, 48, 56, 64 and 72 units.

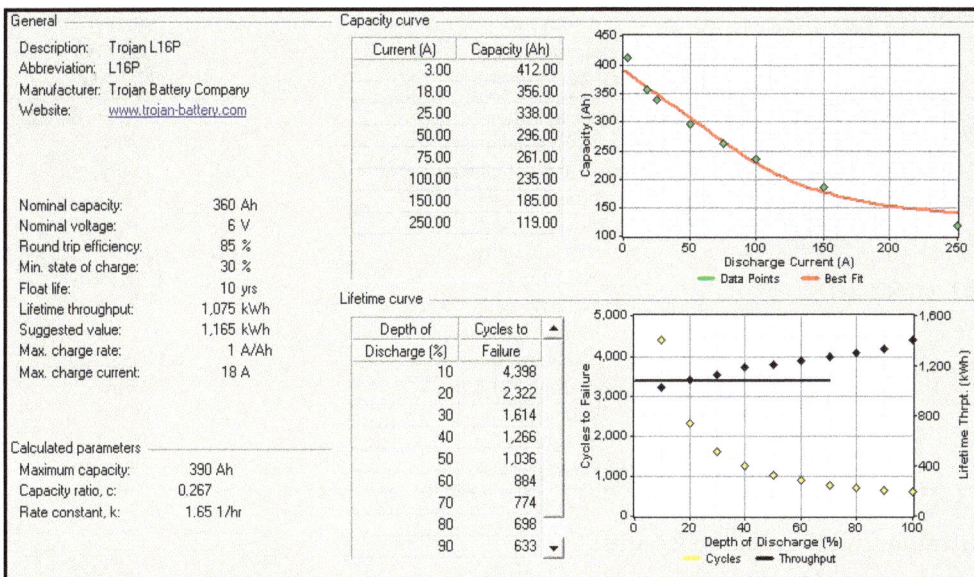

Figure 12. *"Trojan L16P"* battery model characteristics.

5.4.5. Inverter

a. Economic issues: the initial installation costs of an inverter, the replacement costs, and the annual O&M costs per 1 kW are $400, $400, and $10, respectively.

b. Technical issues: the lifetime of the inverter, and the efficiency are set as 15 years, 95%, respectively.

c. Inverter size: the sizes of the simulation-values of the applied inverter are 0.1, 0.15, 0.2, 0.25, and 0.3 kW.

Moreover, the technical specifications, costs, economic parameters, and system constraints that are used in the present study are summarized in Table 2 below.

Table 2. HOMER simulation setup.

System	Parameters	Value
Renewable energy resources	Solar radiation Wind speed	4.0, 4.5, 5.0, 5.5 kWh/m^2/day 4.0 m/s
Control parameters	Annual real interest rate Project lifetime Dispatch strategy Apply setpoint state of charge Operating reserve: as percent of load, hourly load	1.25% 10 years cyclic charging 80% 10%
SPV	Sizes considered Operational lifetime Efficiency System tracking Capital cost Replacement cost O&M cost per year	4, 4.5, 5, 5.5, 6, 6.5, 7 kW 25 years 90% Two axis $1/W $1/W $0.01/W
WTG	Sizes considered Operational lifetime Hub Capital cost Replacement cost O&M cost per year	1, 2, and 3 kW 15 years 50 m $0.6/W $0.6/W $0.05/W
Inverter	Sizes considered Efficiency Operational lifetime Capital cost Replacement cost O&M cost per year	0.1, 0.15, 0.2, 0.25, & 0.3 kW 95% 15 years $0.4/W $0.4/W $0.01/W
Trojan L16P Battery	Number of batteries Round trip efficiency Minimum state of charge Nominal voltage Nominal current Nominal capacity Lifetime throughput Max. charge rate Max. charge current Self-discharge rate Min. operational lifetime Capital cost Replacement cost O&M cost per year	32, 40, 48, 56, 64, 72 85% 30% 6 V 360 Ah at 20 h 6 V × 360 Ah = 2.16 kWh 1075 kWh 1 A/Ah 18 A 0.1% per hour 5 years $300 $300 $10

6. Optimization and Simulation Results

Different average daily solar radiation values of 4.0, 4.5, 5.0 and 5.5 kWh/m^2 are used to simulate the application of solar energy across a wide range of South Korea areas (as shown in Figure 7) with 4.0 m/s wind speed. Additional configuration details are given in Table 2. The energy output, the economic analysis of the proposed hybrid power system, and the related sensitivity analysis are provided in the following paragraphs.

6.1. Optimisation Criteria

Table 3 includes a summary of the economic and technical criteria for an optimal design of the SPV/WTG system based on different values of solar radiation and a 4.0 m/s wind speed.

Table 3. SPV/WTG system: optimal sizing and economic criteria.

Resources		Optimum Sizing				Costs Factor		
Wind Speed (m/s)	Radiation (kWh/m²/day)	SPV (kW)	WTG (kW)	Battery (unit)	Inverter (kW)	Initial Cost (IC) ($)	Annual O&M ($)	NPC ($)
4.0	4.0	6.0	1	64	0.20	25,880	752	29,528
	4.5	5.0	1	64	0.20	24,880	742	28,965
	5.0	4.5	1	64	0.20	24,380	737	28,683
	5.5	4.0	1	64	0.20	23,880	732	28,401

Table 3 indicates that the size of the SPV array decreases with increasing solar radiation. In contrast, the size of the WTG is 1 kW for all solar radiation cases because the wind speed is constant at 4.0 m/s. The optimal number of batteries, which was found by the HOMER simulation for the system, is 64 batteries. In addition, the inverter (DC/AC) needed must be capable of handling 0.2 kW.

The SPV array, which is a 20 Sharp ND-F4Q300 module (polycrystalline), is rated at 6.0 kW with a voltage V_{pm} of 35.20 V_{dc}, a current I_{pm} of 8.52 A, and a power P_{pm} of 300 W. A 20 Sharp ND-F4Q300 module will be connected with four in series and five in parallel to be compatible with the specifications of the solar control regulator chosen in this study, the Solarcon SPT-4830 [35]. This requires that the open circuit voltage for a SCR $V_{oc}^{SPT-4830}$, 192 V_{dc}, be higher than the open circuit voltage of the SPV panel, 180.4 V_{dc} (four SPV modules in series $\times$ $V_{oc}^{ND-F4Q300}$, 45.10 V_{dc}). In addition, the current for a SCR $I^{SPT-4830}$ must be higher than the short circuit current of the SPV panel $I_{sc}^{ND-F4Q300}$, 8.94 A $\times$ 1.3 (safety factor). The optimal size of the WTG was found by the HOMER simulation to be 1 kW; the FT-1000L WTG model is a good choice. The nominal voltage of the *Trojn L16P* battery [31] is 6 V_{dc}, and the nominal capacity is 360 Ah. Thus, the optimal number of batteries (64 batteries) found by the HOMER simulation for the system will be connected with eight in series and eight in parallel because the DC bus-bar is 48 V_{dc}. A typical daily AC load is (air conditioner 91 W + lamps 40 W). Therefore, the inverter must be capable of handling 0.2 kW. *SU200P*, which has specifications of 200 W, an input voltage of 12/24/48 V_{dc}, an output voltage of 220/110 V_{ac}, an AC output frequency of 50 Hz/60 Hz, and a pure sine wave can be chosen.

The system costs consist of the following: (i) the IC cost is paid at the beginning of the project and decreases with decreasing size of the elements of the project; (ii) The operating cost is paid annually, and most of this cost goes towards operating and maintaining the battery bank; (iii) The NPC represents all costs that occur within the project lifetime, including IC costs, component replacements within the project lifetime and O&M costs. Table 3 indicates that the IC and O&M costs decrease with increasing solar radiation because the SPV array size decreases, reducing the total cost (NPC) of the hybrid SPV/WTG system.

6.2. Energy Yield Analysis

Figure 13 summarises the annual energy contributions of different sources for different average daily solar radiation values at a wind speed of 4.0 m/s. The energy delivered to the SPV array load depends on the value of solar radiation and the size of the SPV array. However, the energy delivered to the load of the WTG is the same because the optimal WTG size is fixed.

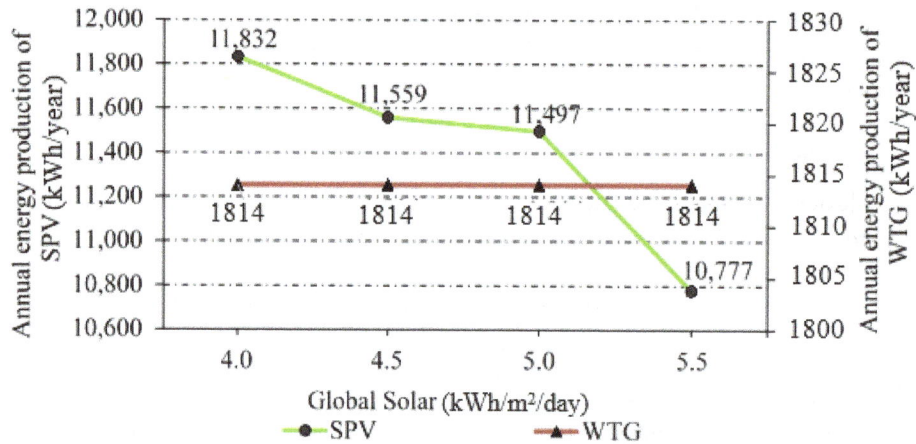

Figure 13. Annual energy contribution of different sources with different average solar radiation values and 4.0 m/s wind speed.

The following statistical analysis discusses the energy production based on the average daily solar radiation (4.0 kWh/m^2) and wind speed (4.0 m/s) for South Korea as a case study. However, this analysis can be extended to other cases, yielding a slight difference in daily peak solar hours per case depending on the average daily solar radiation.

The annual energy contribution of the SPV array is computed based on Equation (2), SPV rated capacity is 6.0 kW × PSH 4.01 h × SPV derating factor 0.9 × 365 days/year, which equals 7904 kWh. However, the tracking system plays a role in increasing the total amount of energy produced by a SPV array. The present simulation adopted a dual axis tracker that increased the total amount of energy to 11,832 kWh. Each hour, HOMER calculates the power output of the wind turbine in a four-step process. First, it determines the average wind speed for the hour at the anemometer height by referring to the wind resource data. Second, it calculates the corresponding wind speed at the turbine's hub height using either the logarithmic law or the power law. Third, it refers to the turbine's power curve to calculate the turbine's power output at that wind speed assuming standard air density. Fourth, it multiplies that power output value by the air density ratio, which is the ratio of the actual air density to the standard air density [21]. The annual energy contribution of the WTG is 1814 kWh. In addition, Figure 14 shows the monthly statistics of the output power for the WTG. It is clear that the maximum energy contribution of the WTG occurred at the first three months of the year (January–March) and at the end two months of the year (November - December), due to decreased the sunshine hours and solar radiation (as shown in Figure 6).

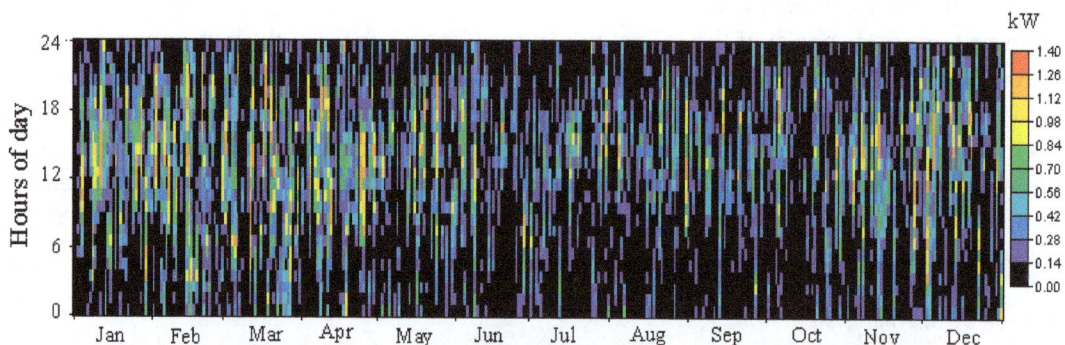

Figure 14. Monthly statistics of the WTG output power.

The total annual energy production of the hybrid SPV/WTG system is 13,646 kWh (87% from the SPV array and 13% from the WTG), while the total annual energy needed by an LTE-macro BS is 8840 kWh, the AC load (air conditioner 91 W + lamps 40 W, operating 7 PM–7 AM) plus the DC load (BS 819 W + microwave link 80 W) multiplied by (24 h × 365 days/year). The difference between electrical production and consumption is equal to the excess electricity of 4004 kWh/year, plus the battery losses of 753 kWh/year, plus inverter losses of 50 kWh/year.

Figure 15 shows the monthly average electric production of the different power sources. The maximum energy contribution of the SPV array occurred in April and May. Meanwhile, the minimum energy contribution occurred in July. These results are attributed to the differences in the average solar radiation rate, as shown in Figure 6. While, the maximum energy contribution of the WTG occurred in February; and the minimum energy contribution occurred in September.

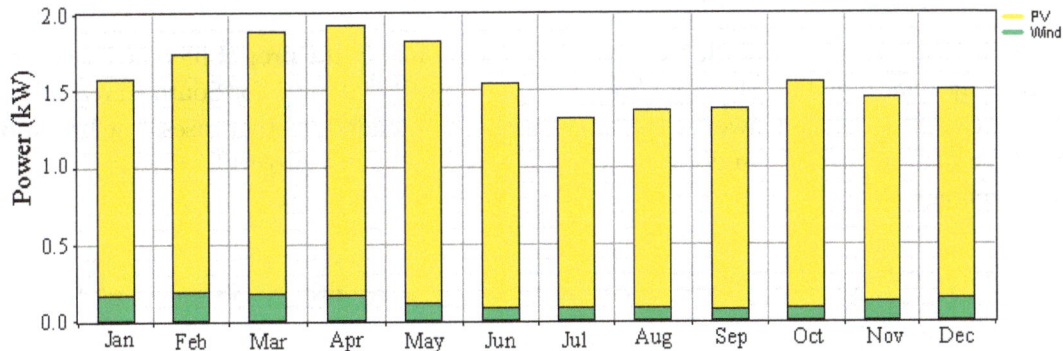

Figure 15. Monthly power contribution of various sources at an average solar irradiation 4.0 kW/m^2/day and average wind speed 4.0 m/s.

The optimal number of batteries, which was found by the HOMER simulation for the system is 64 battery. The battery annual energy-in is 5162 kWh, while the annual energy-out is 4409 kWh, where the roundtrip efficiency was 85%. Batteries can supply LTE-macro BS load autonomy for 95.9 h (3 days and 23.9 h), which is computed based on Equation (5), (number of the batteries is 64 × nominal voltage of a single battery 6 V × nominal capacity of a single battery 360 Ah × DOD 0.7 × 24 h) divided by (daily average LTE-macro BS load 24.22 kWh). However, one battery can supply LTE-macro BS load autonomy 1.5 h. The battery expected life is 10 years, which is computed based on Equation (6). In addition, the seasonal statistics of the maximum and minimum states of charge (SOCs) for the battery are given in Figure 16 and reveal that the highest energy contribution of the battery bank occurred at the end of July because the minimum energy contribution of the hybrid power system occurred in July, as shown in Figure 15.

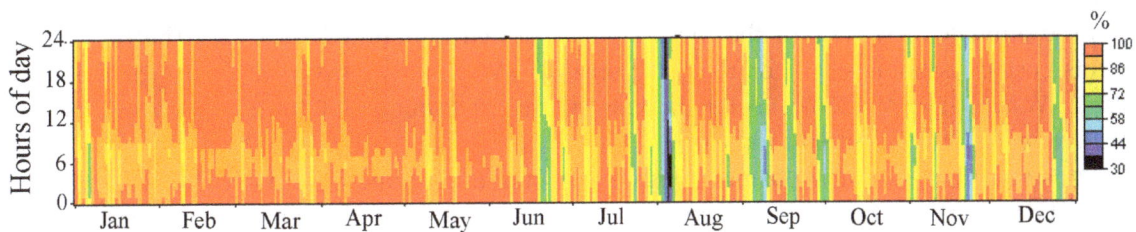

Figure 16. Monthly statistics of the SOC for the battery.

The inverter annual energy-in is 1005 kWh, while the annual energy-out is 955 kWh, based on the daily AC load needs 2.62 kWh × 365 days/year, with 95% efficiency and 8759 h/year operation (operating hours 24 h × 365 days/year). Figure 17 shown the monthly statistics of the output power for the inverter, which indicate that the inverter is operating normally.

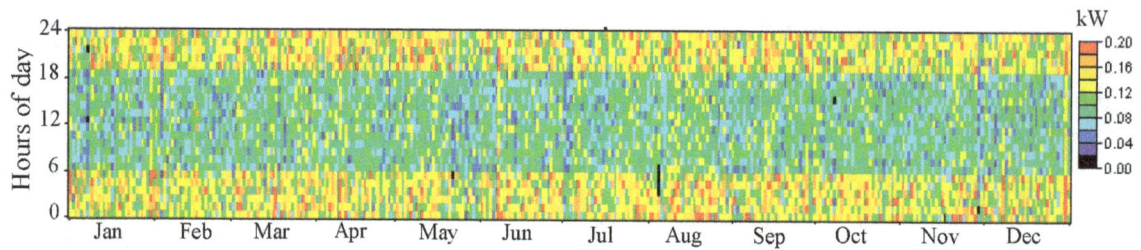

Figure 17. Monthly statistics of the inverter output power.

6.3. Economic Analysis

The cash flow summary of the hybrid power system within the project lifetime based on the average daily solar radiation (4.0 kWh/m^2) and wind speed (4.0 m/s) for South Korea as a case study is given in Figure 18. However, this analysis can be extended to other cases. The breakdown for the IC, replacement, O&M and salvage costs incurred within the project lifetime is given in the following paragraphs.

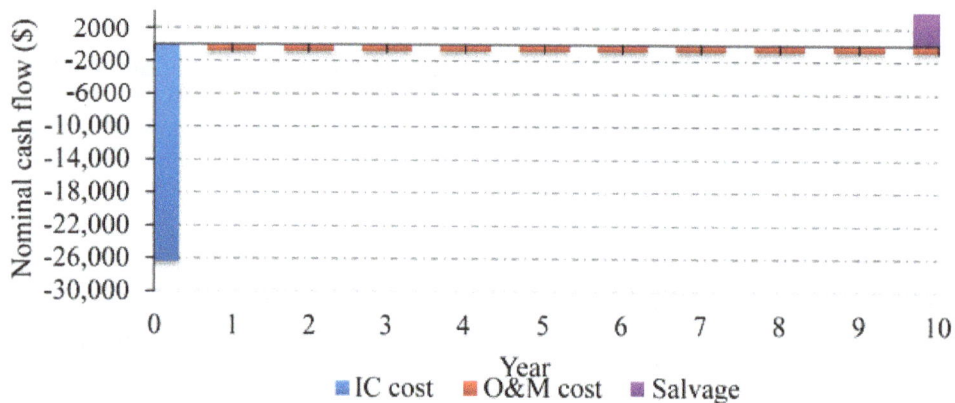

Figure 18. Cash flow summary within the project lifetime.

The *IC cost*, paid once at the beginning of the project, is directly proportional to the size of the system. From Table 3, the IC cost is $25,880. The breakdown of this cost is as follows: (i) For the SPV (size 6.0 kW × cost $1000/1 kW = $6,000); (ii) for the WTG (size 1 kW × cost $600/1 kW = $600); (iii) for the battery units (64 unit × cost $300/unit = $19,200); and (iv) for the inverter (size 0.2 kW × cost $400/1 kW = $80).

The annual *O&M costs* of the hybrid power system amounted to $752. The breakdown of this cost is as follows: (i) for the SPV (size 6.0 kW × $10/1 kW = $60); (ii) for the WTG (size 1 kW × $50/1 kW = $50); (iii) for the battery units (64 unit × $10/unit = $640); and (iv) for the inverter (0.2 kW × cost $10/1 kW = $2).

It is clear that the battery bank represents the bulk of both the IC and O&M costs. However, this cost depended on the number of batteries in the system. Herein, the optimal number of batteries was determined by HOMER to be 64 battery; the number of batteries can be reduced. However, the load autonomy decreases, which is considered to be an important issue, especially in remote rural areas.

The batteries have a lifetime of 10 years, which is the same as the project lifetime; the SPV array has a lifetime of 25 years, and the WTG has a lifetime of 15 years. Thus, neither requires replacement.

Each component has a *salvage value* at the end of the project lifetime. The SPV array salvage value is $3600, the highest in the system, which was computed based on Equation (10), (SPV array remaining lifetime (15)/SPV array lifetime (25)) multiplied by the replacement cost of the SPV array, $6000. While, the WTG salvage value is $200 the inverter salvage value is $27. Thus, the total salvage value at the end project lifetime is $3827.

The economic analysis described above has been conducted on the basis of the nominal system. However, Figure 19 showed the discount factor for each year of the project lifetime, which computed based on Equation (9).

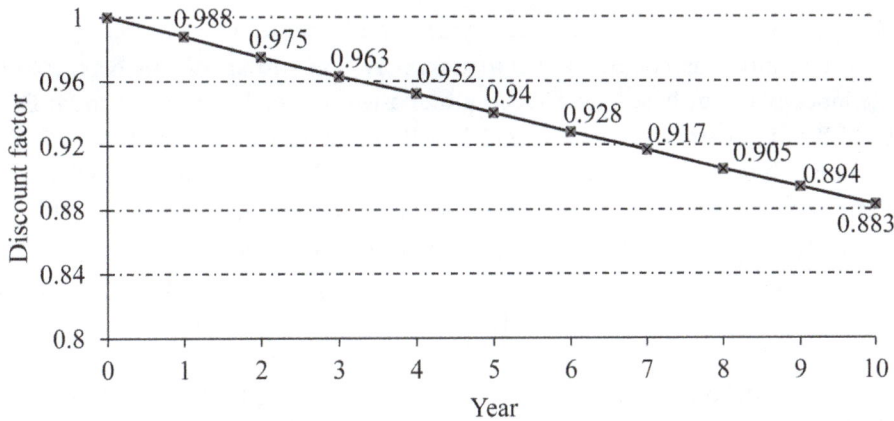

Figure 19. Discount factor for each year of the project lifetime.

The total NPC calculates by summing up the total discounted cash flows in each year of the project lifetime, as follows: include capital costs $25,880 + O&M costs $7075 − salvage $3645, equal $29,810. This analysis can be extended to other cases in which the cost of the system depends on the elements and the size of the components in the system. Figure 20 summarized the total discounted costs occurring within the project lifetime for the hybrid SPV/WTG system.

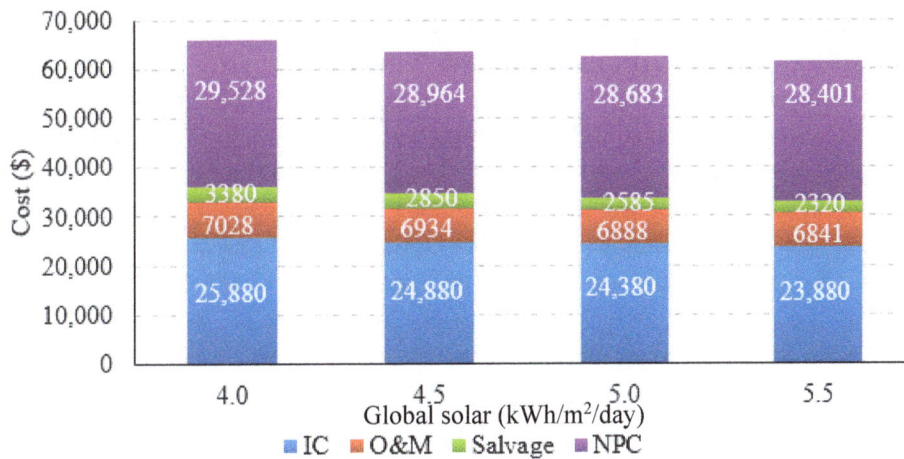

Figure 20. Summary of all costs occurring within the project lifetime for the SPV/WTG hybrid system.

7. Comparison of the Feasibility of Using a Hybrid SPV/WTG system vs. a Diesel Generator

The comparison between the traditional power system DG and the proposed SPV/WTG system is summarized into two important key aspects: (i) economic feasibility and (ii) greenhouse gases (GHG) emissions. The following discussion is based on an average daily solar radiation for South Korea of 4.0 kWh/m^2 and a wind speed of 4.0 m/s as a case study. However, this discussion can be extended to include other cases of solar radiation, with a slight difference in the IC, O&M, and salvage costs.

7.1. Diesel Generator

The DG needed approximately is 4 kW, which computed as (maximum LTE-macro BS, AC and DC load 1.1 kW) divided by (DG efficiency 30% [7] × converter efficiency 95%):

i.   IC cost: the DG IC cost is $2640 (size 4 kW × cost $660/1 kW). However, the fossil fuels not sustainable and expensive, and the price go up continuously.

ii.   O&M cost: the annual cost for the maintenance and operation of the DG amounted to $4680 (without counting the cost of fuel transport). A breakdown of this cost, (i) $438 for DG maintenance per year based on a DG maintenance cost of $0.05/h × annual DG operating hours 8760 h; and (ii) a fuel cost of $4242, based on the diesel price of $1.04/L [36]) multiplied by the a total diesel consumption of 4079 L per year, which computed based on a specific fuel consumption of 0.388 L/kWh × annual electrical production of the DG of 10,512 kWh/year (DG capacity size 4 kW × DG efficiency 0.3 × 24 h × 365 days/year).

iii. Replacement cost: mobile operator may need to change the DG every 3 years, which means at least three times during the life of the project. Thus, the total DG replacement cost is 3 × (size 4 kW × cost $660/1 kW), equal $7920 at least.

iv.   NPC cost: the total NPC include IC cost $2640 + O&M cost $46,800 + Replacement cost $7920, equal $57,360 over the project lifetime (10 years); without counting the cost of fuel transport, which adds further to this cost.

v.   GHG emissions: According to [37], the CO_2 emissions of diesel fuel are 2.68 kg/L. Hence, the total annual CO_2 emissions are 10,931 kg, computed based on the specific annual diesel consumption of 4079 L multiplied by 2.68 kg CO_2/L.

7.2. Hybrid SPV/WTG System

i.   IC cost: the hybrid SPV/WTG system IC cost is $25,880. The IC cost of the SPV/WTG system is high, due to that the components of the system are expensive comparing with DG. However, the global price of SPV/WTG system go down continuously.

ii.   O&M cost: by applying the proposed SPV/WTG system; a large benefits for mobile operators can be achieved for long term. Since the annual O&M cost can be decreasing to $752, which mean saving amounted 83.83% comparing with O&M cost for DG. However, savings rate will increase more and more in the future, due to the continuing rise in fuel prices.

iii. Replacement cost: by applying the proposed SPV/WTG system, the batteries have a lifetime of 10 years, which is the same as the project lifetime, the SPV array has a lifetime of 25 years, and WTG and inverter have has a lifetime of 10 years, so neither requires replacement.

iv.   NPC cost: the total NPC include capital costs $25,880 + O&M costs $7028 − salvage $3380, equal $29,528 over the project lifetime (10 years).

v.   GHG emissions: Renewable energy systems (RESs) are considered an icon of all that is green. Hence, the trend towards RESs is increasing all over the world to eliminate GHG emissions and minimise the effect on both the wallet and the environment.

By applying the proposed SPV/WTG system, the total NPC that can be saving amounted up to 48.52%. Figure 21 summarized OPEX of the traditional power system (DG only) compare with the proposed SPV/WTG system over the project lifetime (10 years).

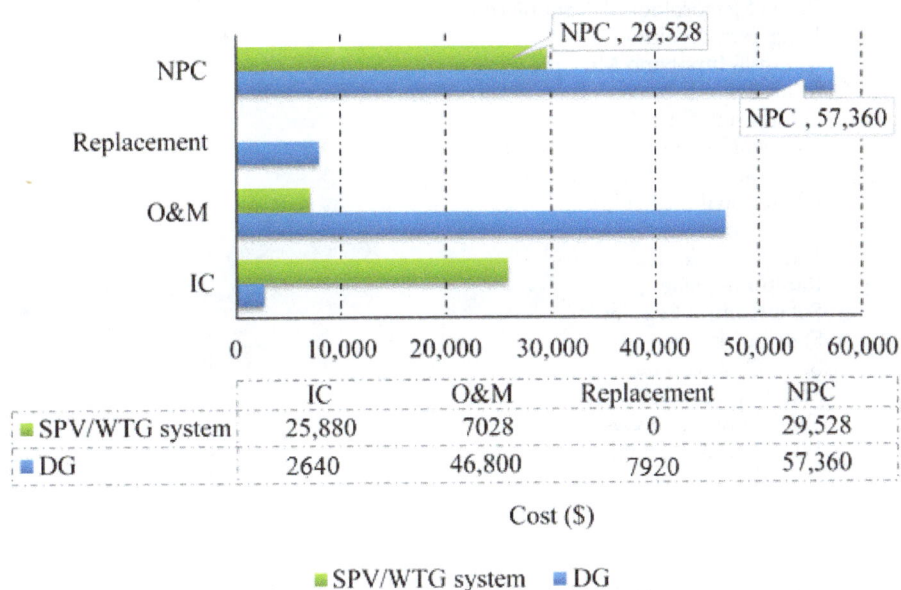

	IC	O&M	Replacement	NPC
■ SPV/WTG system	25,880	7028	0	29,528
■ DG	2640	46,800	7920	57,360

Cost ($)

■ SPV/WTG system ■ DG

Figure 21. Summary an OPEX of the SPV/WTG system vs. DG over the project lifetime.

8. Conclusions

Providing a reliable, secure power and energy system is one of the main issues in cellular networks. Accordingly, this study examined suppling LTE-macro BSs the needed energy by a hybrid SPV with a WTG power system. Three key aspects have been discussed: (i) optimal system architecture; (ii) energy yield analysis; and (iii) economic analysis. The optimum criteria, including economic and technical feasibility parameters, were analysed using the HOMER software package developed by the U.S. National Renewable Energy Laboratory (NREL).

Moreover, the comparison between the proposed hybrid SPV with WTG power system and the classical solution "diesel generator (DG)" was evaluated in terms of: (i) OPEX savings, where the result showed that the OPEX savings is up to $27,832 (representing 48.52%); and (ii) GHG emissions, with the CO_2 emissions decreasing to zero.

Finally, many contributions have been provided, such as decreasing OPEX and gas emissions, ensuring sustainability with better cellular network planning and providing cleaner energy, with an emphasis on the application of optimization methods in power system operation and planning.

Acknowledgments: This work was supported by the faculty research fund of Sejong University in 2016. We thank the reviewers for the fruitful suggestions, which helped us to improve the quality of this work.

Author Contributions: Mohammed H. Alsharif analyzed the data and completed the first draft, and Jeong Kim wrote the solar radiation and wind speed in South Korea and revised the final version of paper.

Conflicts of Interest: The authors declare that they have no competing interests.

Abbreviations

BB	BaseBand
BS	Base Station
CDMA	Code Division Multiple Access
CO_2, NO_x, SO_2	Carbon dioxide, Nitrogen oxides, Sulfur dioxide
CRF	Capital Recovery Factor
DG	Diesel Generator
DOD	Depth of Discharge
FC	Fuel Cell
GHG	GreenHouse Gas
GSM	Global System for Mobile Communication
HOMER	Hybrid Optimisation Model for Electric Renewables
IC	Initial Costs

KMA	Korea Meteorological Administration
LTE	Long Term Evolution
MIMO	Multiple-Input and Multiple-Output
MSW	Modified Sine-Wave
NPC	Net Present Cost
NREL	National Renewable Energy Laboratory
O&M	Operation and Maintenance Costs
OPEX	OPerational EXpenditure
PA	Power Amplifier
PSH	Peak Solar Hours
RF	Radio-Frequency
SCR	Solar Control Regulator
SOC	State of Charge
SPV	Solar Photovoltaic
STC	Standard Test Conditions
TAC	Total Annualised Cost
WTG	Wind Turbine Generator

Symbols

P_{op}	BS operating power
P_{PA}^{DC}	Power consumed by power amplifier
P_{RF}^{DC}	Radio frequency power
P_{BB}^{DC}	Baseband power
P_{tx}^{max}	BS transmission power
P_o	Normalized BS transmission power
P_{BS}	Total Power Consumption for the BS
P_{mc}	Microwave backhaul power
P_{lm}	Auxiliary equipment power
σ_{DC}	DC-DC power supply losses
σ_{cool}	Cooling losses
N_{TRX}	Number of transceivers
η_{PA}	PA efficiency
α	Path loss coefficient
R_o	Coverage radius
E_{PV}	Energy output of the PV array
Y_{PV}	Rated capacity of the PV array
f_{PV}	PV derating factor
A_{batt}	Battery bank autonomy
N_{batt}	Number of batteries
V_{nom}	Nominal voltage of a single battery
Q_{nom}	Nominal capacity of a single battery
$L_{prim,ave}$	Average daily LTE-macro BS load
R_{batt}	Battery bank lifetime
$Q_{lifetime}$	Lifetime throughput of a single battery
Q_{thrpt}	Annual battery throughput
$R_{batt,f}$	Battery float life

References

1. 3GPP System Standards. Available online: http://www.3gpp.org/news-events/3gpp-news/1614-sa_5g (accessed on 30 July 2016).
2. Netmanias Report, LTE in Korea. Available online: http://www.netmanias.com/en/post/reports/6060/kt-korea-lg-u-lte-lte-a-sk-telecom-wideband-lte/lte-in-korea-2013 (accessed on 30 July 2016).
3. Open Signal. Available online: http://opensignal.com/reports/2016/02/state-of-lte-q4-2015/ (accessed on 30 July 2016).
4. Alsharif, M.H.; Nordin, R.; Ismail, M. Survey of Green Radio Communications Networks: Techniques and Recent Advances. *J. Comput. Netw. Commun.* **2013**, *2013*, 453893. [CrossRef]
5. Alsharif, M.H.; Nordin, R.; Ismail, M. Classification, recent advances and research challenges in energy efficient cellular networks. *Wirel. Pers. Commun.* **2014**, *77*, 1249–1269. [CrossRef]
6. Aris, A.M.; Shabani, B. Sustainable power supply solutions for off-grid base stations. *Energies* **2015**, *8*, 10904–10941. [CrossRef]

7. Kusakana, K.; Vermaak, H.J. Hybrid Renewable Power Systems for Mobile Telephony Base Stations in Developing Countries. *Renew. Energy* **2013**, *51*, 419–425. [CrossRef]

8. Seoin, B.; Heetae, K.; Hyun, J.C. Optimal Hybrid Renewable Power System for an Emerging Island of South Korea: The Case of Yeongjong Island. *Sustainability* **2015**, *7*, 13985–14001.

9. Ng, D.W.K.; Lo, E.S.; Schober, R. Energy-Efficient Resource Allocation in OFDMA Systems with Hybrid Energy Harvesting Base Station. *IEEE Trans. Wirel. Commun.* **2013**, *12*, 3412–3427. [CrossRef]

10. Ahmed, I.; Ikhlef, A.; Ng, D.W.K.; Schober, R. Power Allocation for an Energy Harvesting Transmitter with Hybrid Energy Sources. *IEEE Trans. Wirel. Commun.* **2013**, *12*, 6255–6267. [CrossRef]

11. Martínez-Díaz, M.; Villafáfila-Robles, R.; Montesinos-Miracle, D.; Sudrià-Andreu, A. Study of Optimization Design Criteria for Stand-Alone Hybrid Renewable Power Systems. In Proceedings of the International Conference on Renewable Energies and Power Quality (ICREPQ'13), Bilbao, Spain, 20–22 March 2013; pp. 1–5.

12. Kaldellis, J. Optimum hybrid photovoltaic-based solution for remote telecommunication stations. *Renew. Energy* **2010**, *35*, 2307–2315. [CrossRef]

13. Serincan, M.F. Reliability considerations of a fuel cell backup power system for telecom applications. *J. Power Sources* **2016**, *309*, 66–75. [CrossRef]

14. Belkhiri, S.; Chaker, A. Optimization of Hybrid PV/Wind System for Remote Telecom Station, a Case Study of Different Sites in Algeria. In Proceedings of the Chemical, Biological and Environmental Engineering, Ho Chi Minh City, Vietnam, 23–25 March 2016; pp. 1–7.

15. Hossam, K.; Mikhail, A.R.; Hafez, I.M.; Anis, W.R. Optimum Design of PV Systems for BTS in Remote and Urban Areas. *Int. J. Sci. Technol. Res.* **2016**, *5*, 1–9.

16. Salih, T.; Wang, Y.; Adam, M.A.A. Renewable micro hybrid system of solar panel and wind turbine for telecommunication equipment in remote areas in Sudan. *Energy Procedia* **2014**, *61*, 80–83. [CrossRef]

17. Imtiaz, A.W.; Hafeez, K. Stand Alone PV System for Remote Cell Site in Swat Valley. In Proceedings of the 1st International Conference on Technology and Business Management, Peshawar, Pakistan, 2–4 April 2013; pp. 1–5.

18. Nema, P.; Nema, R.; Rangnekar, S. Minimization of Green House Gases Emission by Using Hybrid Energy System for Telephony Base Station Site Application. *Renew. Sustain. Energy Rev.* **2010**, *14*, 1635–1639. [CrossRef]

19. Moury, S.; Khandoker, N.M.; Haider, M.S. Feasibility Study of Solar PV Arrays in Grid Connected Cellular BTS Sites. In Proceedings of the 2012 IEEE International Conference on Advances in Power Conversion and Energy Technologies (APCET), Mylavaram, India, 2–4 August 2012; pp. 1–5.

20. Groupe Speciale Mobile Association. *Green Power for Mobile bi Annual Report 2014*; Groupe Speciale Mobile Association (GSMA): London, UK, 2014.

21. Imran, M.; Katranaras, E.; Auer, G.; Blume, O.; Giannini, V.; Godor, I.; Jading, Y.; Olsson, M.; Sabella, D.; Skillermark, P. *Energy Efficiency Analysis of the Reference Systems, Areas of Improvements and Target Breakdown*; Technical Report, ICT-EARTH Deliverable D2.3; EC-IST Office: Brussels, Belgium, 2011.

22. Auer, G.; Giannini, V.; Desset, C.; Godor, I.; Skillermark, P.; Olsson, M.; Imran, M.A.; Sabella, D.; Gonzalez, M.J.; Blume, O. How much energy is needed to run a wireless network? *IEEE Wirel. Commun.* **2011**, *18*, 40–49. [CrossRef]

23. Schmitt, G. The Green Base Station. In Proceedings of the 4th International Conference on Telecommunication—Energy Special Conference (TELESCON), Frankfurt, Germany, 10–13 May 2009; pp. 1–6.

24. Lambert, T.; Gilman, P.; Lilienthal, P. Micropower System Modeling with HOMER. 2006. Available online: http://homerenergy.com/documents/MicropowerSystemModelingWithHOMER.pdf (accessed on 30 July 2016).

25. Alsharif, M.H.; Nordin, R.; Ismail, M. Energy optimisation of hybrid off-grid system for remote telecommunication base station deployment in Malaysia. *EURASIP J. Wirel. Commun. Netw.* **2015**, *2015*, 1–15. [CrossRef]

26. KMA, Annual Climatological Report 2013, Korea Meteorological Administration. Available online: http://www.kma.go.kr/weather/observation/data_monthly.jsp (accessed on 30 July 2016).

27. NASA. Available online: https://eosweb.larc.nasa.gov/cgi-bin/sse/homer.cgi?email=skip%40larc.nasa.gov&step=1&lat=37.499&lon=126.54958&submit=Submit&ms=1&ds=1&ys=1998&me=12&de=31&ye=1998&daily=swv_dwn (accessed on 30 July 2016).

28. Alsharif, M.H.; Kim, J. Optimal Solar Power System for Remote Telecommunication Base Stations: A Case Study Based on the Characteristics of South Korea's Solar Radiation Exposure. *Sustainability* **2016**, *8*, 942. [CrossRef]

29. NIMS. Available online: http://www.greenmap.go.kr/02_data/data01.do#2#1#1 (accessed on 19 August 2016).

30. The Bank of Korea Monetary Policy. Available online: http://www.bok.or.kr/baserate/baserateList.action?%20menuNaviId=33 (accessed on 30 July 2016).

31. Ge, X.; Cheng, H.; Guizani, M.; Han, T. 5G wireless backhaul networks: Challenges and research advances. *IEEE Netw.* **2014**, *28*, 6–11. [CrossRef]

32. Sharp Solar Electricity Incorporation. Available online: http://www.sharp-world.com/solar/en/solutions/index.html (accessed on 30 July 2016).

33. FT-1000L Wind Turbine Generator Model. Available online: http://www.chinaseniorsupplier.com/Electrical_Equipment_Supplies/Generators/60423002878/Home_use_1000w_wind_turbine_generator_manufacturer.html (accessed on 30 July 2016).

34. Trojan Battery Incorporation. Available online: http://www.trojanbattery.com/ (accessed on 30 July 2016).

35. Leonics Incorporation, SolarCon SPT-Series. Available online: http://www.leonics.com/product/renewable/solar_charge_controller/dl/spt-074.pdf (accessed on 30 July 2016).

36. Global Petrol Prices. Available online: http://www.globalpetrolprices.com/South-Korea/diesel_prices/ (accessed on 30 July 2016).

37. Calculation of CO_2 Emissions. Available online: http://people.exeter.ac.uk/TWDavies/energy_conversion/Calculation%20of%20CO2%20emissions%20from%20fuels.htm (accessed on 30 July 2016).

4

A Feedback Passivation Design for DC Microgrid and Its DC/DC Converters

Feifan Ji [1], Ji Xiang [1,*], Wuhua Li [1] and Quanming Yue [2]

[1] College of Electrical Engineering, Zhejiang University, Hangzhou 310027, China; unusual@zju.edu.cn (F.J.); woohualee@zju.edu.cn (W.L.)

[2] State Grid Zhejiang Electric Power Company, Hangzhou 310007, China; yqm8341@sohu.com

* Correspondence: jxiang@zju.edu.cn

Academic Editor: Josep M. Guerrero

Abstract: There are difficulties in analyzing the stability of microgrids since they are located on various network structures. However, considering that the network often consists of passive elements, the passivity theory is applied in this paper to solve the above-mentioned problem. It has been formerly shown that when the network is weakly strictly positive real (WSPR), the DC microgrid is stable if all interfaces between the microgrid and converters are made to be passive, which is called interface passivity. Then, the feedback passivation method is proposed for the controller design of various DC–DC converters to achieve the interface passivity. The interface passivity is different from the passivity of closed-loop systems on which the passivity based control (PBC) concentrates. The feedback passivation design is detailed for typical buck converters and boost converters in terms of conditions that the controller parameters should satisfy. The theoretical results are verified by a hardware-in-loop real-time labotray (RTLab) simulation of a DC microgrid with four generators.

Keywords: DC microgrid; feedback passivation; distributed control

1. Introduction

With the deterioration of the environment and the reduction of fossil reserves, applications of renewable energy and energy storage systems (ESSs) are rapidly increasing.Most of these new components have a DC feature in essence, which may pose technical and operational challenges for the integration into the existing AC systems. Meanwhile, significant advances in power electronics technology have produced a number of power converters for DC voltage transformation into various levels for different applications. In this context, DC microgrids are emerging as an attractive solution for the integration of renewable-based distributed generators (DGs) and ESSs [1].

The stability of DC microgrids is a critical problem. Its studies have been performed by small or large signal analysis [1–9], in which the DC microgrid is analyzed collectively, based on a comprehensive model of the whole microgrid. The comprehensive model is sensitive to the change of every unit, particularly to the fluctuation of the microgrids' network. The results based on such a high entropy model may have limited effectiveness in providing valuable information for stable design [10]. This paper follows a decomposition approach in which the DC microgrid is analyzed as a network with two parts: a node system consisting of converters and a dynamical edge system consisting of loads and transmission lines to connect the converters and loads. The nodes and edges division is compared to results of a work on a multi-agent systems [11]. According to [11], the outputs of the edge dynamic systems form the external inputs of the node dynamic systems, which are termed "neighboring inputs" representing the coupling actions between nodes. The outputs of the node dynamic systems are the inputs of the edge dynamic systems. As explored in [12], an edge system consisting of resistor-Inductor-capacitor (RLC) components is positive real (PR), or weakly strictly positive real (WSPR) if there exists at least one resistance inside. As it is well-known that the passivity is preserved when two passivity systems are connected properly, this paper makes a passivity design for the converters in a DC microgrid. When each converter has a passive interface, then the DC microgrid with a WSPR (weakly strictly positive real) edge system is guaranteed to be stable.

The passivity-based control (PBC), as a kind of passivation design, has been extensively utilized on the control of converters. In [13], passivity-based feedback controllers are derived for the indirect stabilization of the average output voltage in pulse-width-modulation (PWM) controlled DC–DC power converters of the "boost", "buck-boost", and "buck" types. In [14], the authors present an overview of passivity-based stability assessment, including techniques for space-vector modeling of VSCs (voltage source converters) whereby expressions for the input admittance can be derived. In [15], the author investigate the DC-bus voltage regulation problem for a three-phase boost type PWM AC–DC converter using PBC theory. The models are shown to be Euler–Lagrange (EL) systems corresponding to a suitable set of average EL parameters. The PBC aims to make the controllers themselves be passive, which together with a passive plant stabilizes the closed-loop system. In contrast, the passivity design made by this paper focuses on the interface passivity instead of the passivity of controllers. The interface passivity is a kind of passivity equivalent to the PRness of transfer functions from injecting currents of microgrid i_T to output voltages of converters u_c. i_T and u_c are the output and the input of edge systems, respectively.

In passivity theory, the Kalman–Yakubovich–Popov (KYP) Lemma establishes an equivalence between the conditions in the frequency domain for a system to be positive real, an input–output relationship of the system in the time domain, and conditions on the matrices describing the state-space representation of the system. With the KYP Lemma, many significant results in the control field have been presented for the direct passivation regulating the input–output relationship [16,17], while a few results exist for the indirect passivation regulating the relationship between external input and output [18,19]. The interface passivity addressed in this paper is a kind of indirect passivation that is more difficult than direct passivation since external input can not be controlled. On the contrary, the PBC control is a kind of direct passivation. In [19], necessary and sufficient conditions for the feasibility of indirect passivation of strictly positive real (SPR) was presented for a class of single-input and single-output systems. It should be pointed out that typical DC–DC converters, including "buck" and "boost", do not satisfy the necessary and sufficient condition. This paper proposes a feedback passivation design for DC–DC converters to make them have interface passivity. The resulted transfer functions are PR rather than SPR in the penalty of requiring edge systems to be WSPR rather than PR.

Making a converter's interface be passive is not new. Gu et al. [18] have addressed this subject recently by using a feedforward control method of SISO (single input single output) systems in the frequency domain. In this paper, the interface passivity is discussed in the time domain. In contrast to the frequency domain analysis, the time domain analysis utilizing state-space representation is possible to confirm the state of the system parameters and not merely input–output relations. Moreover, it fits with in the multiple-input and multiple-output essence of two parts of DC microgrids, while the SISO method of frequency domain omitting the off-diagonal coupling term is not rigorous for microgrids. The main contributions of this paper are: (1) a decoupling viewpoint of DC microgrids by which the stability is guaranteed by the interface passivity of node system and the WSPR of the edge system, independent of the network structure; and (2) a feedback passivation method is proposed for controlling typical converters to be of the interface passivity.

The remainder of this manuscript is organized as follows. Section 2 relates the stability of the DC microgrid to the interface passivity. Section 3 details the feedback passivation design of typical DC converters. Section 4 makes a hardware-in-loop real-time laboratory (RTLab) (version 10.7, OPAL-RT Technologies Inc., Montréal, QC, Canada) simulation to verify the analytic results, followed by conclusions in Section 5.

2. Passivity Criterion of Converters in DC Microgrids

In this section, the DC microgrid is analyzed as a network with two parts: dynamical edges and nodes. The structure is shown in Figure 1. The first part is composed of transmission lines and loads, and another part is composed of converters. According to [11], a multi-agent system is often described by a graph, where nodes represent the dynamic subsystems and edges the interactions between these subsystems. One benefit of the division is that we can take advantage of the the passivity of transmission lines more intuitively. That is, the feedback interconnected system of two passive subsystems is passive then stable. From the viewpoint of power electronics, the division of the nodes and dynamical edges makes it easy to depict the requirement of the interfaces, due to the fact that, in a microgrid system, the nodes (converters) under control are connected via dynamical edges (transmission lines and loads). For a certain structure of the dynamical edges, a common requirement can be put forward to every interface. As a result, it is convenient for applying distributed control law for the nodes.

Denote the vectors that collect the current injections and output voltages of converters and loads by i_T and u_c, respectively. We then have:

$$u_c = Z_N i_T \tag{1}$$

where Z_N is a diagonal transfer function matrix between i_T and u_c, in which the controller going to be designed is included.

Meanwhile, denote Y as the admittance transfer function matrix between the output voltage u_c and the injecting current $-i_T$ of the network composed by transmission lines, that is:

$$- i_T = Y u_c \tag{2}$$

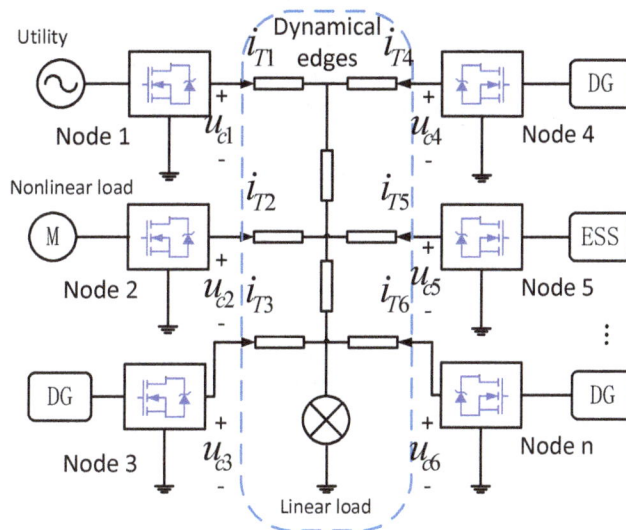

Figure 1. Structure of DC microgrid in this paper. ESS: energy storage system; and DG: distributed generator.

The system composed of Equation (1) and Equation (2) admits the compact block-diagram representation in Figure 2. The converters, loads and the transmission lines are connected via a feedback interconnection. Stable conditions are given based on the feedback interconnection.

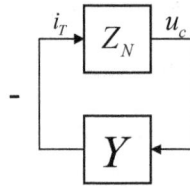

Figure 2. Feedback interconnection of Z_N and Y.

Theorem 1. *If all elements in Z_N are PR, when Y is weakly SPR, the microgrid is asymptotically stable.*

Proof. When Z_N and Y are PR and weakly SPR, respectively, from [20] (Lemma 3.37), the feedback interconnection in Figure 2 is asymptotically stable. □

Remark 1. *The theorem that a system composed of two PR subsystems via feedback interconnection is PR is well-known in passsivity theory. However, it should be mentioned here that passivity yields Lyapunov stability but not asymptotical stability. For instance, $\frac{1}{s}$ is passive but not stable. In [18,21,22], etc., the authors assert the stability only based on the interconnection of PR systems. Even though similar assertions are always true in circuits' analysis, it is essential to explain the fact theoretically. From Definition 2.53 of [20], a PR system is weakly SPR when it is Hurwitz. The assumption of Y is reasonable for when the transmission lines contain resistive elements. Then, Y is Hurwitz, while the edge system consisting of RLC components is passive itself according to [12].*

In fact, Z_N acts like an "interface impedance" of the nodes in a sense. The nodes can also be expressed in an "admittance" form, that is, a transfer function matrix (denote as Y_N) with the input u_c and the output i_T. Similarly, the dynamical edges also have the form of an impedance matrix (denoted as Z). To construct the interconnection feedback structure according to the passivity theory, we have two ways to describe the whole system, using (Z_N, Y) shown in Figure 2 or (Y_N, Z) shown in Figure 3. (Z_N, Z) and (Y_N, Y) are not proposed for the reason that, in these cases, we may encounter the problem as follows: firstly, Z^{-1} or Y^{-1} may not exist. Secondly, when Z or Y is WSPR, the inverse of them may not be WSPR. For example, $\frac{s}{s+1}$ is WSPR while $\frac{s+1}{s}$ is not. In several past works, the analysis are based on admittances or impedances only, and risks will be taken.

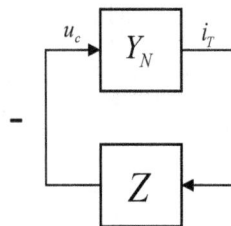

Figure 3. Feedback interconnection of Z and Y_N.

Without loss of generality, the following analysis is based on the structure shown in Figure 2. As Figure 2 shows, in fact, Z_N and Y represent the influence on the stability from converters, loads and electrical network, respectively. Due to the feedback interconnection and the weak SPRness of Y, making every transfer function in Z_N be PR is a convenient way to stabilize the closed loop. It should be mentioned that the loads are passive according to Definition 2.1 in [20]. Ths is to make the grid-connected interfaces of converters be passive and then Z_N is passive.

By the way, the result in Theorem 1 is always true, regardless of the structure changing—for example, the plug and play of distributed generators (DGs) or ESSs and the fluctuations of loads and transmission lines. The remainder of the paper is focused on the interfaces' passivation problem for converters.

3. Feedback Passivation

This section focuses on designing a controller to comply with the interface passive criteria described above.

3.1. Preliminary

The PBC theory offers a systematical method for system passivation, which is successfully applied to controlling many kinds of converters, including buck, boost, buck-boost and three-phase DC–AC VSCs [13,15,21–23]. However, these approaches concentrate on the passivity seeking between the control input and output but not the one with respect to external input and output, required by the interface passivity.

In [19], the authors derive necessary and sufficient conditions for the existence of a control law that renders the closed-loop system SPR with respect to an external input. However, most DC–DC converters, including those considered in this paper, do not satisfy the conditions. Considering that the edge system is WSPR, the goal of feedback passivation on DC–DC converters is relaxed to seek the PR, weaker than the SPR. The necessary and sufficient conditions for the feasibility of rendering a system PR with respect to an external input is still an open problem. This section does not challenge it but presents sufficient designs to realize it for some typical DC–DC converters, which is still challenging and new.

Another difficulty arises from the requirement of unbiased tracking, which is often realized by a proportion integration (PI) controller. In contrast to the static feedback control, PI control is a dynamic control introducing extra state variables. This poses some structure limitations on the feedback passivation design.

The state-space equation of a converter with dynamic controller can be written as:

$$\begin{aligned} \dot{x} &= Ax + Bu + G\omega \\ z &= Hx \end{aligned} \tag{3}$$

where x represents the state variables from the converters and the associated dynamic controller, u, is the control input, with z the output and ω the external inputs. The goal of feedback passivation now is to find a control law:

$$u = Kx \tag{4}$$

to make the transfer function from ω to z be PR, i.e., interface passivity.

For the convenience of the analysis below, review the following lemma firstly, derived from the KYP lemma.

Lemma 1. *Consider the closed system composed by system Equation (3) and the control law Equation (4), and denote $A_c = A + BK$. The interface, i.e., transfer function from ω to z is passive if there exists a P that satisfies:*

$$\begin{aligned} P &= P^T \\ P &> 0 \\ PG &= H^T \end{aligned} \tag{5}$$

This makes:

$$PA_c + A_c^T P \leq 0 \tag{6}$$

3.2. Feedback Passivation for Buck Converters

The buck converter operating in the voltage source mode is shown in Figure 4. With a classical averaging method, the state-space model is given by:

$$
\begin{aligned}
L\dot{i}_L &= d\left(V_g - i_L R_L - u_c\right) + d'\left(-i_L R_L - u_c\right) \\
&= -R_L i_L - u_c + V_g d \\
C\dot{u}_c &= i_L + i_T
\end{aligned}
\tag{7}
$$

When the converter operates in the voltage source mode, the compensator containing an integrator is:

$$
\begin{aligned}
\dot{\xi} &= u_{ref} - u_c \\
d &= u = k_u u_c + k_i \xi + k_L i_L - k_u u_{ref}
\end{aligned}
\tag{8}
$$

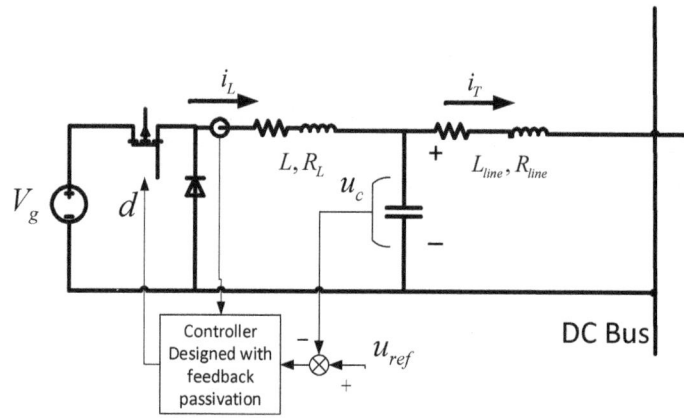

Figure 4. A grid-connected buck converter, operating in the voltage source mode.

In the analysis below, u_{ref} will be ignored, due to the fact that, according to the internal model principle in output regulation theory. When the closed loop system is stable, the asymptotically tracking of u_{ref} is naturally realized. Write Equation (7) and Equation (8) together, and the state- space equation of the open-loop system is given in the form of Equation (3), where:

$$
x = \begin{bmatrix} i_L \\ u_c \\ \xi \end{bmatrix} \quad u = d \quad \omega = i_T \quad z = u_c
$$

$$
A = \begin{bmatrix} -\frac{R_L}{L} & -\frac{1}{L} & 0 \\ \frac{1}{C} & 0 & 0 \\ 0 & -1 & 0 \end{bmatrix} \quad B = \begin{bmatrix} \frac{V_g}{L} \\ 0 \\ 0 \end{bmatrix}
\tag{9}
$$

$$
G = \begin{bmatrix} 0 \\ \frac{1}{C} \\ 0 \end{bmatrix} \quad H = \begin{bmatrix} 0 & 1 & 0 \end{bmatrix}
$$

and the state feedback is:

$$
u = Kx
\tag{10}
$$

Theorem 2. *For the buck circuit shown in Figure 4, when:*

$$K = [k_1, k_2, k_3] \tag{11}$$

satisfies:

$$\begin{cases} k_1 < 0 \\ k_2 < 0 \\ 0 < k_3 < \frac{R_L}{LV_g} \end{cases} \tag{12}$$

the interface, i.e., the transfer function from i_T to u_c, is passive.

Proof. Under the control law Equation (11), the closed loop $A + BK$ can be denoted as:

$$A_c = \begin{bmatrix} -\frac{R_L - k_1 V_g}{L} & -\frac{1 - k_2 V_g}{L} & \frac{k_3 V_g}{L} \\ \frac{1}{C} & 0 & 0 \\ 0 & -1 & 0 \end{bmatrix} \tag{13}$$

When P^{-1} is selected as:

$$P^{-1} = \begin{bmatrix} -(k_2 - \frac{1}{V_g})\frac{V_g}{L} & 0 & 1 \\ 0 & \frac{1}{C} & 0 \\ 1 & 0 & -\frac{V_g k_1 - R_L}{k_3 V_g} \end{bmatrix} \tag{14}$$

as a result:

$$A_c P^{-1} + P^{-1} A_c^T$$

$$= \begin{bmatrix} k_3 - \frac{V_g}{L}\left(k_2 - \frac{1}{V_g}\right)\left(k_1 - \frac{R_L}{V_g}\right) & 0 & 0 \\ 0 & 0 & 0 \\ 0 & 0 & 0 \end{bmatrix} \leqslant 0 \tag{15}$$

Considering Lemma 1, Equation (6) is equivalent to Equation (15); thus, interface passivity is achieved. ☐

Remark 2. *In Theorem 2, an intuitive design method is given, and it is easy to use for engineering practice. Because the proposed method can be treated as a classical dual-loop control law with an inner current loop and an outer voltage loop, the gain of the inner loop is $-k_1$ and the PI parameters of the outer loop are $\frac{k_2}{k_1}$ and $-\frac{k_3}{k_1}$. From the design point of view, we just need to consider Equation (12) as an additional requirement of the dual-loop design method as Figure 5 shows, which can be applied intuitively for engineering practice. From the physical sense, k_1 and k_2 reflect the damping of the inner loop and the outer loop respectively. From Equation (15), it is found that k_3 can not be too large to keep the semi-negativeness of $A_c P^{-1} + P^{-1} A_c^T$. Meanwhile, k_3 should be positive to keep $\|A_c\| = -\frac{k_3 V_g}{C} < 0$, which represents the product of three eigenvalues.*

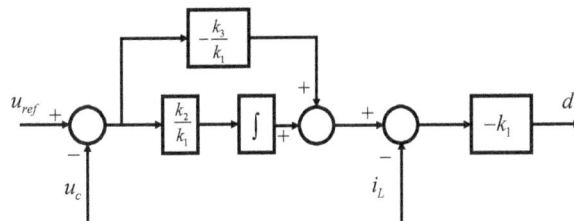

Figure 5. The proposed method can be treated as a classical dual-loop control law with an inner current loop and an outer voltage loop. The gain of the inner loop is $-k_1$ and the PI parameters of the outer loop are $\frac{k_2}{k_1}$ and $-\frac{k_3}{k_1}$.

3.3. Feedback Passivation for Boost Converters

For boost converters shown in Figure 6, the large-signal model can be obtained, through a classical modeling method of averaging, that:

$$
\begin{aligned}
L\dot{i}_L &= d\left(V_g - i_L R_L\right) + d'\left(V_g - i_L R_L - u_c\right) \\
&= -R_L i_L - d' u_c + V_g \\
C\dot{u}_c &= d i_T + d'\left(i_L + i_T\right) = d' i_L + i_T
\end{aligned}
\tag{16}
$$

and the compensator here is:

$$
\begin{aligned}
\dot{\xi} &= I_{ref} - i_L, \\
d &= u = k_u u_c + k_i \xi + k_L i_L
\end{aligned}
\tag{17}
$$

In this case, the converter operates as a current source. The small-signal model of the open loop system can be obtained by linearization in the form of Equation (3), where:

$$
x = \begin{bmatrix} \hat{i}_L \\ \hat{u}_c \\ \hat{\xi} \end{bmatrix} \quad u = \hat{u} \quad \omega = \hat{i}_T \quad z = \hat{u}_c
$$

$$
A = \begin{bmatrix} -\frac{R_L}{L} & \frac{D-1}{L} & 0 \\ \frac{1-D}{C} & 0 & 0 \\ -1 & 0 & 0 \end{bmatrix} \quad B = \begin{bmatrix} \frac{U_c}{L} \\ -\frac{I_L}{C} \\ 0 \end{bmatrix}
\tag{18}
$$

$$
G = \begin{bmatrix} 0 \\ \frac{1}{C} \\ 0 \end{bmatrix} \quad H = \begin{bmatrix} 0 & 1 & 0 \end{bmatrix}
$$

and the state feedback is:

$$
u = Kx
\tag{19}
$$

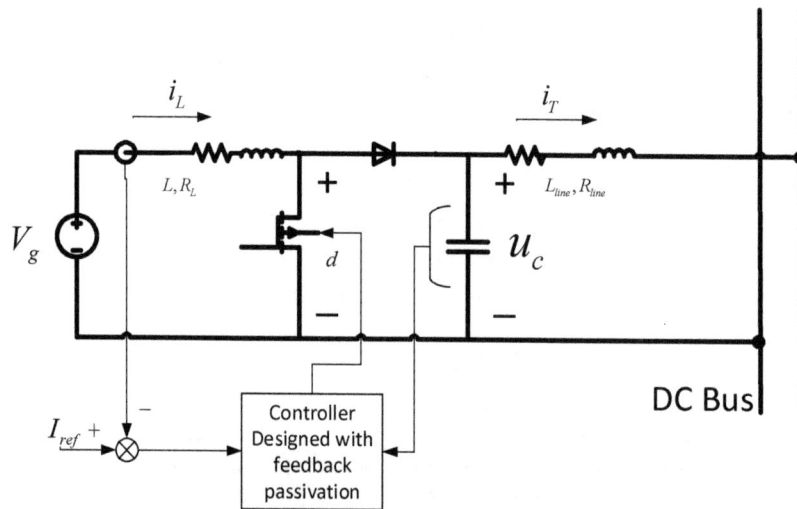

Figure 6. A grid-connected boost converter, operating in the current source mode.

In Equation (18), D, U_c and I_L are the parameters of the equilibrium point. Different from the aforementioned buck circuit, whose regulated output is the interface's output voltage itself, for the boost circuit here, the regulated output is not the interface's output voltage but the inductor's current. As a result, another equation representing the regulated output is:

$$\hat{i}_L = \begin{bmatrix} 1 & 0 & 0 \end{bmatrix} \begin{bmatrix} \hat{i}_L \\ \hat{u}_c \\ \hat{\xi} \end{bmatrix} \tag{20}$$

Familiar with the analysis of the buck converter, when the stability is achieved, the asymptotically tracking of $\hat{I}_{ref}$ is realized, so $\hat{I}_{ref}$ and the output of Equation (20) is ignored in the following analysis.

Theorem 3. *For the boost circuit shown in Figure 6, when:*

$$K = \begin{bmatrix} k_1 & 0 & k_3 \end{bmatrix} \tag{21}$$

where $k_1 < 0, 0 < k_3 \leq \frac{(U_c k_1 - R_L)(k_1 I_L + D - 1)}{I_L L}$, the interface, i.e., the transfer function between i_T and u_c, is passive.

Proof. With feedback gain K, denote $A + BK = A_c$, where:

$$A_c = \begin{bmatrix} \frac{U_c k_1 - R_L}{L} & \frac{D-1}{L} & \frac{U_c k_3}{L} \\ \frac{1-D-I_L k_1}{C} & 0 & -\frac{I_L k_3}{C} \\ -1 & 0 & 0 \end{bmatrix} \tag{22}$$

According to Lemma 1, the interface is passive if there exists a P that satisfies Equations (5) and (6). Here, select P as:

$$P = \begin{bmatrix} La_1 & 0 & La_2 \\ 0 & C & 0 \\ La_2 & 0 & a_3 \end{bmatrix} > 0 \tag{23}$$

where:

$$a_1 = \frac{k_1 I_L}{D-1} + 1$$
$$-\frac{R_0}{L} a_1 \leq a_2 < 0 \tag{24}$$
$$a_3 = -a_1 a_2 R_z - a_2 R_0$$

with:

$$R_0 = R_L - U_c k_1$$

$$R_z = \frac{(1-D)U_c}{I_L}$$

As a result:

$$PA_c + A_c^T P$$
$$= \begin{bmatrix} -2a_1 R_0 - 2La_2 & 0 & 0 \\ 0 & 0 & 0 \\ 0 & 0 & -2a_2^2 R_z \end{bmatrix} \leqslant 0 \tag{25}$$

and the interface passivity is achieved. $\square$

Remark 3. *In Theorem 3, the physical meaning of R_0 and R_z is obvious. The former parameter is equivalent to increase the resistance of the inductor, while the later one can be treated as an equivalent load of the circuit. The proposed control law can be understood as a modified single current loop PI controller, with PI parameters $-k_1$ and k_3, as Figure 7 shows. The control law is also applicable in practice. $k_1 < 0$ keeps the trace of A_c*

(sum of eigenvalues) negative. k_3 should have an upper bound to keep the semi-negativeness of $PA_c + A_c^T P$, and $k_3 > 0$ keeps $\|A_c\| = \frac{I_L k_3 (D-1)}{C}$ (product of eigenvalues) negative.

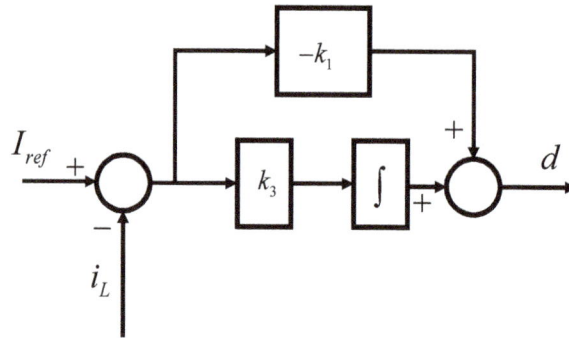

Figure 7. The proposed control law can be understood as a modified single current loop PI controller, with PI parameters $-k_1$ and k_3.

The interface passivity and analysis procedures for buck circuits operating as current sources and boost circuit operating as voltage sources are similar to the two conditions discussed above and not presented here.

4. Real-Time Laboratory Experimental Results

In order to verify the proposed feedback passivation method, a DC microgrid test system is built in the RTLab 10.7. Figure 8 shows the structure of the hardware in loop platform. The system consists of DGs, battery ESSs, local loads and transmission lines.

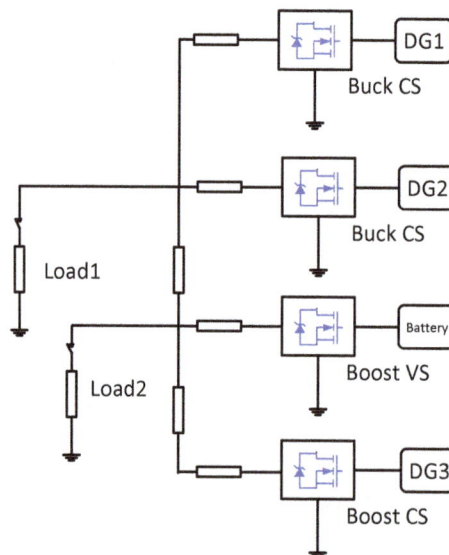

Figure 8. DC microgrid platform in real-time laboratory (RTlab) layout, where CS and VS mean the operating mode of current source and voltage source, respectively.

Three DGs work in current source mode, and the ESS operates in the voltage source mode to maintain the bus voltage at 560 *V*. Buck and boost circuits are adopted for DC–DC conversion. The detailed converter configurations are listed in Table 1.

Table 1. Converter configurations. CS: current source; and VS: voltage source.

Items	DG1	DG2	DG3	ESS
Converter Type	Buck	Buck	Boost	Boost
Mode	CS	CS	CS	VS
Rated Power (kW)	14	5	14	40
Switching Frequency (kHz)	10	10	10	10
Input Voltage (V)	800	800	200	800
Inductor (mH)	2.4	2.4	1.2	1.2
Output Capacitor (μF)	470	470	470	470

In this paper, through the direct code generation method using Matlab (2013b, Mathworks Inc. Natick, MA, USA) and code composer studio (CCS), the proposed passive control algorithm is compiled and downloaded to digital signal processor (DSP) (TMS28335). The control loop in the experiment is realized by external DSP chips, and only the main circuits are simulated in software.

4.1. Test on Feedback Passivation

Firstly, the original passivity criterion is validated. According to the methods proposed in Section 3, based on the control diagrams in Figures 5 and 7, if all converters are controlled under the feedback passivation method to achieve passivity, then the system is stable. As Figure 9a shows, when DG1 and DG2 converters are switched to the conventional dual-loop control to become non-passive, the system experiences a transition first, and then becomes stable with more seriously fluctuating waveforms. In Figure 9b, the two converters are switched back to controllers with feedback passivation and the waveforms restored to be smooth. When three DG converters are controlled to be non-passive, then the system is unstable, as Figure 9c shows. The results above prove the effectiveness of the feedback passivation.

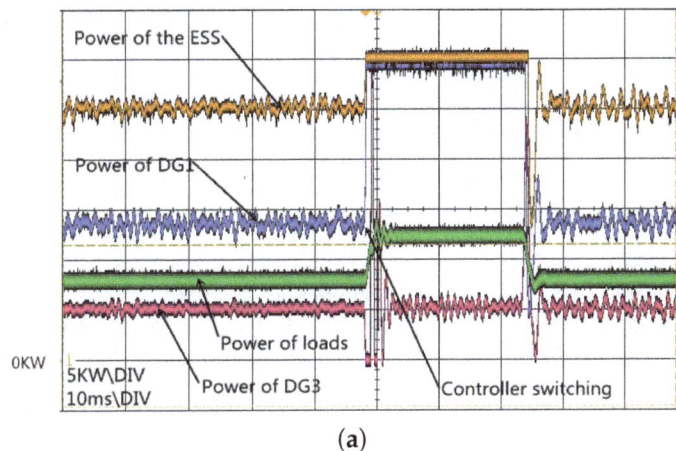

(a)

Figure 9. *Cont.*

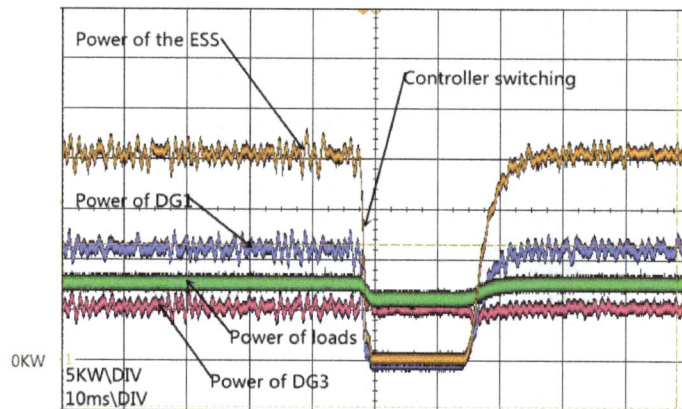

(b)

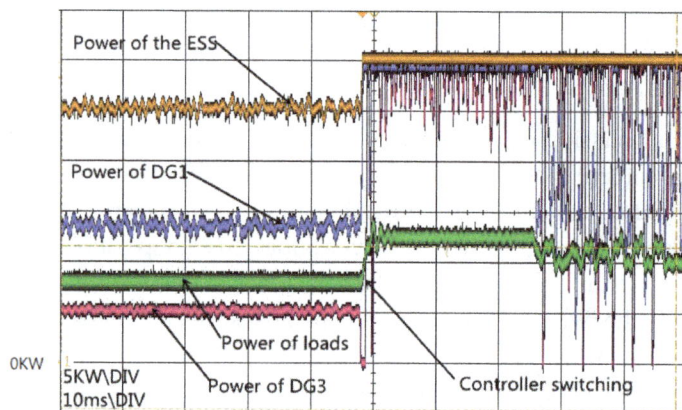

(c)

Figure 9. (**a**,**b**) Two non-passive converters: stable but seriously oscillate; and (**c**) three non-passive converters: unstable.

4.2. Test on Stability in Varying Microgrid Structures

The most obvious advantage of the feedback passivation method is that the stability is always guaranteed no matter to what structure the microgrid changes—for instance, the plug-and-play issue of distributed resources.

To verify this benefit, DGs and loads are switching in this experiment. Three DGs, ESSs and one load (50 Ω, 0.1 mH) are connected to the microgrid initially. Figure 10a shows the transition of the connecting of another load (50 Ω, 0.1 mH). Then, as Figure 10b shows, DG2s are disconnected from the system.

As a comparison, the same procedure is implemented in the scenario while converters of DG1 and DG2 are controlled to be non-passive by the conventional dual-loop controllers and the system is stable. Figure 11a,b show the load connection and DG2 disconnection, respectively.

From the figures above, one can find that the overshoot of the control group is obviously more serious than the group with feedback passivation. For instance, in Figure 10a, no overshoot exists in the power waveform of the ESS, while for the control group, the overshoot is about 2 KW, as Figure 11a shows. The comparison of Figures 10b and 11b confirms the same phenomenon. It should be mentioned that, in Figure 11b, the waveforms' fluctuation becomes smoother after DG2 disconnection. It is reasonable that the removal of a non-passive part from the system is benefit to stability.

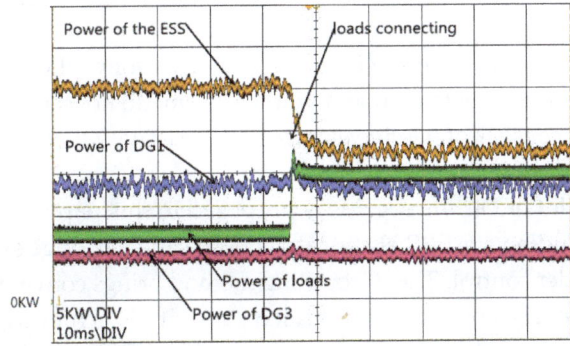

(a)

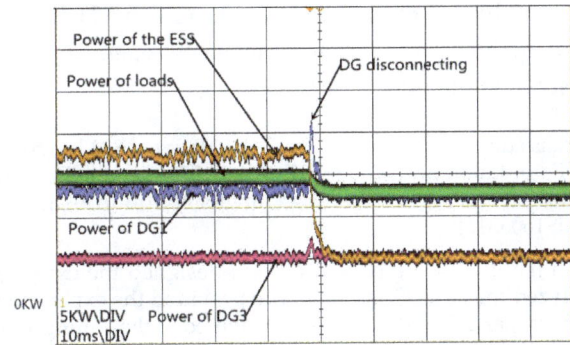

(b)

Figure 10. (**a**) Load connection and (**b**) DG disconnection of converters with feedback passivation.

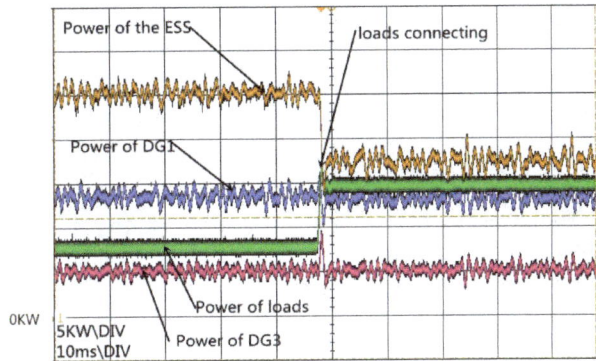

(a)

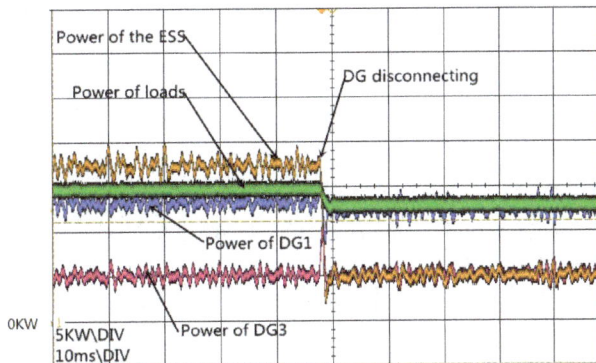

(b)

Figure 11. (**a**) Load connection and (**b**) DG disconnection of the control group.

5. Conclusions

Due to the various network structures of DC microgrids, the conventional methods for the stability analysis, by which the microgrids are treated as a whole system, do not seem suitable enough. Because the electrical network often consists of passive elements, the passivity theory is applied to the interface passivity in this paper.

A decoupling viewpoint of DC microgrids is proposed in this paper, by which the microgrid is divided into two parts, including a WSPR electrical network called an edge system and a node system consisting of converters under control. The decoupling method brings convenience to stability analysis due to the fact that when every converter's interface passivity is achieved, the microgrid is stable. In this paper, the feedback passivation method for typical DC–DC converters is proposed in the time domain, and the proposed method is used in practical applications.

The major contributions of this paper are verified through an RTlab experiment, compared with the typical dual-loop control method, which proves the effectiveness, applicability and robustness of the proposed feedback passivation method.

Acknowledgments: This work is supported by grants from the National 863 Program of China (2015AA050104), the National Natural Science Foundation of China (61374174, 61573314), the Research Project of the State Key Laboratory of Industrial Control Technology of Zhejiang University (ICT1606) and the Science and Technology Project of SGCC (SGZJ0000BGJS1600312).

Author Contributions: Feifan Ji and Ji Xiang conceived and designed the experiments; Feifan Ji performed the experiments; Feifan Ji analyzed the data; Quanming Yue provided the experimental environment; Feifan Ji, Ji Xiang and Wuhua Li wrote the paper. All authors have contributed to the editing and proofreading of this paper.

Conflicts of Interest: The authors declare no conflicts of interest.

Abbreviations

AC	Alternating current
CCS	Code composer studio
DC	Direct current
DSP	Digital signal processor
DG	Distributed generators
EL	Euler–Lagrange
ESS	Energy storage systems
KYP	Kalman–Yakubovich–Popov
PBC	Passivity-based control
PI	Proportion integration
PR	Positive real
PWM	pulse-width-modulation
RLC	resistor-lnductor-capacitor
SISO	Single input single output
SPR	Strictly positive real
VSC	Voltage source converter
WSPR	Weakly strictly positive real

References

1. Justo, J.; Mwasilu, F.; Lee, J.; Jung, J.W. AC-Microgrids versus DC-microgrids with distributed energy resources: A review. *Renew. Sustain. Energy Rev.* **2013**, *24*, 387–405.

2. Kahrobaeian, A.; Mohamed, Y.R. Analysis and mitigation of low-frequency instabilities in autonomous medium-voltage converter-based microgrids with dynamic loads. *IEEE Trans. Ind. Electron.* **2014**, *61*, 1643–1658.

3. Kundur, P.; Balu, N.J.; Lauby, M.G. *Power System Stability and Control (The EPRI Power System Engineering)*; McGraw-Hill: New York, NY, USA, 1994.

4. Kundur, P.; Lauby, M.G. *Analysis of Electric Machinery and Drive Systems*; IEEE Press: Piscataway, NJ, USA, 2002.

5. Katiraei, F.; Iravani, M.; Lehn, P. Micro-grid autonomous operation during and subsequent to islanding process. *IEEE Trans. Power Deliv.* **2005**, *20*, 248–257.

6. Kabalan, M.; Singh, P.; Niebur, D. Large signal lyapunov-based stability studies in microgrids: A Review. *IEEE Trans. Smart Grid* **2016**, doi:10.1109/TSG.2016.2521652.

7. Lu, X.; Sun, K.; Guerrero, J.; Vasquez, J.; Huang, L.; Wang, J. Stability enhancement based on virtual impedance for DC microgrids with constant power loads. *IEEE Trans. Smart Grid* **2015**, *6*, 2770–2783.

8. Shamsi, P.; Fahimi, B. Stability assessment of a DC distribution network in a hybrid micro-grid application. *IEEE Trans. Smart Grid* **2014**, *5*, 2527–2534.

9. Lu, X.; Wan, J. Modeling and control of the distributed power converters in a standalone DC microgrid. *Energies* **2016**, *9*, 217.

10. Middlebrook, R.D. The general feedback theorem: A final solution for feedback systems. *IEEE Microw. Mag.* **2006**, *7*, 50–63.

11. Xiang, J.; Li, Y.; Hill, D. Cooperative output regulation of multi-agent systems coupled by dynamic edges. *IFAC Proc. Vol.* **2014**, *19*, 1813–1818.

12. Anderson, B.; Vongpanitlerd, S. *Network Analysis and Synthesis: A Modern Systems Theory Approach*; Prentice Hall: Upper Saddle River, NJ, USA, 1973.

13. Sira-Ramirez, H.; Perez-Moreno, R.; Ortega, R.; Garcia-Esteban, M. Passivity-based controllers for the stabilization of DC-to-DC power converters. *Automatica* **1997**, *33*, 499–513.

14. Harnefors, L.; Wang, X.; Yepes, A.; Blaabjerg, F. Passivity-based stability assessment of grid-connected VSCs–An overview. *IEEE J. Emerg. Sel. Top. Power Electron.* **2016**, *4*, 116–125.

15. Lee, T.S. Lagrangian modeling and passivity-based control of three-phase AC/DC voltage-source converters. *IEEE Trans. Ind. Electron.* **2004**, *51*, 892–902.

16. Hoang, H.; Tuan, H.; Nguyen, T. Frequency-selective KYP lemma, IIR filter, and filter bank design. *IEEE Trans. Signal Process.* **2009**, *57*, 956–965.

17. Hoang, H.; Tuan, H.; Apkarian, P. A Lyapunov variable-free KYP lemma for SISO continuous systems. *IEEE Trans. Autom. Control* **2008**, *53*, 2669–2673.

18. Gu, Y.; Li, W.; He, X. Passivity-based control of DC microgrid for self-disciplined stabilization. *IEEE Trans. Power Syst.* **2015**, *30*, 2623–2632.

19. Arcak, M.; Kokotovi, P. Feasibility conditions for circle criterion designs. *Syst. Control Lett.* **2001**, *42*, 405–412.

20. Brogliato, B.; Lozano, R.; Maschke, B.; Egeland, O. *Dissipative Systems Analysis and Control*; Springer: Berlin/Heidelberg, Germany, 2007.

21. Hernandez-Gomez, M.; Ortega, R.; Lamnabhi-Lagarrigue, F.; Escobar, G. Adaptive PI stabilization of switched power converters. *IEEE Trans. Control Syst. Technol.* **2010**, *18*, 688–698.

22. Jeltsema, D.; Scherpen, J. A power-based perspective in modeling and control of switched power converters [Past and Present]. *IEEE Ind. Electron. Mag.* **2007**, *1*, 7–54.

23. Leyva, R.; Cid-Pastor, A.; Alonso, C.; Queinnec, I.; Tarbouriech, S.; Martinez-Salamero, L. Passivity-based integral control of a boost converter for large-signal stability. *IEE Proc. Control Theory Appl.* **2006**, *153*, 139–146.

5

Operation Optimization of Steam Accumulators as Thermal Energy Storage and Buffer Units

Wenqiang Sun [1,2,*], Yuhao Hong [1,3] and Yanhui Wang [1]

[1] Department of Thermal Engineering, School of Metallurgy, Northeastern University, Shenyang 110819, China; hong.yuhao@chinaboilers.com (Y.H.); 13080876162@163.com (Y.W.)

[2] State Environmental Protection Key Laboratory of Eco-Industry, Northeastern University, Shenyang 110819, China

[3] Department of Technology, Hangzhou Boiler Group Co., Ltd., Hangzhou 310021, China

* Correspondence: neu20031542@163.com or sunwq@mail.neu.edu.cn

Academic Editor: Brian Agnew

Abstract: Although steam is widely used in industrial production, there is often an imbalance between steam supply and demand, which ultimately results in steam waste. To solve this problem, steam accumulators (SAs) can be used as thermal energy storage and buffer units. However, it is difficult to promote the application of SAs due to high investment costs, which directly depend on the usage volume. Thus, the operation of SAs should be optimized to reduce initial investment through volume minimization. In this work, steam sources (SSs) are classified into two types: controllable steam sources (CSSs) and uncontrollable steam sources (UCSSs). A basic oxygen furnace (BOF) was selected as an example of a UCSS to study the optimal operation of an SA with a single BOF and sets of parallel-operating BOFs. In another case, a new method whereby CSSs cooperate with SAs is reported, and the mathematical model of the minimum necessary thermal energy storage capacity (NTESC) is established. A solving program for this mathematical model is also designed. The results show that for UCSSs, applying an SA in two parallel-operating SSs requires less capacity than that required between a single SS and its consumer. For CSSs, the proposed minimum NTESC method can effectively find the optimal operation and the minimum volume of an SA. The optimized volume of an SA is smaller than that used in practice, which results in a better steam storage effect.

Keywords: steam accumulator (SA); optimal operation; minimum volume; steam source (SS); necessary thermal energy storage capacity (NTESC)

---

1. Introduction

Steam has been used as a heating or power source in various industries, including chemical, dyeing, pharmaceutical, and electrical industries [1–4]. However, some steam is discharged directly into the atmosphere due to underdeveloped equipment and recovery techniques. The Chinese steel industry can be used as an example. The waste heat recovery rate of large- and medium-scale steel plants in China is approximately 25.8%. This leaves room for further recycling and recovery [5].

The boiler is a major device in a steam system used in industrial production and residential heating. In countries like China, the actual operating thermal efficiency of a boiler is only approximately 57%, which is much lower than its designed thermal efficiency [6]. For steam supply systems with boilers as a steam source (SS), there generally exists an imbalance between steam generation and steam demand. Because of the intermittently operated steam users, the amount of steam consumed shifts frequently, and thus, the steam demand often fluctuates. Therefore, the boiler combustion or feedwater capacity is usually adjusted to keep the boiler pressure stable. It has been shown by Tanton et al. [7] that the thermal efficiency of a boiler decreases with increasing load fluctuation frequency. In addition, Elman [8] concluded that the thermal efficiency of a boiler reduces with the increase in load disturbance amplitude.

To balance the steam load between SSs and consumers, steam accumulators (SAs) are used as thermal energy storage and buffer units [9,10], which improves the operating condition and supplied steam quality of boilers, thus saving large amounts of energy. With the development of energy-saving technologies and devices, SAs are employed in various fields outside of boiler systems. This widespread use can be attributed to the main advantages of higher energy storage performance, rapid steam discharge rate, and elimination of steam load fluctuation. In a basic oxygen furnace (BOF) waste heat recovery system, SAs store the intermittently generated steam and then transform the intermittent steam supply into a continuous and stable SS for a turbine. Additionally, as an important part of the steam catapult of a warship, marine SAs have the additional characteristics of short charging and discharging time, large instantaneous steam consuming, etc. In a solar thermal power system, SAs can generate a continuous flow of saturated steam when solar radiation fluctuates in order to make the thermal power system run smoothly without interruptions [11–13].

Previous studies focused on the thermodynamic performance of SAs. Some models [10,11,14] are based on the thermal equilibrium between the liquid and steam phase, although a substantial deviation from thermal equilibrium could exist in SAs during rapid steam storage or release. To overcome the drawbacks of the equilibrium SA model, the non-equilibrium model of SA operation was developed by Studovic and Stevanovic [15]. Additionly, Stevanovic et al. [16,17] developed a non-equilibrium method to evaluate the SA, and compared the results with equilibrium situations.

In SA modelling, Maklakov et al. [18] studied the factors influencing the stable operation of an SA. Walter and Linzer [19] analyzed the stable mass flow and energy flow in an SA. Liu [20] established a theoretical model to discuss the change in the wall temperature of SAs, concluding that the imbalance of the temperature field inside the SA leads to the resultant irreversible energy loss. To improve the thermal efficiency of SAs, Yang and Manning [21], Steinmann and Eck [22], and Su et al. [23] investigated SAs in industrial plants, and the results of these industrial experiments provided theoretical bases for SA design and optimization.

Engelhardt et al. [24] developed an approximate, yet accurate method for calculating average volume concentrations of impurities and corrosion products in an SA. Cao [25] analyzed the necessary thermal energy storage capacity (NETSC) and established a calculation model from an economic perspective. Valenzuela et al. [26] discussed the control scheme of the direct steam generation in solar boilers. All of these previous studies contributed greatly to the design and application of SAs.

However, as an effective and efficient energy storage and buffer unit, their true application has not yet been fully satisfied. Currently, even in situations that SAs can meet steam demands, boilers are more common than SAs because the initial investment and land occupied are similar but SAs do not generate steam directly. It is expected that a minimum, yet reasonable volume may promote the application of SAs, as the investment required for an SA depends largely on its volume. In addition, the characteristics of SS affect its function in the supply system, and a wide variety of SSs could supply steam for SAs. Thus, in this work, SSs are divided into two types: controllable and uncontrollable, while the optimal operation and minimum volume of an SA are studied.

2. Principle of SA and Classification of SSs

2.1. SA Operation Principle

The operating status of an SA consists of a steam storage process and a steam release process, as shown in Figure 1. The two processes and their operation principles are as follows:

- Steam storage process: The pressure of steam from a high-temperature SS is higher than that inside the SA. When the steam inlet valve is open, steam flows into the SA automatically. The saturated water and saturated steam with relatively low temperature and pressure is stored in the SA, to which high-temperature steam is added. Through rapid heat exchange, the high-temperature steam is cooled, and the previously mentioned saturated water and steam are heated to a higher temperature to form a new equilibrium state. As the steam storage process continues, it sees a steady rise in the temperature of the water and steam inside the SA, followed by an increasing pressure and water level. Once the internal pressure reaches the specified maximum value, the steam inlet valve closes, and the steam storage process ends. As shown in Figure 2, during the steam storage process, the pressure rises from the specified minimum value p_2 to the maximum p_1, and the steam is condensed, leading to an increase in the water enthalpy along the saturated liquid line at the left side of Figure 2.

- Steam release process: This is an inverse process of the steam storage process detailed above. When the SA provides steam for users, the valve at the steam outlet is open. Since the pressure in the steam pipe is lower than that in the SA, the saturated steam in the SA releases automatically under the pressure difference, resulting in a lower pressure inside the SA. When the internal pressure is lower than the saturation pressure corresponding to the temperature of water stored, the saturated water becomes superheated, and the water boils immediately to evaporate as saturated steam. At this point the water temperature and water level in the SA drop until they have decreased to the specified minimum value, at which point the steam outlet valve closes, and the steam release process ends. During this process, the pressure decreases from p_1 to p_2, as shown in Figure 2, meaning that the superheated water converts to saturated water, and thus, the water enthalpy reduces along the saturated liquid line.

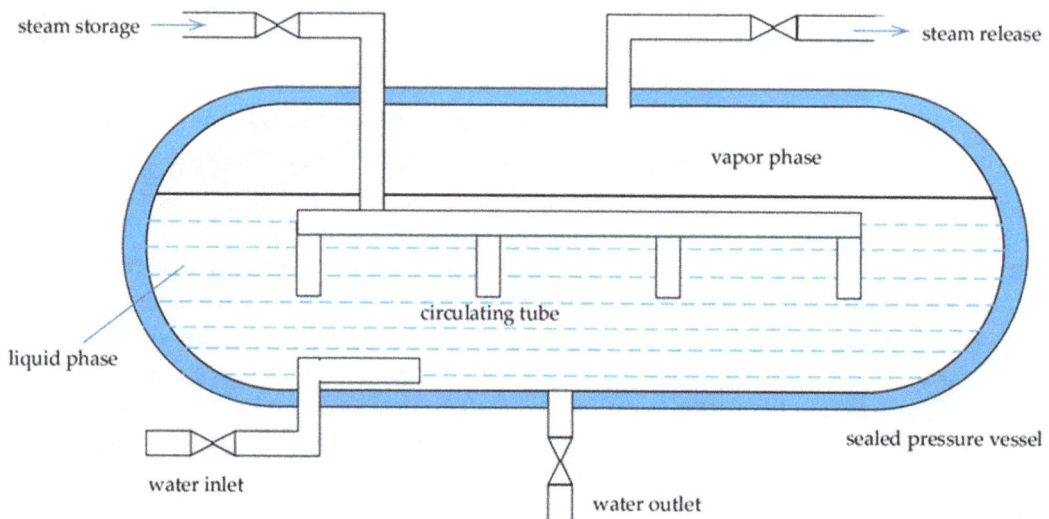

Figure 1. Schematic layout of a steam accumulator (SA).

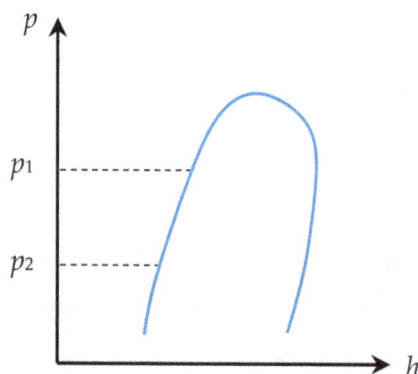

Figure 2. Pressure-enthalpy diagram of steam storage and release processes. Pressure varies within p_1 to p_2 in the whole SA operating process.

2.2. SS Classification

Usually, SSs are divided into high-, medium-, and low-pressure according to their pressure. In this study, another classification system is used according to the controllability of SS:

- Controllable steam source (CSS): The steam generation process is of considerable controllability, and therefore, the steam's flow rate, pressure, and temperature are steady, change periodically, or can be controlled easily, e.g., coal-fired boiler and other station boilers.

- Uncontrollable steam source (UCSS): This type generally has the following characteristics: (a) The generation of the steam is influenced by other factors, even randomly intermittent factors; (b) the flow rate of the steam supplied fluctuates frequently; and (c) the temperature and pressure of the steam supplied vary significantly over a large range. It is hard to control the flow rate, pressure, and temperature of steam from UCSSs. Examples include solar thermal plants, incinerators fueled by municipal solid wastes, and basic oxygen furnace (BOF) waste heat power generation.

In both CSSs and UCSSs, SAs are installed between SSs and steam consumers. They can not only exponentially increase the instant steam supply capacity but also stabilize the flow rate, pressure, and temperature of steam supplied at an expected range. Thus, they work both as a thermal energy storage unit and as a buffer unit in the steam supply system, especially for intermittent and fluctuating SSs or steam consumers.

3. Mathematical Model of Steam Accumulator (SA) Operation Optimization

3.1. SA Operation Optimization for UCSS

As mentioned above, the steam generation of UCSSs is easily affected by other random factors, is usually intermittent and frequently fluctuates. Thus, it is of great importance to optimize the operation of SA for UCSSs.

The operating pressure of SAs change within the specified range. The specified maximum value, called charging pressure (p_1 in Figure 2), is the highest pressure at the end of the steam storage process. The specified minimum value, called discharging pressure (p_2 in Figure 2), is the lowest pressure at the end of steam release process.

The thermal energy storage capacity (TESC) of an SA is the amount of steam mass, or steam thermal energy, generated from the SA from p_1 down to p_2. For a fixed SA and SS with certain parameters, TESC is determined by the difference between the charging pressure and the discharging pressure. TESC can be increased by increasing the difference between the charging and discharging pressures. For a specified pressure difference, TESC decreases with the increase in discharging pressure. However, the charging pressure cannot be raised freely due to limited pressure of SS and steam pipe resistance. Additionally, discharging pressure is limited by the minimum steam pressure demanded by steam consumers as well as the pipe resistance.

Charging pressure is set as p_1, and discharging pressure is set as p_2. According to the definition mentioned above,

$$p_1 = p_{SS} - \Delta p_1, \tag{1}$$

$$p_2 = p_{SC} + \Delta p_2, \tag{2}$$

where p_{SS} and p_{SC} are the pressures of SS and steam consumers respectively, measured in MPa; Δp_1 is the pressure drop from SS to SA, due to the pipe resistance, measured in MPa; and Δp_2 is the pressure drop from SA to steam consumers, measured in MPa.

The TESCs of the metal part of the SA and vapor phase are neglected because they are much lower than in the water phase in SA. When calculating, only the TESC of the water phase is considered; thus, the volume of an SA can be calculated by:

$$V = \frac{G}{g \times \eta \times \varphi} \tag{3}$$

where V is the volume of SA, measured in m^3; G is the necessary thermal energy storage capacity (NTESC) of SA, measured in kJ (in caloric) or kg (in mass); η is the SA efficiency and is between 0.98 and 0.99; φ is water filling coefficient, which is defined as the ratio of water filling volume to the total SA volume and is approximately 0.9; and g is specific thermal energy storage capacity (STESC), which is defined as the TESC of 1 m^3 water from the full storage state to the completely released state, measured in kJ/m^3 (in caloric) or kg/m^3 (in mass).

The STESC can be determined by two methods: one based on the principle of energy balance, and the other by integrating the steam release equation. In regards to the energy balance method, it is assumed that x kg steam will be generated when 1 kg saturated water changes from p_1 to p_2, and thus the energy balance equation of steam release process is:

Enthalpy of saturated water (before pressure drop) = Enthalpy of released steam
(after pressure drop) + Enthalpy of remaining saturated water (after pressure drop).

In this study, enthalpies are measured by average values. Then,

$$h_1' = x \left(\frac{h_1'' + h_2''}{2} \right) + (1 - x)h_2' \tag{4}$$

where x is the amount of steam generated per unit water at p_2, measured in kg-steam/kg-water; h_1' and h_2' are the enthalpy of saturated water at p_1 and p_2, respectively, measured in kJ/kg; h_1'' and h_2'' are the enthalpy of saturated steam at p_1 and p_2, respectively, measured in kJ/kg.

According to Equation (4), STESC can be expressed as:

$$g = x \times \rho_1 = \rho_1 \times \frac{h_1' - h_2'}{\left(\frac{h_1'' + h_2''}{2} \right) - h_2'}, \tag{5}$$

where ρ_1 is the density of saturated water at p_1, measured in kg/m^3.

As for the integral method, it is assumed that X kg water is contained in the SA. The thermal energy of dq (in kJ/kg) will be released when the pressure inside the SA decreases from p to $(p - \Delta p)$. In addition, the released thermal energy makes dX kg water flash to vapor. The evaporation latent heat of water is r, then

$$X dq = r dX \tag{6}$$

thereby

$$\frac{dX}{X} = \frac{dq}{r} = \frac{dq/T}{r/T} = \frac{ds'}{r/T} \tag{7}$$

where $ds' = dq/T$ is the entropy change of water, kJ/(kg·K), and T is the absolute temperature of water, K.

Integrating Equation (7) from p_1 to p_2 yields

$$\ln \frac{X_2}{X_1} = \int_{p_1}^{p_2} \frac{ds'}{r/T} \tag{8}$$

where X_1 and X_2 are the amount of water at pressure p_1 and p_2, respectively.

Thus, the STESC can be calculated as

$$g = \frac{X_1 - X_2}{X_1} \times \rho_1 = \rho_1 \times \left(1 - e^{\int_{p_1}^{p_2} \frac{ds'}{r/T}} \right) \tag{9}$$

According to Equation (9), the STESCs under various charging and discharging pressures are listed in Table 1.

Table 1. Specific thermal energy storage capacities (STESCs) under different charging and discharging pressures (in kg/m^3).

Disharing Pressure (MPa)	Charging Pressure (MPa)											
	0.7	0.8	0.9	1.0	1.2	1.4	1.5	1.6	1.8	2.0	2.2	2.4
0.2	66	74	81	87	99	110	115	119	127	136	143	149
0.3	48	57	65	71	84	95	99	104	113	121	127	134
0.4	33	42	50	57	69	81	86	91	100	108	116	122
0.5	22	31	39	46	59	70	76	80	90	97	106	112
0.6	-	-	28	34	47	59	65	69	78	87	95	102
0.7	-	-	-	-	38	50	56	61	70	78	86	92
0.8	-	-	-	-	-	-	47	53	63	71	78	84
0.9	-	-	-	-	-	-	-	44	55	63	70	76
1.0	-	-	-	-	-	-	-	-	47	56	64	70
1.1	-	-	-	-	-	-	-	-	-	49	57	63
1.2	-	-	-	-	-	-	-	-	-	43	50	56

NTESC should be analyzed and calculated based on the real SS condition, the fluctuation regulation of steam consumption load, and the structure of the steam supply system. Initially, it is recommended to work out the average load line according to the real-time steam consumption load curve. Figure 3 is a steam consumption load curve during a cycle. The function of the curve is:

$$Q = f(t) \tag{10}$$

where Q is the real-time steam consumption load, t/h; t is time, in h; f is the function that presents the relationship between Q and t.

The average steam load can be expressed as:

$$\overline{Q} = \frac{1}{t_2 - t_1} \int_{t_1}^{t_2} f(t)\mathrm{d}t \tag{11}$$

where $\overline{Q}$ is the average steam consumption load, measured in t/h; t_1 and t_2 are beginning time and end time, respectively, measured in h.

Then, the difference between real-time load curve and average load line is calculated, and the difference integrated to obtain the integral curve, as shown in Figure 4. The difference between the maximum and minimum points in Figure 4 is the NTESC:

$$G = \Delta Q_{\max} - \Delta Q_{\min} \tag{12}$$

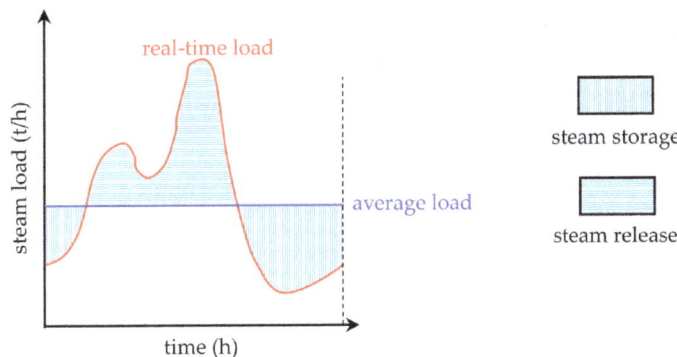

Figure 3. Steam consumption load curve during a cycle. The curve denotes the real-time load; the straight line denotes the average load; the shadow in horizontal line denotes the steam storage; and the shadow in vertical line denotes the steam release.

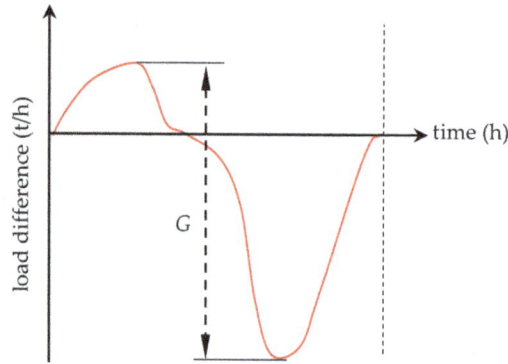

Figure 4. Integral curve of load difference. It is the integral of the difference between real-time load and average load. G is the difference between the maximum and minimum values on the integral curve.

3.2. SA Operation Optimization for CSS

The steam generation process of a CSS has considerable controllability, and thus, the flow rate, pressure, and temperature of the generated steam can be controlled. It is of great significance to optimize the cooperation of a CSS and an SA. However, SA designers routinely determine the volume of an SA according to steam users without considering its cooperation with a CSS. If it is considered, the operation of an SA and a CSS such as a boiler will be regulated to interact with each other, leading to a minimum SA volume and investment cost.

In a cycle, the steam consumption load is $Q(t)$, and the boiler's operating load is $M(t)$; thus, the amount of steam that should be stored or released is $C(t) = M(t) - Q(t)$. If $C(t) > 0$, the SA works in steam storage case, whereas if $C(t) < 0$, the SA works in steam release case. By integrating $C(t)$, the NTESC can be obtained.

Given that a boiler is generally described by its daily load curve, one day is chosen as a cycle with a time length of 24 h. The number of load records in one hour is R, so the number of total load records is $24R$, labelled $S = 1, 2, \ldots, 24R$. $Q(t)$ can be determined by fitting the $24R$ load records. The allowed maximum number of changing boiler loads in a cycle is assumed as N, i.e., the allowed maximum segments is N. According to practical experience for a boiler, a load change per 2–3 h has very little influence on its thermal efficiency. The time span between two adjacent load changes is set as L, which is no less than 2 or 3 h.

There are several feasible segmentation schemes satisfying the requirements of N, L, R, and S mentioned above. The optimal segmentation scheme is the one with the lowest NTESC among the feasible schemes. Assume the load record labelled s is the starting time of a cycle. For the k-th segmentation scheme, the beginning time of segment i is $b_i = t_{i-1}$, and the end time is $e_i = t_i$. In segment i, the operating load of the boiler at time t is:

$$M(t) = \frac{t - t_{i-1}}{t_i - t_{i-1}} \int_{t_{i-1}}^{t_i} Q(t)\mathrm{d}t. \tag{13}$$

A set of b_i and e_i ($i = 1, 2, \ldots, N$) specifies a segmentation scheme k. If the value of S changes from 1 to $24R$, the total number of segments is K:

$$K = \begin{cases} 1, & (N = 1) \\ 24R - N \times R \times L + 1, & (N = 2) \\ \frac{1}{2} \sum_{P_{N-2}=1}^{24R-NRL+1} \sum_{P_{N-3}=1}^{P_{N-2}} \cdots \sum_{P_2=1}^{P_3} \sum_{P_1=1}^{P_2} P_1(P_1 + 1), & (N \geq 3) \end{cases} \tag{14}$$

where P denotes the permissible value of a cut-point in a segment.

For specified L and N, the minimization of NTESC can be expressed as:

$$\begin{cases} \min G = C_{max} - C_{min}; \\ \text{s.t. } b_i = 1, \ e_N = 24R, \\ LR(i-1)+1 \leq b_i \leq 24R - LR(N-i+1), \ for \ i = 2, 3, \ldots, N, \\ b_{i+1} - b_i \geq LR, \ for \ i = 1, 2, \ldots, N-1, \\ e_i = b_{i+1} - 1, \ for \ i = 1, 2, \ldots, N-1, \\ S = 1, 2, \ldots, 24R, \\ k = 1, 2, \ldots, K. \end{cases} \qquad (15)$$

To solve the minimization problem (15) to obtain the minimum NTESC and optimal segmentation scheme from all feasible schemes, an optimization algorithm should be developed. The basic process for solving this problem is shown in Figure 5, where a cycle is divided into two segments to be integrated. For each segment, $M(t)$ and $C(t)$ can be calculated by Equation (13) from beginning time b_i to end time e_i. Starting from scheme A, e_1 and b_2 is changed gradually and recalculated until scheme B. Then, the starting time of a cycle should change in turn within sets of S, i.e., b_1 changes from 1 to 2 to 3, all the way until to $24R$; while e_2 changes from $24R$ to 1, to 2, until $(24R - 1)$. The above process is repeated to calculate all feasible two-segment schemes.

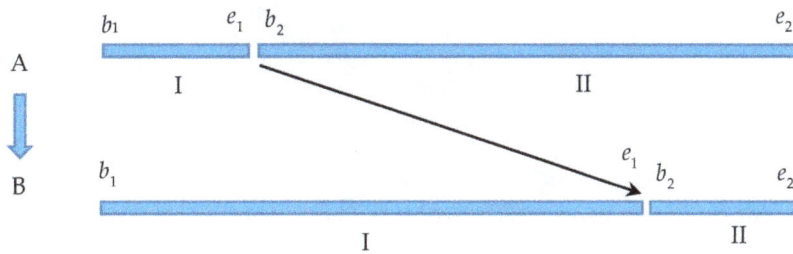

Figure 5. Basic process of model solving.

When the number of segments reaches N, the exhaustive segmentation schemes can be determined via multiple recursive calls of a basic process. Finally, the optimal scheme, i.e., the minimum NTESC, can be screened out. Note that the beginning time and end time of segment i (see Figure 6) are within the ranges:

$$\begin{cases} e_{i-1} + LR + 1 \leq b_i \leq 24R - LR(N-i+1), \\ b_i + LR \leq e_i \leq 24R - LR(N-i). \end{cases} \qquad (16)$$

The whole process for solving this problem is illustrated in Figure 7.

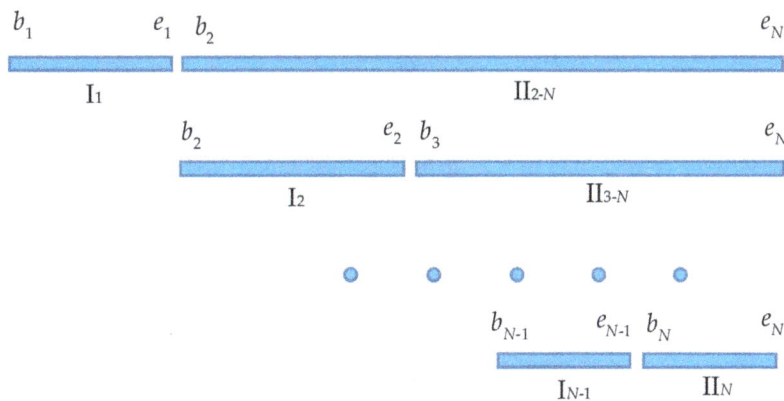

Figure 6. Recursive call of basic process for segment number at N.

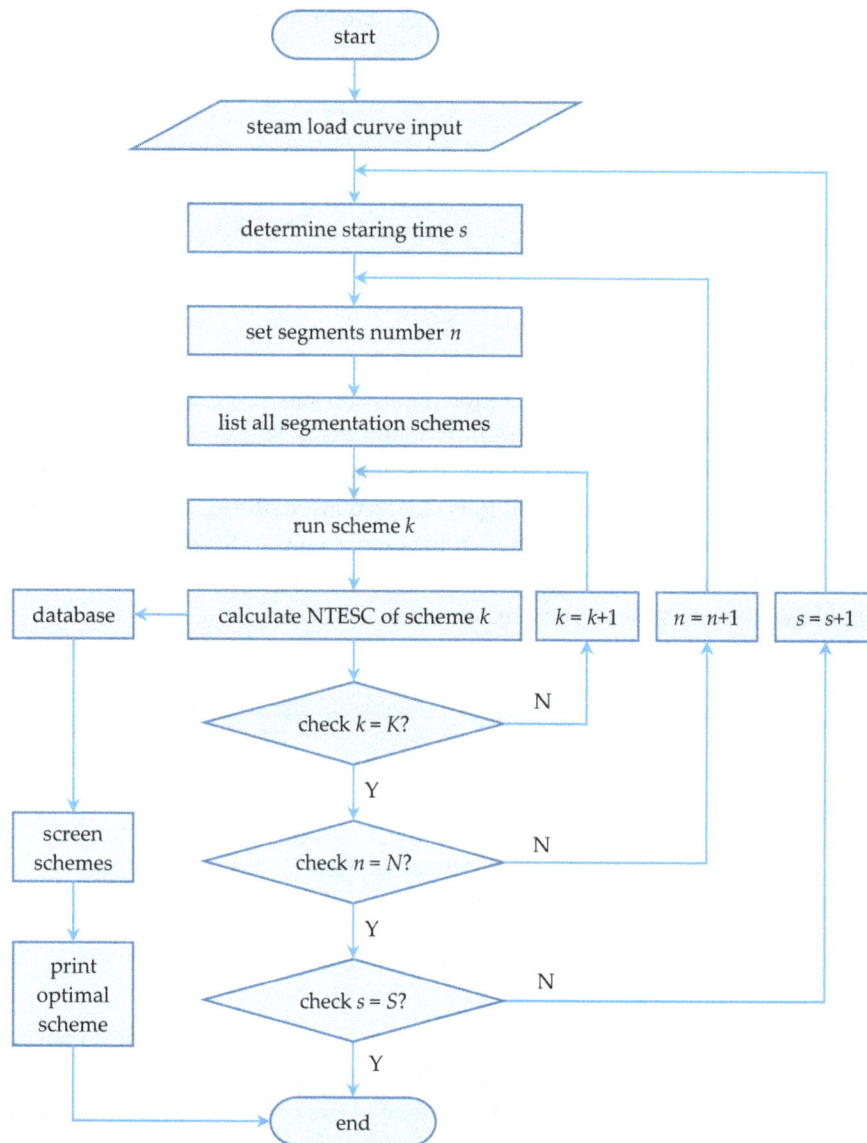

Figure 7. Problem solving flow chart. NTESC: necessary thermal energy storage capacity.

4. Case Results and Discussion

4.1. SA Cooperating with a UCSS

There is a marked steam fluctuation in a BOF waste heat recovery system. The smelting cycle is approximately 30–40 min, of which the oxygen blowing period is approximately 10–16 min. Steam is recovered only during the blowing period. SAs gather the fluctuating and intermittently generated steam and supply that steam to users in a continuous and steady manner.

In a steel-making branch plant of Ansteel, China, there are two BOFs with a nominal capacity of 260 t. Table 2 lists the parameters and Figure 8 shows the steam generation curve of a 260 t BOF. It can be seen from Figure 8 that the oxygen blowing period starts from 9'30" and ends at 24'30". From Equation (11), the average steam generation during the oxygen blowing period and the whole smelting cycle can be obtained as follows: the average value of blowing period is 120.64 t/h, and the average value of smelting cycle is 51.70 t/h. The amount of generated steam in one smelting cycle is 30.16 t. Note that there are two peaks. This is because a second oxygen blowing period is usually conducted after the first one, according to the liquid steel detection result of the first blowing period, with the purpose of controlling the carbon content in the liquid steel.

Table 2. Parameters of basic oxygen furnaces (BOFs), steam accumulators (SAs), and generated steam.

Item	Value	Unit
nominal capacity of BOF	260	t
smelting cycle of BOF	35	min
oxygen blowing period of BOF	15	min
pressure of generated steam	2.45	MPa
temperature of generated steam	223	°C
charging pressure of SA	2.40	MPa
discharging pressure of SA	1.05	MPa
water filling coefficient of SA	0.90	-
thermal efficiency of SA	0.99	-

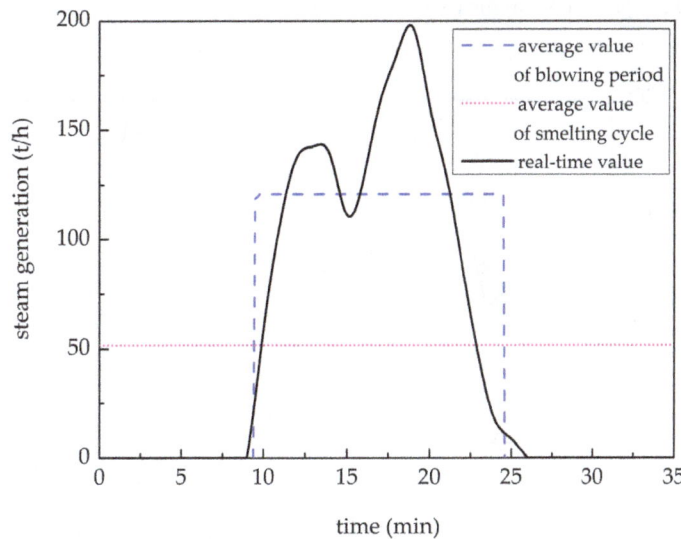

Figure 8. Steam generation of a 260 t BOF. The solid line denotes the real-time load value; the dashed line denotes the average value of the blowing period; and the short dot denotes the average value of smelting cycle.

According to the real-time steam generation curve and average steam generation line shown in Figure 8, the NTESC is 17.24 t from Equation (12), as found via integral calculation.

The charging pressure of the SA is the difference between the pressure of steam from the BOF and pipe resistance and has a value of 2.40 MPa. In addition, the discharging pressure is the sum of the pressure of the steam entering the turbines and the pipe resistance and has a value of 1.05 MPa.

The enthalpy of steam can be obtained once the charging pressure and discharging pressure are calculated. From Equation (5) and Table 1, the STESC is 74.6 kg/m^3. The water filling coefficient is 0.9, and the thermal efficiency of the SA is 0.99. The volume of the SA for each BOF can be calculated from Equation (3) as 259.30 m^3.

With the development of BOF smelting and energy recovery technologies, BOF upgrades towards the large-scale direction, and the amount of steam generated increases. Therefore, it is almost a certainty that the amount of steam needing to be stored or released will rise in the near future. However, the cost of SAs is currently too high to be employed in these cases. An SA cooperating with sets of parallel-operating BOFs may be a more optimized solution that can be economically employed in future steam systems.

To minimize the steam generation peak value, two BOFs blow oxygen in turn, as shown in Figure 9. A smelting cycle is schematically depicted in a rectangle around a dashed line. The starting time is set at 2′30″ later than when BOF #B ends its oxygen blowing. The oxygen blowing period of BOF #A starts from 1′30″ to 16′30″, while BOF #B blows from 18′30″ to 33′30″. Therefore, the case where two BOFs generate steam at the same time can be effectively avoided. The steam generation is shown in Figure 10. The average steam generation during the oxygen blowing period and the whole smelting cycle can be obtained by Equation (11).

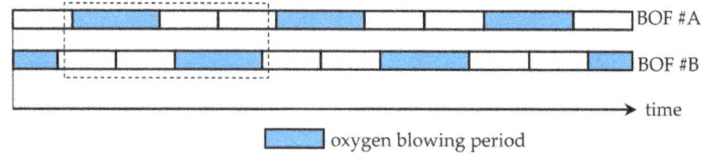

Figure 9. Schematic diagram of alternated operation of BOFs. The shadowed zones denote the oxygen blowing periods. The dashed rectangle denotes a melting cycle.

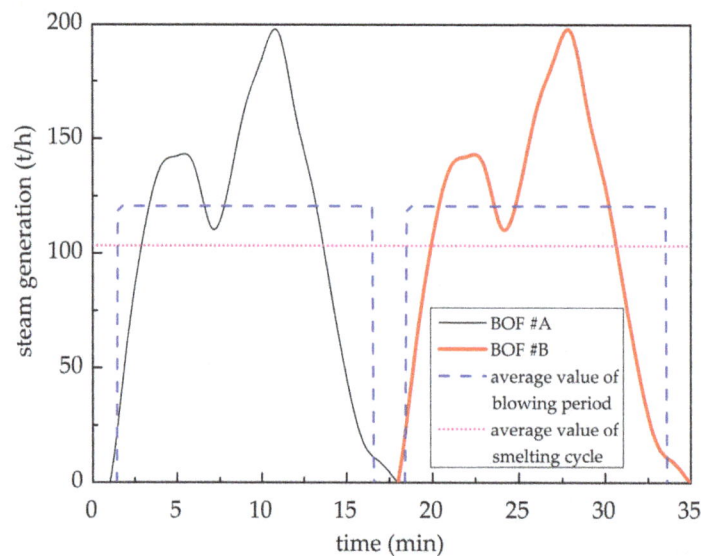

Figure 10. Steam generation of two 260 t parallel-operating BOFs. The fine line denotes the real-time load of BOF #A; the heavy line denotes the real-time load of BOF #B; the dashed line denotes the average value of the two BOFs of blowing period; and the short dot denotes the average value of the two BOFs of smelting cycle.

It can be seen from Figure 10 that for two parallel-operating BOFs, the average steam generation during oxygen blowing period is 120.64 t/h, and the average value during the whole smelting cycle is 103.40 t/h. According to the real-time steam generation curve and average steam generation line of two parallel-operating BOFs shown in Figure 10, the NTESC is 8.62 t from Equation (12) via integral calculation. The charging pressure and discharging pressure of the SA cooperating with two parallel-operating BOFs are also 2.40 MPa and 1.05 MPa respectively, based on which the enthalpy of steam can be obtained. The STESC is 74.6 kg/m^3 from Equation (5). Additionally, the water filling coefficient is 0.9, and the thermal efficiency of the SA is 0.99. Thus, the volume of the SA for the two parallel-operating BOFs is 129.69 m^3 according to Equation (3). Compared with the SA cooperating with a single BOF, the volume of SA cooperating with two parallel-operating BOFs is reduced remarkably by 49.98%, with the advantages of less initial investment and less area of land occupied. Note that the optimization results show that two steam sources cooperating with one SA performed even better. It should be noted that use of more than two accumulators in a single system is not discussed in this study.

4.2. SA Cooperating with a CSS

Shanghai Heavy Machinery Factory was selected as an example of an SA cooperating with a CSS. In this factory, an SA with the volume of 155 m^3 is used in the steam supply system. The relevant parameters are listed in Table 3, and the steam consumption load is shown in Figure 11. It can be seen from Figure 11 that the steam is generated at a constant rate. However, the steam demand of steam consumers varies with time.

Table 3. Parameters of SA.

Item	Value	Unit
charging pressure of SA	1.50	MPa
discharging pressure of SA	0.40	MPa
water filling coefficient of SA	0.90	-
thermal efficiency of SA	0.99	-

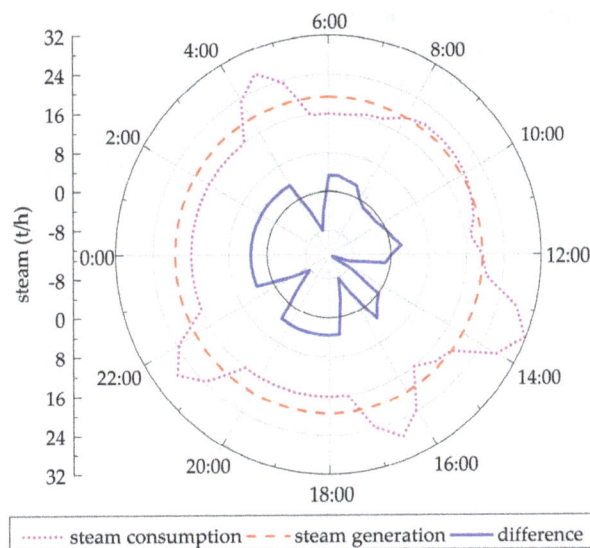

Figure 11. Daily steam consumption load curve. The dashed line denotes the steam generation; the short dot denotes the steam consumption; and the solid line denotes the difference between steam generation and steam consumption.

The record of steam load in this factory is $R = 1$. To ensure a relatively high thermal efficiency of the boiler, the time span between two adjacent load changes is set as $L \geq 3$. Thus, the number of segments is between 1 and 8. The optimal operation scheme and minimum NTESC can be achieved based on Figure 7. The optimized scheme has five segments: 6:00–9:00, 9:00–12:00, 12:00–15:00, 15:00–18:00, and 18:00–6:00. The steam load curve is presented in Figure 12. Compared with the current operation scheme shown in Figure 11, the steam generation is no longer a constant value throughout the day.

The NTESC of the optimal scheme is 9.064 t from Equation (12). The charging pressure and discharging pressure of the SA are 1.5 MPa and 0.4 MPa, respectively, from which the STESC can be calculated to be 86 kg/m^3 via either Equation (5) or Table 1. The water filling coefficient is 0.9, and the thermal efficiency of the SA is 0.99. Thus, according to Equation (3), the volume of the SA is 116 m^3. Compared with the current 155 m^3 SA, the volume of the optimized operational setup is only 116 m^3, which also satisfies the steam consumption load of steam users. The volume reduction rate is 25.16%.

In addition, by comparing Figures 11 and 12, it can be seen that the load difference between steam generation and steam consumption of the optimal setup is between −8.94 t/h and 6.76 t/h, while the current operation is between −12.33 t/h and 3.42 t/h. The maximum load difference of the optimal operation is reduced 27.49% compared to that of the current setup.

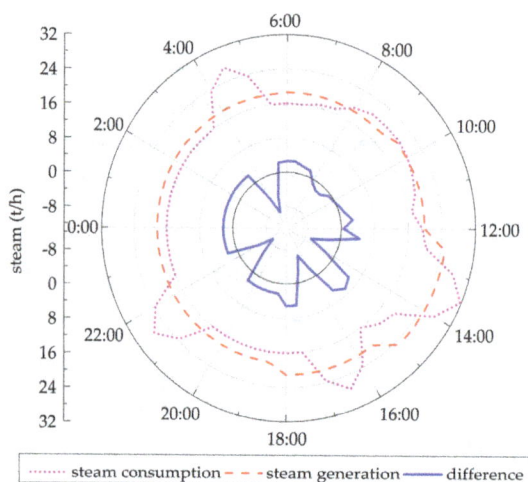

Figure 12. Load diagram of the optimal operation scheme. The dashed line denotes the steam generation; the short dot denotes the steam consumption; and the solid line denotes the difference between steam generation and steam consumption.

The steam is stored in the SA for a cycle time of 8 t. The hourly changes accumulated in the TESC of the optimal operation setup and the current setup are compared in Figure 13. It can be found that the accumulated TESC of the optimized operation setup is smoother than that of the current setup. In the optimal setup, the steam storage and release processes appear in turn, and the levels of transient steam stored and released are equally matched, resulting in a smoother accumulated TESC change. The difference between the maximum and minimum accumulated TESCs in the optimal setup is 9.66 t, rather than 12.91 t in the current setup, with a reduction in fluctuating margin of 25.17%.

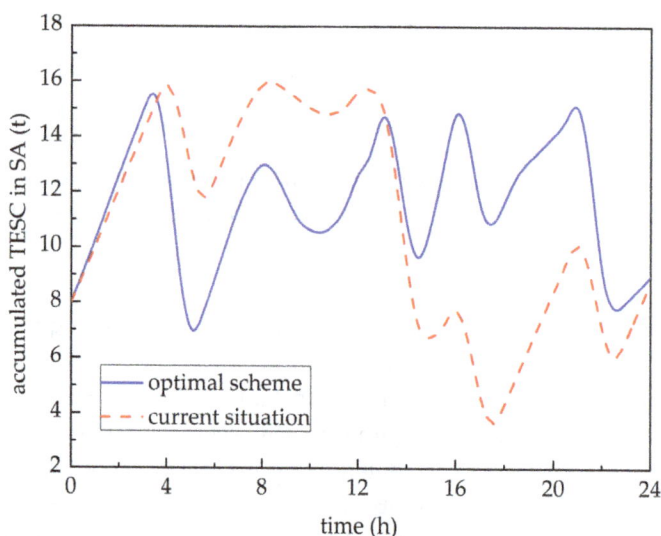

Figure 13. Comparison of accumulated thermal energy storage capacity (TESC) of the optimal and current operational setups. The dashed line denotes the current setup; and the solid line denotes the optimal setup.

Figure 14 compares the steam storage rate and release of the optimal operation setup and the current setup. The variance of the current setup is 797.35, and the variance of the optimal operational setup is 614.24, with an increase of 22.96% in stability. According to [18], there is a tendency towards a non-equilibrium with the increase of steam storage and release rates. Therefore, the optimal operational setup has a more uniform temperature field inside the SA, lower thermal loss and exergy loss, and thus, a better steam storage effect.

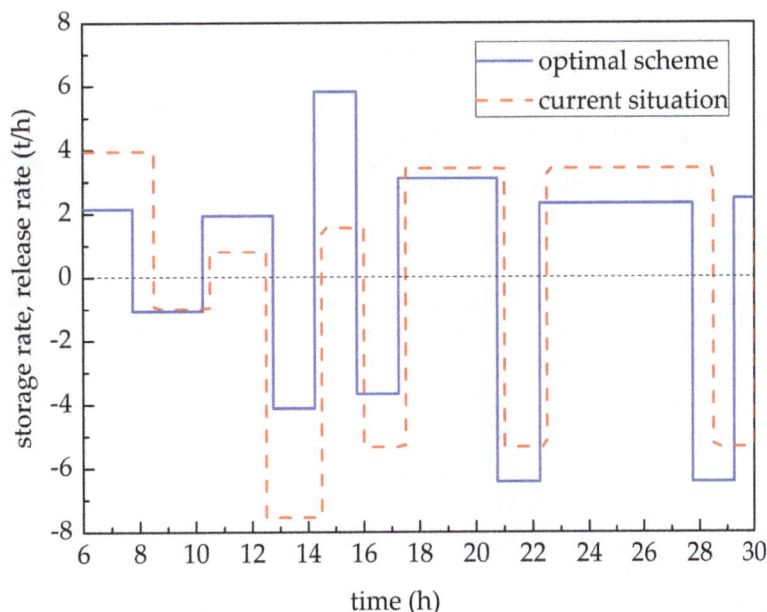

Figure 14. Comparison of steam storage rate and release rate of the two setups. The dashed line denotes the current setup; and the solid line denotes the optimal setup.

5. Conclusions

Steam accumulators (SAs) are used in different applications. Based on the operation principle of an SA and the controllability of steam sources (SSs), SSs are classified in a new way into controllable steam source (CSS) and uncontrollable steam sources (UCSS). When used as a thermal storage unit, thermal energy storage capacity (TESC) is the most important factor that should be considered, while when used as a steam buffer unit, the most important issue is the balance of steam generation and consumption.

For UCSS, the main functions of the SAs are to store emergency steam, to stabilize the fluctuant steam generation, and to make the intermittently supplied steam continuous. The operation of an SA cooperating with a basic oxygen furnace (BOF) as a UCSS has been studied and optimized. It is shown that the optimal operational setup, an SA cooperating with two parallel-operating BOFs, saves 49.98% of the SA volume in the investigated application.

For a CSS, the main function of an SA is to buffer and balance the steam generation and consumption. The mathematical model of the minimum necessary thermal energy storage capacity (NTESC) has been established and solved. Additionally, the optimal operation of an SA in a heavy machinery factory is studied by optimizing the minimum NTESC. The optimized volume of the SA is 116 m^3, with a volume reduction rate of 25.16%, a maximum transient load difference reduction of 27.49%, a fluctuating margin reduction of 25.17%, and an increase in steam storage and release stability of 22.96, respectively.

Acknowledgments: This work was supported by the Fundamental Research Funds for the Central Universities, China (N140203002), and the National Natural Science Foundation of China (21561122001).

Author Contributions: Wenqiang Sun designed the overall research; Yuhao Hong developed the mathematical models and retrofitted the configuration; Yanhui Wang analyzed the data; Wenqiang Sun wrote the paper.

Conflicts of Interest: The authors declare no conflict of interest.

Abbreviations

The following abbreviations are used in this manuscript:

BOF	basic oxygen furnace
CSS	controllable steam source
NTESC	necessary thermal energy storage capacity
SA	steam accumulator
SS	steam source
STESC	specific thermal energy storage capacity
TESC	thermal energy storage capacity
UCSS	uncontrollable steam source

References

1. Castro-Dominguez, B.; Mardilovich, I.P.; Ma, L.C.; Ma, R.; Dixon, A.G.; Kazantzis, N.K.; Ma, Y.H. Integration of methane steam reforming and water gas shift reaction in a Pd/Au/Pd-based catalytic membrane reactor for process intensification. *Membranes* **2016**, *6*, 44. [CrossRef] [PubMed]

2. Kessy, H.N.E.; Hu, Z.; Zhao, L.; Zhou, M. Effect of steam blanching and drying on phenolic compounds of litchi pericarp. *Molecules* **2016**, *21*, 729. [CrossRef] [PubMed]

3. Shamsi, S.; Omidkhah, M.R. Optimization of steam pressure levels in a total site using a thermoeconomic method. *Energies* **2012**, *5*, 702–717. [CrossRef]

4. Sun, W.; Zhao, Y.; Wang, Y. Electro- or turbo-driven?—Analysis of different blast processes of blast furnace. *Processes* **2016**, *4*, 28. [CrossRef]

5. Ma, G.; Cai, J.; Zhang, L.; Sun, W. Influence of steam recovery and consumption on energy consumption per ton of steel. *Energy Procedia* **2012**, *14*, 566–571. [CrossRef]

6. Yu, J. Status and transformation measures of industrial coal-fired boiler in China. *Clean Coal Technol.* **2012**, *18*, 89–91. (In Chinese)

7. Tanton, D.M.; Cohen, R.R.; Probert, S.D. Improving the effectiveness of a domestic central-heating boiler by the use of heat storage. *Appl. Energy* **1987**, *27*, 53–82. [CrossRef]

8. Elman, J.L. Finding structure in time. *Cogn. Sci.* **1990**, *14*, 179–211. [CrossRef]

9. Sabihuddin, S.; Kiprakis, A.E.; Mueller, M. A numerical and graphical review of energy storage technologies. *Energies* **2015**, *8*, 172–216. [CrossRef]

10. Shnaider, D.A.; Divnich, P.N.; Vakhromeev, I.E. Modeling the dynamic mode of steam accumulator. *Auto Remote Control* **2010**, *71*, 1994–1998. [CrossRef]

11. Bai, F.; Xu, C. Performance analysis of a two-stage thermal energy storage system using concrete and steam accumulator. *Appl. Therm. Eng.* **2011**, *31*, 2764–2771. [CrossRef]

12. Xu, E.; Yu, Q.; Wang, Z.; Yang, C. Modeling and simulation of 1 MW DAHAN solar thermal power tower plant. *Renew. Energy* **2011**, *36*, 848–857. [CrossRef]

13. Chen, G.Q.; Yang, Q.; Zhao, Y.H.; Wang, Z.F. Nonrenewable energy cost and greenhouse emissions of a 1.5 MW solar power tower plant in China. *Renew. Sustain. Energy Rev.* **2011**, *15*, 1961–1967. [CrossRef]

14. Price, N. Steam accumulators provide uniform loads on boilers. *Chem. Eng.* **1982**, *89*, 131–135.

15. Studovic, M.; Stevanovic, V. Non-equilibrium approach to the analysis of steam accumulator operation. *Thermophys. Aeromech.* **1994**, *1*, 53–60.

16. Stevanovic, V.D.; Maslovaric, B.; Prica, S. Dynamics of steam accumulation. *Appl. Therm. Eng.* **2012**, *37*, 73–79. [CrossRef]

17. Stevanovic, V.D.; Petrovic, M.M.; Milivojevic, S.; Maslovaric, B. Prediction and control of steam accumulation. *Heat Transf. Eng.* **2015**, *36*, 498–510. [CrossRef]

18. Maklakov, N.N.; Khramov, S.M. Application of a heat hydraulic accumulator to thermal stabilization of the evaporation zone of a heat pipe. *J. Eng. Phys. Thermophys.* **2003**, *76*, 1–5. [CrossRef]

19. Walter, H.; Linzer, W. Flow stability of heat recovery steam generators. *J. Eng. Gas Turbines Power* **2006**, *128*, 840–848. [CrossRef]

20. Liu, X.; Gong, C.; Liu, S.; Nie, Q. Analytical calculation and experimental study on temperature rising process of steam accumulator. *Acta Energiae Solaris Sin.* **1998**, *19*, 102–104.

21. Yang, S.; Manning, B.W. Fitness for service evaluation of cracked divider plate bolt locking tabs for nuclear steam generators. *Nucl. Eng. Des.* **2009**, *239*, 2242–2264. [CrossRef]

22.　Steinmann, W.D.; Eck, M. Buffer storage for direct steam generation. *Sol. Energy* **2006**, *80*, 1277–1282. [CrossRef]

23.　Su, S.; Huang, S.Y.; Wang, X.M. Experiments and homogeneous turbulence model of boiling flow in narrow channels. *Heat Mass Transf.* **2005**, *41*, 773–779.

24.　Engelhardt, G.R.; Macdonald, D.D.; Millett, P.J. Transport processes in steam generator crevices. II. A simplified method for estimating impurity accumulation rates. *Corros. Sci.* **1999**, *41*, 2191–2211. [CrossRef]

25.　Cao, J. Optimization of thermal storage based on load graph of thermal energy system. *Int. J. Thermodyn.* **2000**, *3*, 91–97.

26.　Valenzuela, L.; Zarza, E.; Berenguel, M.; Camacho, E.F. Direct steam generation in solar boilers. *IEEE Control Syst. Mag.* **2004**, *24*, 15–29. [CrossRef]

6

Performance Study on a Single-Screw Expander for a Small-Scale Pressure Recovery System

Guoqiang Li [1], Yuting Wu [1],*, Yeqiang Zhang [2], Ruiping Zhi [1], Jingfu Wang [1] and Chongfang Ma [1]

[1] Key Laboratory of Enhanced Heat Transfer and Energy Conservation,
Ministry of Education and Key Laboratory of Heat Transfer and Energy Conversion, Beijing Municipality,
College of Environmental and Energy Engineering, Beijing University of Technology, Beijing 100124, China;
guoqiang121913@126.com (G.L.); zhiruiping@gmail.com (R.Z.); jfwang@bjut.edu.cn (J.W.);
machf@bjut.edu.cn (C.M.)

[2] School of Energy and Power Engineering, Zhengzhou University of Light Industry, No. 5 Dongfeng Road,
Zhengzhou 450002, China; zhangyeqiang@zzuli.edu.cn

* Correspondence: wuyuting@bjut.edu.cn

Academic Editor: K.T. Chau

Abstract: A single-screw expander with 195 mm diameter is developed to recover pressure energy in letdown stations. An experiment system is established using compressed air as a working fluid instead of natural gas. Experiments are conducted via measurements for important parameters, such as inlet and outlet temperature and pressure, volume flow rate and power output. The influence of inlet pressure and rotational speed on the performance are also analyzed. Results indicate that the single-screw expander achieved good output characteristics, in which 2800 rpm is considered the best working speed. The maximum volumetric efficiency, isentropic efficiency, overall efficiency, and the lowest air-consumption are 51.1 kW, 83.5%, 66.4%, 62.2%, and 44.1 kg/(kW·h), respectively. If a single-screw expander is adopted in a pressure energy recovery system applied in a certain domestic natural gas letdown station, the isentropic efficiency of the single-screw expander and overall efficiency of the system are found to be 66.4% and 62.2%, respectively. Then the system performances are predicted, in which the lowest methane consumption is 27.3 kg/(kW·h). The installed capacity is estimated as 204.7 kW, and the annual power generation is 43.3 MWh. In the next stage, a pressure energy recovery demonstration project that recycles natural gas will be established within China, with the single-screw expander serving as the power machine.

Keywords: natural gas; pressure energy recovery; single-screw expander

---

Highlights

- A single-screw expander is applied in a natural gas pressure energy recycle system.
- Performance tests of the single screw expander are carried out by using compressed air.
- The maximum expander total efficiency of 62.2% is achieved at approximately 2800 rpm
- The installed capacity is estimated as 204.7 kW in a domestic natural gas letdown station

1. Introduction

With the development of the national economy, the demand on energy supply is also increasing by 10% annually [1]. Natural gas is currently extensively used all over the world because it is an important, clean, and secure fossil fuel. Natural gas is conveyed at high pressure with the assistance of pipelines (up to 12 MPa). The natural gas pressure must be reduced through a regulator or valves in letdown stations prior to safe utilization. Currently, the pressure energy in this process is wasted, becoming more serious when the natural gas has high expansion ratios and lower flow rates.

With the use of a commercial software to simulate power generation system [2], on the basis of the annually averaged natural gas properties and flow rates, the electricity generation from the available energy can meet all electrical demand from the refinery [3]. According to relevant data with regard to the properties and daily flow rates of the natural gas through Khorasan province (Iran) City Gate Stations in an entire year, a total of 762 MWh of electrical power could be generated [4]. A potential power generation system was proposed in Bangladesh [5], and this system was simulated with the thermo flow software. Results indicate that the exergy recovery is 96%, and the payback period is estimated as two years or so [6]. If turbo expanders were furnished for every pressure reduction station in Egypt, then the potential electricity generation capacity from the gas distribution network would amount to 110 MW [7]. Energy recovery potential has been analyzed by using energy and exergy in natural gas transport systems [8]. A turbo expander can be used to recycle the gas pressure energy. The annual recovered electrical power on City Gate Stations at Takestan, Iran was calculated to reach 1.1 GWh or so for a flow rate of 20,000 m^3/h [9]. Recycling natural gas pressure energy using turbine is feasible for a system. The use the pressure exergy of natural gas for the compression of air was suggested [10]. A pilot project using a turbo expander was established at a pressure reducing station in the city of Onesti, to recover waste pressure energy [11]. Advanced numerical simulation was proposed for the thermodynamic modeling of natural gas single acting reciprocating expansion engine under various working conditions for high pressure cases. Results show that the output power increases with increasing crank radius and inlet port diameter, which also increases with increasing motor speed too [12].

A turbo expander is not only used to obtain clean electricity but is also applied in chemical processes, such as gas transportation, ethylene plants, refineries, and air separation facilities. A few pressure energy recycling projects have been recently reported and hybrid applications have also been mentioned [13].

A combined cooling, heating and power system based on an internal combustion engine was proposed for power generation, refrigeration, and domestic hot water production. Results show that primary energy efficiency, energy saving ratio, and cost saving ratio of the system can reach 0.944, 0.304, and 0.417, respectively [14]. A new approach was also presented to improve boiler thermal efficiency by integrating absorption heat pumps with natural gas boilers for waste heat recovery. This approach provided a pathway to achieve realistic high-efficiency natural gas boilers for applications [15]. A new hybrid turbo expander-fuel cell system has also been considered. In this system, natural gas was utilized to preheat the gas [16]. The cost-effectiveness of the hybrid turbo expander-fuel cell system depends on the turbo expander and fuel cell investment costs, fuel cost and electrical energy price. The increase fuel cells and turbo expander investment costs ratio increase ensures that the hybrid system can be acceptably applied in a few projects in the future [17]. The cited paper quantified the energy that could be extracted from various pressure reduction facilities using an expander coupled to an electric generator. If the coupled technologies operate at their assumed peak efficiencies, then electricity can be extracted from the pressure reduction with 75% exergetic efficiency, and hydrogen can be produced with 45% exergetic efficiency [18]. A simplified and novel method of the molten carbonate fuel cell hybrid system using a turbo expander was also proposed. An exergy analysis of the hybrid system demonstrates that an overall efficiency of up to 60% is achievable. This new technology could provide solutions and be a substitute in future energy supply systems [19]. The combined heat and power system was proposed by using a turbo expander. Results show that if a gas turbine was utilized along with the turbo expander, then the amount of electrical energy would double compared with the case in which only the turbo expander was used. The results also show that the discounted payback period is approximately 3.2 years [20]. A combined heat and power system for pressure reducing stations was investigated via modeling and optimization. Results show that the shortest payback period is 1.23 years [21].

Thermoeconomic analysis of an expansion system was applied in the natural gas transportation process. The system contains a hybrid energy generation unit that generates electricity partially from the physical exergy of pressurized natural gas, and partially from the primary energy of fuel [22]. A new approach was proposed to improve the performance of a gas turbine. The approach has been applied in one of the Khangiran refinery gas turbines. Results show that the gas turbine inlet air temperature could be reduced by 4~25 K and the performance could be improved by 1.5%~5% in ten months [23].

Expanders have been used in most waste heat and pressure energy recovery systems. The two main types of expanders can be classified as velocity-type, such as the axial turbine expander, and volume-type, such as the sliding vane expander, scroll expander, screw expander and reciprocating piston expander [24]. Normal applications include natural gas processing, petrochemical processing, air separation, pressure letdown, waste heat recovery, geothermal power generation, cryogenic refrigeration, dew point control and simulation of high altitude atmospheric conditions for turbine engine test cells [25]. Turbine expander has numerous advantages [26]; however it is generally applied to power cycles with power output greater than 50 kW, because its efficiency would be unacceptable in small scale power cycles [27]. Turbine expander has a faster rotational speed, and an excess gear box is indispensable if it is utilized in waste heat and large scale pressure energy recovery systems. Compared with velocity-type expanders, volume-type expanders are more suitable for small scale pressure and low temperature energy recovery systems, because they are characterized by lower flow rates, higher expansion ratios, and a much lower rotational speed.

Recently, the sliding vane expander has been used in the energy recovery system, mechanical power recovered was up to 1.9 kW and 4.4% of the engine shaft power [28,29]. The experimental activity on the sliding vane rotary expander as the device to convert the enthalpy of the working fluid, the overall cycle efficiency and mechanical output power achieved was close to 8% and 2 kW, respectively [30]. The scroll expander has been gaining interest as a power machine in small scale systems. This device does not require inlet or exhaust valves, and thus, the noise is reduced, and the durability of the unit is improved. Another advantage is that the rolling contacts provide a sealing effect. As a result, using a large volume of oil as a sealant is not required, and the leakage issue is relieved [31]. Compared with other volume-type expanders, scroll expanders have the most complicated geometry and may even be applied to relatively small scale power systems, down to 0.1~2 kW [32,33].

Two types of screw expanders exist: twin-screw expanders and single-screw expanders. The twin-screw expander has been extensively used in Rankine cycle systems, especially for geothermal and waste heat applications. The twin-screw expander depends on precise numerically-controlled machining to achieve a leak-resistant fit. Compared with the twin-screw expander, the single-screw expander has numerous advantages, such as long service life, balanced loading of the main-screw, high volumetric efficiency, low noise, low leakage, low vibration and simple configuration, and so on [34]. The single-screw expander can realize 1~200 kW power output range, and it is more suitable for small scale pressure energy and low temperature recovery systems. Desideri et al. described the experimental results of a small scale ORC system which utilized a single-screw expander modified from a single-screw compressor. In total, 120 steady-state experimental data points have been measured and the isentropic efficiency of the expander ranged from 27.3% to 56.35% [35].

A scheme which utilizes the pressure energy in natural gas pipelines with single-screw expander is shown in Figure 1. Any failure in the pressure power generation can be readily cut off, while the original natural gas pressure system continues to run.

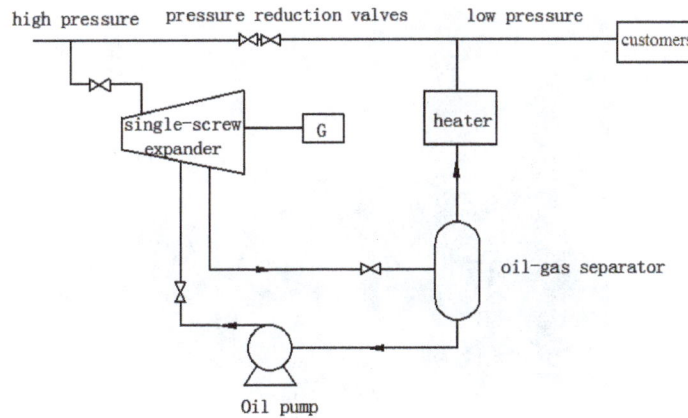

Figure 1. Schematic view of a basic high pressure natural gas pipeline with single-screw expander.

Single-screw expanders with 117 and 175 mm diameter screw have been developed by our team [34,36]. On the basis of our performance experiments on these single-screw expanders, a single-screw expander with a 195 mm diameter screw was developed last year; the in volume ratio of this expander is 4.5. This single-screw expander is of CP (C-cylinder, P-planar) type. The component contains screws, gate rotors, bearings, and other normal parts. Polytetrafluoroethylene (PTFE) is used as the material for the gate rotors to reduce the friction resistance between screw and gate rotors. A balance hole connects the high pressure leakage room with the low pressure discharge chamber of the expander. This hole is drilled on the shell, which is quite different from the balance hole drilled on the screw or main shaft. Moreover, the gaps between the screw and the shell and between the gate rotors and the groove have been optimized.

The main task of this paper is to describe the performance test conducted on this prototype. Meanwhile, the effects of rotational speed and inlet pressure of the working fluid on the prototype performance were also studied. Finally, the scale of the recoverable natural gas pressure energy in a domestic natural letdown station was predicted.

2. Experimental System

The key components of single-screw expanders are one screw and two gate rotors, which create a working chamber with the shell, as shown in Figure 2. As the screw and gate rotors rotate, the working chamber becomes bigger until it connects with the exit chamber. Figure 3 shows a photo of this prototype, and Table 1 illustrates the parameters of the prototype.

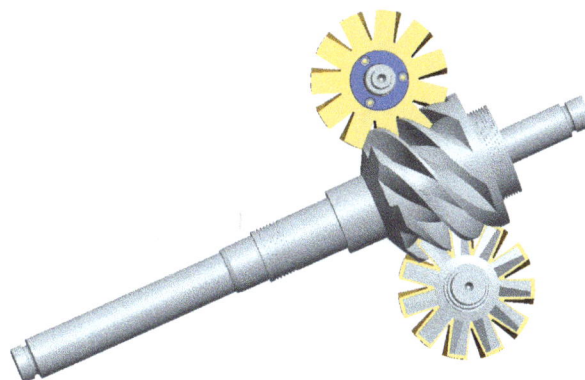

Figure 2. Three-dimensional structure of the single-screw expander.

Figure 3. Photo of the single-screw expander prototype.

Table 1. Parameters of the prototype.

Parameter	Diameter of the Screw (mm)	Groove Number of the Screw	Diameter of the Gate Rotor (mm)	Tooth Number of the Gate Rotor	Center Distance (mm)	Displacement (dm^3/r)
value	195	6	195	11	156	29

Considering the risk in handling natural gas, compressed air is used as the working fluid in the experiment instead of natural gas. The experimental system is illustrated in Figure 4. An air compressor provided high pressure air to the storage tank, which served as the stable gas source in the test. Compressed air flowed at a preset flow rate with different pressures from the storage tank and entered the single-screw expander by adjusting the inlet valve. Exhaust gas was discharged outside of the room. Through an eddy current dynamometer for the load, the power output was converted to heat, which was carried away by cooling water. Oil is injected into the expander and mixed with air to lubricate and seal. In the oil-gas separator, oil is separated from the exhaust gas and stored in the lower part of the separator. The parameters of working fluid, including intake volume flow, inlet and outlet pressure, and inlet and outlet temperature, as well as the performance parameters of the single-screw expander, such as torque, power output, and rotational speed, were obtained. A photo of the experimental platform is shown in Figure 5.

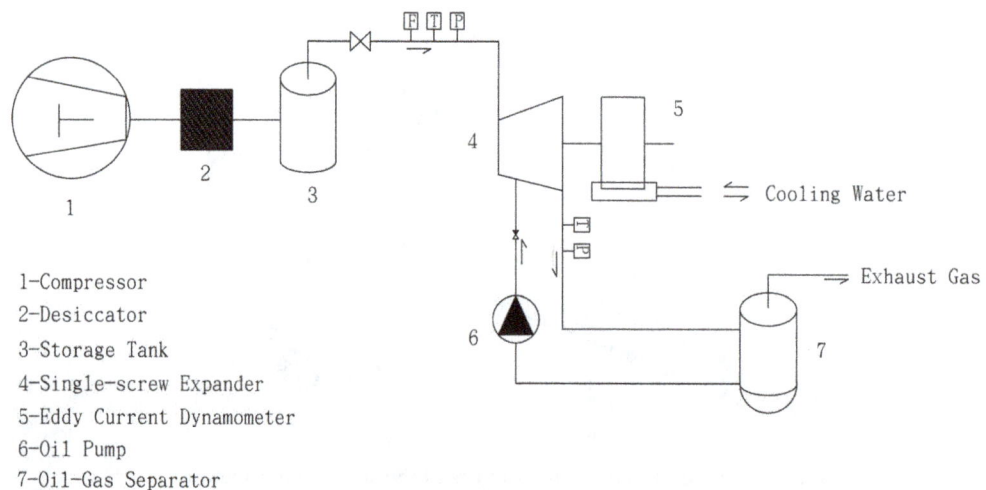

1–Compressor
2–Desiccator
3–Storage Tank
4–Single–screw Expander
5–Eddy Current Dynamometer
6–Oil Pump
7–Oil–Gas Separator

Figure 4. Schematic diagram of the experimental system.

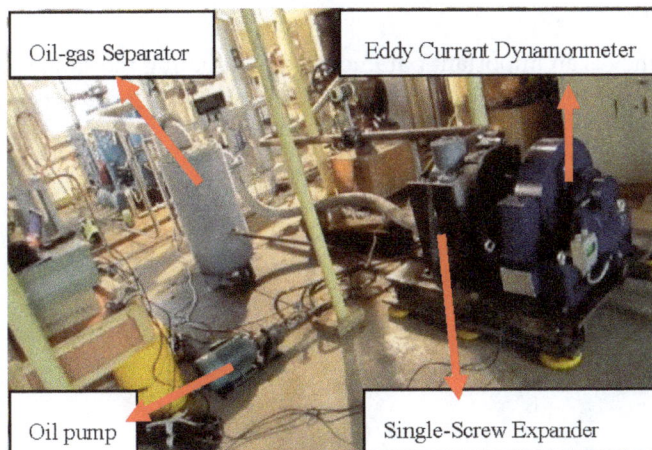

Figure 5. Experimental platform.

During the experiment, two pressure transmitters of the SMP131 type were used with a current output of 0~20 mA. The measurement range of one is 0~1.0 MPa, the other is 0~3.0 MPa and their accuracy is 0.5%. Two temperature transmitters of LG200 were used with a current output of 0~20 mA. Their measurement ranges are 0~50 °C and −100~50 °C respectively, both with an accuracy of 0.5%. The pressure transmitters and temperature transmitters are supplied by Leeg Instruments Co., Ltd. (Shanghai, China). A vortex flow-meter of FFM63 type provided by Floworld Meter Co., Ltd. (Shanghai, China) was used with current output of 4~20 mA. The measurement range is 50~280 m^3/h, and its accuracy level is 1.5%. The torque transducer model used has the measurement ranges of 0~160 N·m and 0~10,000 r/min, which is manufactured by Xiangyi Dynamic Test Instrument Co., Ltd. (Changsha, China) and the measurement accuracy of torque is ±0.2%~0.4% FS, and that of rotational speed is ±1 r/min.

3. Data Processing

The main indexes for evaluating the performance of single-screw expanders are volume efficiency η_V, isentropic efficiency η_s, mechanical efficiency η_M, total efficiency η_t, and air-consumption ratio d.

(1) Volumetric efficiency

Volume efficiency is an important parameter of single-screw expanders. For single-screw expander, the intake pressure is higher than the discharge pressure, and working fluid leaks though gaps from inlet chambers to discharge rooms. The actual flow rate through the expander is thus increased, and it may become larger than the theoretical flow rate. Therefore it is reasonable to define the volume efficiency of expander as

$$\eta_V = q_{Vth}/q_V \times 100\% \tag{1}$$

where, q_{Vth} is theoretical volume flow rate through the expander, and q_V is the measured one. It is obvious that η_V is probably smaller than 100% for positive-displacement expanders. A small value of volume efficiency denotes an expander with serious leakage.

(2) Isentropic efficiency

Isentropic efficiency is an important parameter for evaluating the effect of internal irreversible losing single-screw expanders during the isentropic process, which is defined as the ratio of the actual specific enthalpy drop to the isentropic specific enthalpy drop for the same inlet state and the same outlet pressure.

$$\eta_s = \Delta h/\Delta h_s \times 100\% \tag{2}$$

where Δh is the actual specific enthalpy drop, and Δh_s is the isentropic specific enthalpy drop.

(3) Mechanical efficiency

Mechanical efficiency is an important parameter for evaluating the single-screw expander during the expansion process, which is defined as the ratio of power output drop to the total measured enthalpy drop.

$$\eta_M = Pe/\Delta h \times 100\% \tag{3}$$

where, Pe is the power output of the single-screw expander, and Δh is the actual specific enthalpy drop.

(4) Total efficiency

Total efficiency is the ratio of the power output to the enthalpy drop of the working fluid in an ideal isentropic process, which is defined by

$$\eta_t = Pe/(m\Delta h_s) \times 100\% \tag{4}$$

where, Pe is the power output of the single-screw expander, m is the mass flow, and Δh_s is the specific enthalpy drop.

(5) Air-consumption ratio

Air-consumption ratio is another important parameter for evaluating the economical efficiency of this prototype, and is defined as

$$d = m/Pe \tag{5}$$

4. Experimental Performances Analysis

Performance tests were conducted for different inlet pressures and rotating speeds, ranging from 7 bar to 16 bar and from 1600 rpm to 2800 rpm, respectively.

4.1. Power Output

The power output Pe is an important parameter for evaluating the performance of single-screw expanders. The effects of rotational speed on power output are shown in Figure 6. As seen from this figure, the power output increases with increases of rotational speed and inlet pressure, and the maximum of the power output is 51.1 kW (the maximum error is 0.2 kW) obtained at the condition of 16 bar and 2800 rpm. However, the increase rate of the power output with increase of inlet pressure is faster than that with increase of rotational speed. This shows that, with the increase of rotational speed, the power consumption of the friction loss between the screw and the shell increases gradually, which leads to the slow growth of the output power. Figure 6 illustrates that the influence of inlet pressure is bigger than that of rotational speed.

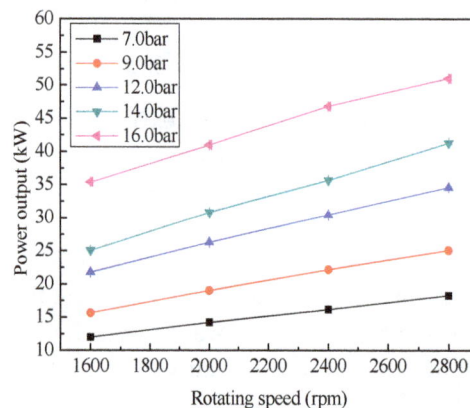

Figure 6. Power output versus rotating speed.

4.2. Volumetric Efficiency

Volume efficiency is another important parameter of single-screw expanders. Figure 7 shows the changes of the volumetric efficiency versus rotational speed. Overall, at a different pressure, the volumetric efficiency reaches a maximum value of 83.5% (the maximum error is 1.4%). It can be seen that in a certain range of pressure, the rotational speed 2800 rpm of a single-screw expander is more conducive to the formation and stability of oil slick, and that the amount of gas leakage maybe reaches the minimum, so that the volume efficiency has the highest value. Theoretically, the volume efficiency should remain unchanged at a certain constant rotational speed no matter what the inlet pressure is, because the basic volume of the single-screw expander is invariant. This conclusion is validated by Figure 8. That is to say, the operation was steady in the case of 2800 rpm. The corresponding parameters can be used as the performance parameters of this prototype.

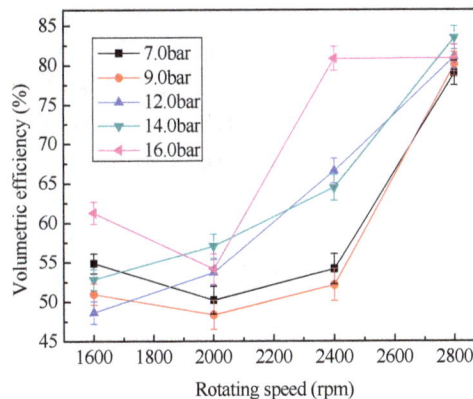

Figure 7. Volumetric efficiency versus rotating speed.

4.3. Isentropic Efficiency

The variation of isentropic efficiency with changes of rotational speed is shown in Figure 8. The influencing factors on isentropic efficiency are the irreversible losses in the system, including internal leakage, eddy current, friction and so on. From Figure 8, under the same inlet pressure, the isentropic efficiency increases with the increase of rotational speed. At the maximum value of 66.4% (the maximum of error is 0.7%) in the case of 2800 rpm, with the increase of rotational speed, gas flow increases, gas leakage ratio decreased, and the friction loss is not sufficient to affect the decline of outlet temperature, thus increasing the adiabatic efficiency.

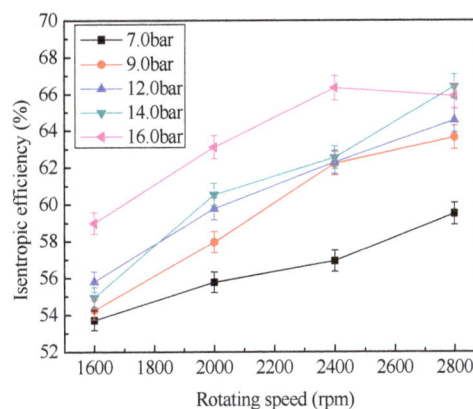

Figure 8. Isentropic efficiency versus rotating speed.

Under the same rotating speed, the isentropic efficiency first increases with the increase of inlet pressure. However, when inlet pressure increased to a certain value, the isentropic efficiency began to decline. This is because the pressure increases, the gas flow gradually increased, but the corresponding reduction in the proportion of gas leakage, and thus increase the adiabatic efficiency. When the pressure exceeds a certain value, since the increase of inlet pressure causes the increase of internal leakage correspondingly, which leads to the increase of outlet temperature, and thus the isentropic efficiency declines. It shows that the best performance of single-screw expander happens at 2800 rpm, at which the irreversible loss achieves at the minimum.

4.4. Mechanical Efficiency

Mechanical efficiency is an important parameter for evaluating the performance of the single-screw expander during the expansion process. The variation of mechanical efficiency with rotational speed is shown in Figure 9. At the same pressure, when the rotational speed is lower than 2000 rpm, with the increase of rotational speed, gas flow increases, gas leakage ratio decreased, but the rate of friction loss is greater than the power output, so that the mechanical efficiency is reduced. When the rotational speed is 2000–2800 rpm, the friction loss rate is less than the rate of the power output, so the mechanical efficiency increases, reaching the maximum value. From Figure 9, under the different inlet pressure, at the rotating speed of 2800 rpm, the mechanical efficiency up to the maximum value of 93.7% (the maximum error is 1.7%), it can illustrate that the best performance of a single-screw expander is achieved at 2800 rpm.

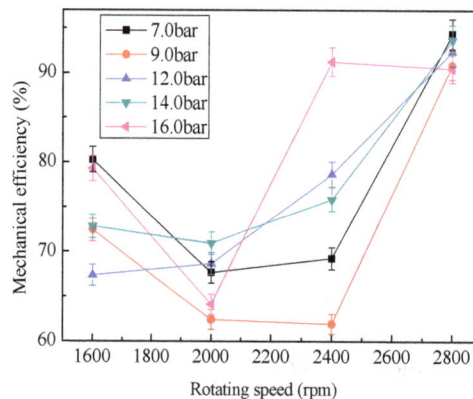

Figure 9. Mechanical efficiency versus rotating speed.

4.5. Total Efficiency

The variations of total efficiency with changing rotational speed are shown in Figure 10, the total efficiency at 2800 rpm is better than that at other speed under the same inlet pressure, and the maximum of total efficiency reaches 62.2% (the maximum error is 1.0%) under the condition of 14.0 bar and 2800 rpm. Figure 10 shows that the total efficiency increases with the increase of rotational speed. The total efficiency reaches a maximum at 2800 rpm under the same inlet pressure. It again illustrates that the best performance of the single-screw expander is achieved at 2800 rpm, and that the sealing performance and friction loss of the expansion machine achieves an optimal value, which makes the total efficiency reach the maximum at a speed of 2800 rpm.

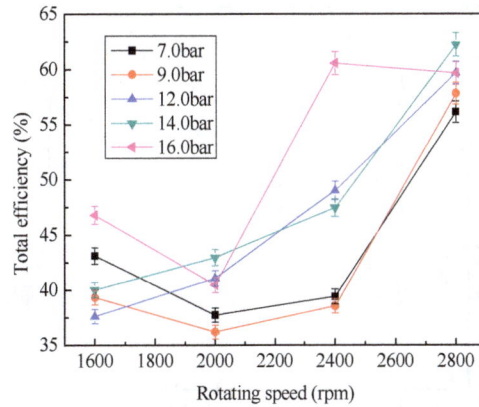

Figure 10. Total efficiency versus rotating speed.

4.6. Air Consumption

Figure 11 shows the variations of air consumption with changing rotational speed. From Figure 11, the air-consumption decreases with the increase of the inlet pressure at 2000 rpm, 2800 rpm and 3200 rpm, and there is the lowest air-consumption when the rotating speed is at 2800 rpm. It proves that the best economic performance of single-screw expander is obtained at 2800 rpm, at which the air consumption is also the lowest, i.e., 44.1 kg/(kW·h).

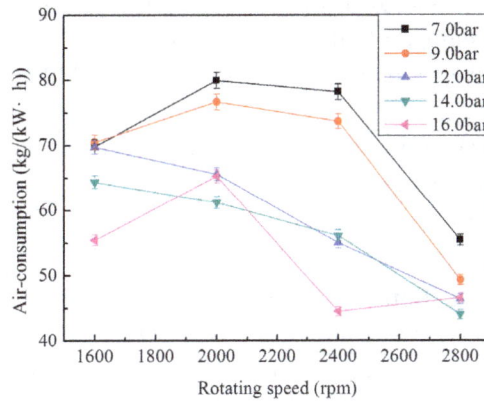

Figure 11. Air-consumption versus rotating speed.

5. Prediction

5.1. Operational Parameters

We then proceed to predict the scale of the recoverable natural gas pressure energy in the Jiyuanzhongyu natural gas letdown station in Henan province in 2014. The single-screw expander with 195 mm diameter is used as the power machine, and the inlet/outlet pressure of natural gas in the letdown station are shown in Figure 12. As seen in Figure 12, the maximum, minimum and average inlet/outlet pressures are 9.04/3.60 MPa, 4.75/1.62 MPa and 7.57/3.12 MPa respectively.

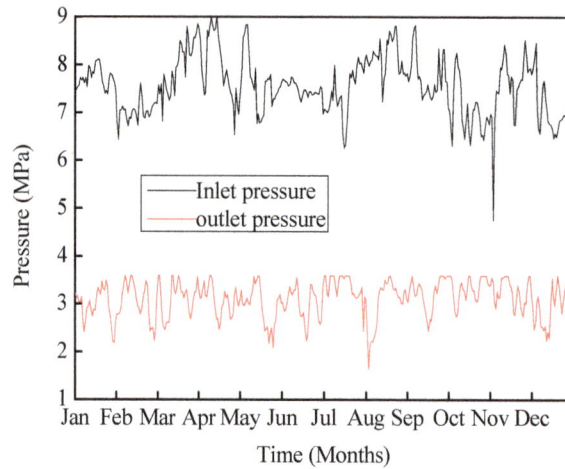

Figure 12. Inlet/outlet pressure of natural gas during 2014.

5.2. Assumptions

The following hypotheses are adopted in the analysis of the natural gas pressure energy recovery system:

(1) The natural gas can be defined as pure methane because of the high concentration of methane in the natural gas components.

(2) The inlet/outlet pressure and the inlet temperature of the Jiyuanzhongyu natural gas letdown station are the inlet/outlet pressure and inlet temperature of the single-screw expander respectively.

(3) The inlet volume flow rate of natural gas is assumed to be 110 m^3/h, which is the average inlet volume flow rate in the experiment.

(4) The isentropic efficiency of the single-screw expander and overall efficiency of the system is taken as 66.4% and 62.2%, respectively.

5.3. Predictions

Figure 13 shows the prediction curves of output power and methane-consumption ratio. The output power of April and August is larger, and that of February, July and November is small can be seen from Figure 13a; the methane consumption of February, July and November is larger, and that of April and August is small can be seen from Figure 13b; The predicted parameters are summarized in Table 2.

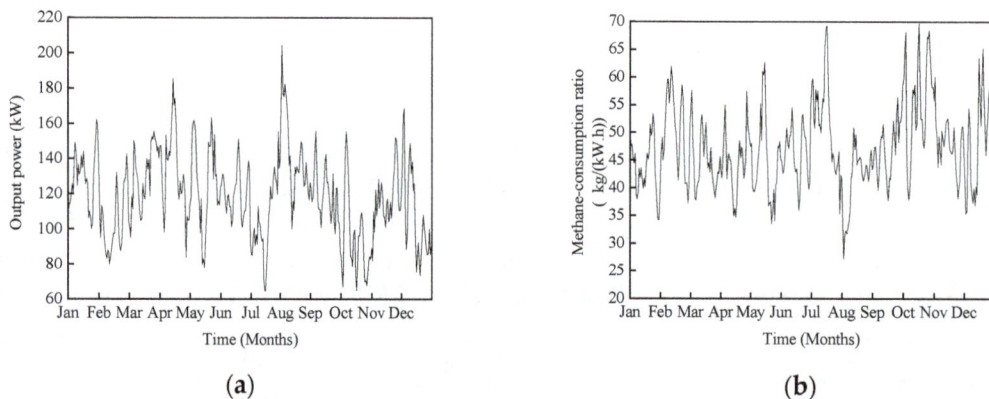

(a)

(b)

Figure 13. Predicted parametric curves of output power (**a**) and methane-consumption ratio (**b**).

Table 2. Predicted results summary.

Value	Output Power (kW)	Methane-Consumption Ratio (kg/(kW·h))
Maximum	204.7	69.6
Mean	118.6	47.4
Minimum	64.7	27.3

Based on the predicted results in Table 2, it is concluded that if the natural gas energy could have been used to generate electricity and the lowest methane consumption was 27.3 kg/(kW·h), the installed capacity can be estimated as 204.7 kW, and the annual power generation would be a total of 43.3 MW·h in 2014.

6. Conclusions

The pressure energy recovery system that will be applied in the Jiyuanzhongyu natural gas letdown station in Henan province was analyzed through experiments on a pressure recovery system using a single-screw expander. The following conclusions can be drawn.

The single-screw expander achieved good output at 2800 rpm. The greatest power output, the maximum volumetric efficiency, the maximum of isentropic efficiency and overall efficiency were 51.1 kW, 83.5%, 66.4% and 62.2% respectively.

When the pressure energy recovery system with a single-screw expander is applied in a domestic natural gas letdown station, the isentropic efficiency of the single-screw expander and overall efficiency of the system are 66.4% and 62.2%, respectively. The lowest methane consumption is 27.3 kg/(kW·h), the installed capacity is estimated as 204.7 kW and the annual power generation would be a total of 43.3 MWh. Thus, the single-screw expander has great prospects in small-scale waste heat pressure energy recovery systems, as well as in new and renewable energy applications.

Acknowledgments: The authors acknowledge the financial support provided by the International Natural science foundation of china (Grant Numbers 51361135702) and the National Basic Research Program of China (Grant Numbers 2013CB228306). Also, the authors gratefully acknowledge the support of data provided by the Compressor Dept. Petro china West East Gas Pipeline Company.

Author Contributions: Guoqiang Li wrote the paper. Guoqiang Li, Yuting Wu and Yeqiang Zhang conceived and designed the experiments; Guoqiang Li and Yeqiang Zhang performed the experiments; Ruiping Zhi, Jingfu Wang and Chongfang Ma revised the paper and provided useful suggestions; all authors read and approved the manuscript.

Conflicts of Interest: The authors declare no conflict of interest.

Nomenclature

d	air-consumption ratio (kg/kW·h)
q_{Vth}	theoretical volume flowrate (m^3/h)
q_V	measured volume flowrate (m^3/h)
Δh	actual specific enthalpy drop (kJ/kg)
Δh_s	isentropic enthalpy drop (kJ/kg)
Pe	power output (kW)
m	mass flow (kg/h)
n	rotate speed (r/min)
V_0	maximum basic volume (m^3)
Z_1	groove number of single-screw rotor

Greek Symbol

η_s	isentropic efficiency (%)
η_M	mechanical efficiency (%)
η_t	total efficiency (%)
η_V	volumetric efficiency (%)

Subscripts

M	mechanical energy
s	isentropic
t	total
V	volume
0	ambient condition

References

1. Lin, W.; Zhang, N.; Gu, A. LNG (liquefied natural gas): A necessary part in China's future energy infrastructure. *Energy* **2010**, *35*, 4383–4391. [CrossRef]

2. Atik, K. Thermoeconomic optimization in the design of thermoelectric cooler. In Proceedings of the 5th International Advanced Technologies Symposium, Karabuk, Turkey, 13–15 May 2009; pp. 1775–1778.

3. Farzaneh-Gord, M.; Deymi-Dashtebayaz, M. Recoverable energy in natural gas pressure drop stations: A case study of the Khangiran gas refinery. *Energy Explor. Exploit.* **2008**, *26*, 71–82. [CrossRef]

4. Farzaneh-Gord, M.; Hashemi, S.; Sadi, M. Energy destruction in Iran's natural gas pipe line network. *Energy Explor. Exploit.* **2007**, *25*, 393–406. [CrossRef]

5. Mansoor, S.A.; Mansoor, A. Power generation opportunities in Bangladesh from gas pressure reducing stations. In Proceedings of the 3rd International Conference on Electrical & Computer Engineering, Dhaka, Bangladesh, 28–30 December 2004; p. 3.

6. Seresht, R.T.; Jalalabadi, H.K.; Rashidian, B. Retrofit of Tehran City Gate Station (CGS No. 2) by Using Turboexpander. In Proceedings of the ASME 2010 Power Conference, American Society of Mechanical Engineers, Chicago, IL, USA, 13–15 July 2010; pp. 207–212.

7. Elsobki, M.S.; El-Salmawy, H.A. Power generation using recovered energy from natural gas networks. In Proceedings of the 17th International Conference on Electricity Distribution, Barcelona, Spain, 12–15 May 2003.

8. Kostowski, W. The possibility of energy generation within the conventional natural gas transport system. *Strojarstvo* **2010**, *52*, 429–440.

9. Rezaie, N.Z.; Saffar-Avval, M. Feasibility Study of Turbo Expander Installation in City Gate Station. In Proceedings of the 25th International Conference on Efficiency, Cost, Optimization and Simulation of Energy Conversion Systems and Processes, Perugia, Italy, 26–29 June 2012; p. 47.

10. Bisio, G. Thermodynamic analysis of the use of pressure exergy of natural gas. *Energy* **1995**, *20*, 161–167. [CrossRef]

11. Andrei, I.; Valentin, T.; Cristina, T.; Niculae, T. Recovery of Wasted Mechanical Energy from the Reduction of Natural Gas Pressure. *Procedia Eng.* **2014**, *69*, 986–990. [CrossRef]

12. Gord, M.F.; Jannatabadi, M. Simulation of single acting natural gas Reciprocating Expansion Engine based on ideal gas model. *J. Nat. Gas Sci. Eng.* **2014**, *21*, 669–679. [CrossRef]

13. Daneshi, H.; Khorashadi Zadeh, H.; Lotfjou Choobari, A. Turboexpander as a distributed generator. In Proceedings of the 2008 IEEE Power and Energy Society General Meeting—Conversion and Delivery of Electrical Energy in the 21st Century, Pittsburgh, PA, USA, 20–24 July 2008; p. 7.

14. Wang, J.; Wu, J.; Zheng, C. Simulation and evaluation of a CCHP system with exhaust gas deep-recovery and thermoelectric generator. *Energy Convers. Manag.* **2014**, *86*, 992–1000. [CrossRef]

15. Qu, M.; Abdelaziz, O.; Yin, H. New configurations of a heat recovery absorption heat pump integrated with a natural gas boiler for boiler efficiency improvement. *Energy Convers. Manag.* **2014**, *87*, 175–184. [CrossRef]

16. Howard, C.; Oosthuizen, P.; Peppley, B. An investigation of the performance of a hybrid turboexpander-fuel cell system for power recovery at natural gas pressure reduction stations. *Appl. Therm. Eng.* **2011**, *31*, 2165–2170. [CrossRef]

17. Darabi, A.; Shariati, A.; Ghanaei, R.; Soleimani, A. Economic assessment of the hybrid turbo expander-fuel cell gas energy extraction plant. *Turk. J. Electr. Eng. Comput. Sci.* **2016**, *24*, 733–745. [CrossRef]

18. Maddaloni, D.J.; Rowe, A.M. Natural gas exergy recovery powering distributed hydrogen production. *Int. J. Hydrogen Energy* **2007**, *32*, 557–566. [CrossRef]

19. Rashidi, R. *Thermodynamic Analysis of Hybrid Molten Carbonate Fuel Cell Systems*; University of Ontario Institute of Technology: Oshawa, ON, Canada, 2008.

20. Eftekhari, H.; Akhlaghi, K.; Farzaneh-Gord, M.; Khatib, M. A Feasibility Study of Employing an Internal Combustion Engine and a Turbo-expander in a CGS. *Int. J. Chem. Environ. Eng.* **2011**, *2*, 343–349.

21. Sanaye, S.; Nasab, A.M. Modeling and optimizing a CHP system for natural gas pressure reduction plant. *Energy* **2012**, *40*, 358–369. [CrossRef]

22. Wojciech, S.U.; Kostowski, J. Thermoeconomic assessment of a natural gas expansion system integrated with a co-generation unit. *Appl. Energy* **2013**, *101*, 58–66.

23. Farzaneh-Gord, M.; Deymi-Dashtebayaz, M. A new approach for enhancing performance of a gas turbine (case study: Khangiran refinery). *Appl. Energy* **2009**, *86*, 2750–2759. [CrossRef]

24. Qiu, G.; Liu, H.; Riffat, S. Expanders for micro-CHP systems with organic Rankine cycle. *Appl. Therm. Eng.* **2011**, *31*, 3301–3307. [CrossRef]

25. Bloch, H.; Soares, C. *Turboexpanders and Process Applications*; Gulf Professional Publishing: Boston, MA, USA, 2001.

26. Pei, G.; Li, Y.; Li, J.; Ji, J. An experimental study of a micro high-speed turbine that applied in Organic Rankine cycle. In Proceedings of the Asia-Pacific Power and Energy Engineering Conference (APPEEC), Chengdu, China, 28–31 March 2010; pp. 1–4.

27. Peterson, R.; Wang, H.; Herron, T. Performance of a small-scale regenerative Rankine power cycle employing a scroll expander. *Proc. Inst. Mech. Eng. Part A* **2008**, *222*, 271–282. [CrossRef]

28. Bianchi, G. Exhaust Waste Heat Recovery in Internal Combustion Engines. Ph.D. Thesis, University of L'Aquila, L'Aquila, Italy, March 2015.

29. Cipollone, R.; Contaldi, G.; Bianchi, G.; Murgia, S. Energy Recovery Using Sliding Vane Rotary Expander. In Proceedings of the 8th International Conference on Compressors and their Systems, London, UK, 9–10 September 2013; pp. 183–194.

30. Cipollone, R.; Bianchi, G.; Battista, D.D.; Contaldi, G.; Murgia, S. Mechanical Energy Recovery from Low Grade Thermal Energy Sources. *Energy Procedia* **2014**, *45*, 121–130. [CrossRef]

31. Wang, H.; Peterson, R.; Herron, T. Experimental performance of a compliant scroll expander for an organic Rankine cycle. *Proc. Inst. Mech. Eng. Part A* **2009**, *223*, 863–872. [CrossRef]

32. Lemort, V.; Quoilin, S.; Cuevas, C.; Lebrun, J. Testing and modeling a scroll expander integrated into an Organic Rankine Cycle. *Appl. Therm. Eng.* **2009**, *29*, 3094–3102. [CrossRef]

33. Lemort, V.; Declaye, S.; Quoilin, S. Experimental characterization of a hermetic scroll expander for use in a micro-scale Rankine cycle. *Proc. Inst. Mech. Eng. Part A* **2012**, *226*, 126–136. [CrossRef]

34. Wang, W.; Wu, Y.; Ma, C.; Liu, L.; Yu, J. Preliminary experimental study of single-screw expander prototype. *Appl. Therm. Eng.* **2011**, *31*, 3684–3688. [CrossRef]

35. Desideri, A.; den Broek, M.; Gusev, S.; Lemort, V.; Quoilin, S. Experimental Campaign and Modeling of a Low-capacity Waste Heat Recovery System Based on a Single-screw Expander. In Proceedings of the 22nd International Compressor Engineering Conference, West Lafayette, IN, USA, 14–17 July 2014.

36. He, W.; Wu, Y.; Peng, Y.; Zhang, Y.; Ma, C.; Ma, G. Influence of intake pressure on the performance of single-screw expander working with compressed air. *Appl. Therm. Eng.* **2013**, *51*, 662–669. [CrossRef]

7

Lobatto-Milstein Numerical Method in Application of Uncertainty Investment of Solar Power Projects

Mahmoud A. Eissa [1,2,*] **and Boping Tian** [1,*]

1   Department of Mathematics, Harbin Institute of Technology, Harbin 150001, China

2   Department of Mathematics, Faculty of Science, Menoufia University, Menoufia 32511, Egypt

*   Correspondence: mahmoud.eisa@science.menofia.edu.eg (M.A.E.); bopingt361147@hit.edu.cn (B.T.)

Academic Editor: Senthilarasu Sundaram

Abstract: Recently, there has been a growing interest in the production of electricity from renewable energy sources (RES). The RES investment is characterized by uncertainty, which is long-term, costly and depends on feed-in tariff and support schemes. In this paper, we address the real option valuation (ROV) of a solar power plant investment. The real option framework is investigated. This framework considers the renewable certificate price and, further, the cost of delay between establishing and operating the solar power plant. The optimal time of launching the project and assessing the value of the deferred option are discussed. The new three-stage numerical methods are constructed, the Lobatto3C-Milstein (L3CM) methods. The numerical methods are integrated with the concept of Black–Scholes option pricing theory and applied in option valuation for solar energy investment with uncertainty. The numerical results of the L3CM, finite difference and Monte Carlo methods are compared to show the efficiency of our methods. Our dataset refers to the Arab Republic of Egypt.

Keywords: stochastic differential equation; numerical simulation; real option; renewable energy; Egypt

1. Introduction

A great deal of effort is being put into researching and developing renewable energy (RE) technologies. RE can be generated from wind, solar, biomass, sunlight, tides and flowing water. The primary reason for this effort stems from the environmental impact of using fossil fuels, such as nitrogen and sulfur oxides (NOx and SOx), as well as oil spills, similar to the recent major spill in the Gulf of Mexico [1]. In addition, the rising demand for electricity is considered as one of the main reasons that also make RE development to serve to increase energy security by reducing reliability on foreign imports of fossil fuels.

Despite the delay with respect to some countries in the world, we can see the U.S., as well as several other regions, such as Western Europe, East Asia and North Africa, having a massive increase in the construction and operation of renewable power production sites. Particularly, we mean the production of electricity from renewable energy sources (RES). In the IEO2016 [2], long-term global prospects continue to improve for generating electricity from RES. RES are the fastest-growing source of energy for electricity generation, with annual increases averaging 2.9% from 2012 to 2040.

One of the RES is solar energy, which can be converted into electricity using photovoltaic (PV) technology [3,4]. Solar is the world's fastest-growing form of RES, with net solar generation increasing by an average of 8.3%/year. Solar energy shared 859 billion kWh (15%) of the 5.9 trillion kWh of new renewable generation added over the projection period (see [5]).

The main drivers for fast-growing solar have not only been the economic efficiency and technology breakthroughs in renewable power production, but also the favorable government support due to environmental concerns. We can see that currently, such government interest in the support and incentives to private investors, but private investors are driven by profit maximization. There are two major groups of schemes that can pave the way for a wider spread: (1) the scheme of tariff-based capacity (a payment for kWh of energy generated); (2) the scheme of the quota system (the government obliges heavy industries to use a percentage of their electricity consumption from RES). Wiser et al. [6] addressed some of the government support schemes, which are typically in the form of subsidies and incentives that are front-loaded in the construction and early operating years in the U.S. Furthermore, Fagiani et al. [7] discussed the dilemma that arises from certain support schemes, such as the market risk. Regardless of the market risk factor leading to making the optimal use of RES, which implies limiting the cost to society, but the market risk simultaneously deters investors, thus this provides for less RE and a higher price.

In general, the policy instruments aim to keep investors' risks within reasonable limits. In addition, the policies have a strong effect on the price and quantity risks faced by an investor. Therefore, we note that the drivers and investors usually feel major concern in this investment because of uncertain returns. Daim et al. [8] discussed identifying future adoption, products and technologies for residential and industrial consumers in the form of a graphical technology road map for wind energy. Sorsimo et al. [9]. presented the polices used by European nations to stimulate offshore wind development and discuss the impact of similar policies in the U.S. Furthermore, the performance of 'market-based' British renewable obligation and German 'feed-in tariff' systems of RE procurement systems are analyzed by Toke [10].

The RE is an uncertain investment, such that it is long term, costly and depends on a feed-in tariff system. The valuation for RE investment must consider the irreversibility and flexibility enjoyed by decision makers (i.e., the option to delay investment), in addition to the uncertainty. A.Dixit et al. [11] addressed the subject of traditional valuation techniques based on discounted cash flows inferior to real option analysis under these circumstances. Here, we follow the real option approach (ROA) to address the real option valuation (ROV) of an investment in solar energy (SE) projects and the optimal time to invest under a number of different payment settings [12,13]. Fernandes et al. [14] presented a review of the current state of the art in the application of ROA to investment in non-renewables and RES. Abadie et al. [15] provides a literature review of the real option valuation for the operation of a wind farm. According to [14,15], this particular literature in the RE sector is still limited. Therefore, attempts to fill this gap would be welcome.

In this work, we consider the ROV of private potential investment in RE under the energy and environmental policies, as well as the analysis and assessment of the impact of uncertainty sources. In other words, the ROV has a crucial dimension (the option to delay an irreversible investment in RE) under the policies and support schemes, which are provided by drivers; as such, it should be embodied in the total value of RE. A real options framework is modeled for use in RE investment using stochastic differential equation (SDEs).

Following this approach, Abadie et al. [15] addressed the value of an operating wind farm and the real option to investment in it under different support schemes. The model considers up to three sources of uncertainty: the electricity price, the wind load and the renewable obligation certificate (ROC) price. They resorted to a trinomial lattice combined with Monte Carlo simulation, when the analytical solutions are lacking. The authors considered the data referring to the U.K. Gazheli et al. [16] developed a real option model in order to take into account the uncertainty and irreversibility of the farmer deciding to lease agricultural land to a company installing a PV power plant. The uncertainty in the agricultural commodity price in addition to the irreversible science that it is a 20-year commitment from the farmer are considered. Subsidies introduced by the government to increase the investments in the RE sector are discussed. The model is applied to a province in Italy.

Stochastic differential equations (SDEs) are using to model problems in many fields of science [17]. In practice, numerical solutions are becoming increasingly important, because the analytic solutions are usually not available for SDEs. The well-known Euler–Maruyama (EM) method for SDEs was presented with a strong convergence order of 0.5 in [18]. In order to improve the fundamental analysis of numerical approximations, various implicit numerical methods using split-step techniques have been derived based on the Euler method. In 2002, Higham et al. [19] derived the split-step backward Euler (SSBE) method. In addition, the split-step theta (SSθ) methods, which generalize the SSBE method when $\theta = 1$, were discussed in [20,21]. Although, these numerical methods are A-stable for linear SDEs, these methods have a strong convergence order of 0.5. Using the additional term of the Itô–Taylor expansion, the Milstein method was presented with a strong convergence order of 1.0 [18]. Based on the Milstein method, Wang et al. [22] presented the drifting split-step backward Milstein (DSSBM) method. Guo et al. [23] constructed the modified split-step composite θ-Milstein (MSSCTM) methods. In 2015, Voss et al. [24] combined the predictor-corrector method with a Milstein method to investigate the split-step Adams–Moulton–Milstein (SSAMM) method. In 2016, the modified split-step theta Milstein (DSSθM) methods were presented by Tian et al. [25]. Although, these methods improved the convergence order to be 1.0, unfortunately, we can see that the mean-square (MS) stability conditions of these split-step methods have some restrictions for the parameters and step-size h; furthermore, these methods are not A-stable. As far as the authors know, no implicit split-step numerical methods have a strong convergence order of 1.0 and are A-stable for SDEs.

Numerical methods are needed for real option valuation in cases where analytic solutions are either unavailable or not easily comparable. Approximation of the stochastic process for an underlying asset can be applied to real option valuation. There are several candidate models for the stochastic evaluation of the underlying asset (see [26]). An overview of two numerical methods is available in the context of the Black–Scholes–Merton method [27,28]. Brennan et al. [29] considered finite difference methods (FDM). Boyle [30] gave the simulation of the stochastic process using the Monte Carlo (MC) method. The comparative study of FDM and the MC method for pricing European options was considered in [31]. In addition, the methods are typically tailored to fit into a specific problem at hand (see [32,33]).

It is well known that, when the real option can be modeled using a partial differential equation, then FDM are sometimes applied. Despite the large number of research discussed using FDM for ROV, the FDM have become uncommon in use today (particularly amongst practitioners) due to the required mathematical sophistication; these also cannot readily be used for high-dimensional problems [33]. Although the MC method has also developed, is increasing and is especially applied to high-dimensional problems, its convergence to the correct values is still slow, which leads to a significant increase in run-time [34]. Therefore, recently, there has been increasing interest in deriving new numerical methods, which can possibly avoid the shortcomings in the aforementioned methods. In this work, the new classes of split-step numerical methods are constructed, which are A-stable, with convergence with order 1.0. Using Lobatto3C (The Lobatto3C methods are algebraically stable, B-stable and L-stable. Therefore, the Lobatto3C methods are considered excellent for stiff ordinary differential equation (ODE) problems [35].) methods in collusion with the Milstein method, the Lobatto3C-Milstein (L3CM) methods are derived. The new numerical methods L3CM methods are applied to valuing the real options, and the results are compared with those of FDM and MC methods.

In this paper, a real option framework is modeled for use in RE investment. The real option framework considers the volatility in RE price during the project lifetime and the development lag between launching the project and starting the production (since the net production revenue cannot be started instantaneously, a time lag has to be allowed between the decision to establish the RE plant, and the actual production is the cost of delay (if the cash flows are evenly distributed over time and the exclusive rights last n years (20 years), the annual cost of delay can be written as: $\frac{1}{n} = \frac{1}{20} = 5\%$ a year; though, this cost of delay rises each year, to $\frac{1}{19}$ in Year 2, $\frac{1}{18}$ in Year 3, and so on, making the cost of delaying the exercise larger over time)). The real option framework differs from the previous work, since the new numerical methods, L3CM, are integrated with option theory and the four economic elements, cost, value, risk and flexibility, to value a real option. We examine the new L3CM methods with two other commonly-used methods, the FDM and MC methods, in an options valuation for investment with uncertainty in a case study.

The paper is organized as follows. In Section 2, we show the development in the RE sector. In addition, the investment in generating electricity using solar energy is discussed. The situation of the RE sector in Egypt is provided in Section 3. In Section 4, the L3CM methods are derived to apply in a real option framework. A real option framework is designed for use in RE investment in Section 5. In Section 6, a case study of solar thermal energy in Egypt is introduced. Furthermore, a comparison between the L3CM, FDM and MC methods is presented to explain the efficiency of the new numerical methods.

2. Renewable Energy Investment

Other concerns, like the rising demand of electricity and the risks of climate change, increase the importance of RES. In fact, RES are becoming ever more relevant in the generation of electricity. RES account for a rising share of the world's total electricity supply, and they are the fastest growing source of electricity generation in the IEO 2016 [5] (see Figure 1). Total generation from RES increases by 2.9%/year, and the renewable share of world electricity generation grew from 22% in 2012 to 29% in 2040. The generation of electricity from solar is increasing by an average of 8.3%/year. Of the 5.9 trillion kWh of new renewable generation added over the projection period, solar energy accounts for 859 billion kWh.

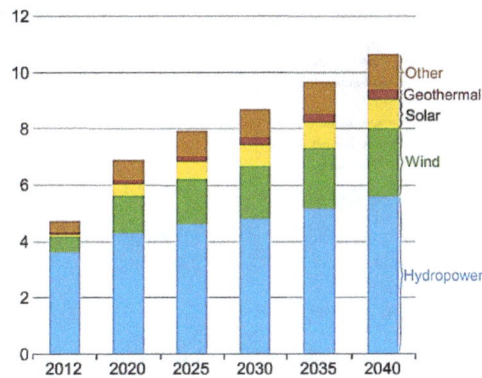

Figure 1. World net electricity generation from renewable power by fuel for trillion kWh [36].

A great deal of effort is being put into researching and developing renewable energy technologies. Bloomberg New Energy Finance tracks deals across the financing continuum, from R&D funding and venture capital for technology and early-stage companies, through to the asset fiance of utility-scale generation projects [36] (see Figure 2).

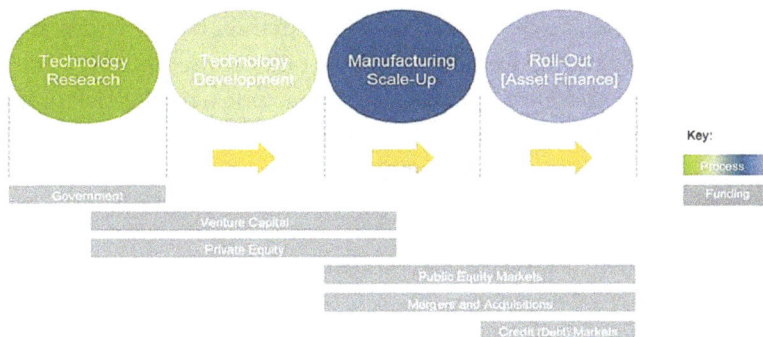

Figure 2. Bloomberg New Energy Finance tracks [36].

RE set new records in 2015. Investments reached nearly $286 billion, six-times more than in 2004. For the first time, more than half of all added power generation capacity came from RES. All of this happened in a year for which the prices of fossil fuel commodities-oil, coal and gas plummeted. So far, the drivers of investment in RE, including climate change policies and improving cost-competitiveness, have been more than sufficient to enable RE to keep growing its share of world electricity generation. Figure 3a shows that investment in RE rose 5% to $285.9 billion, taking it above the previous record of $278.5 billion reached in 2011, and that investment in RE has been running at more than $200 billion per year for six years now. The stand-out contribution to the rise in investment from the new record came from China, which lifted its outlays by 17% to $102.9 billion, some 36% of the global total. Investment also increased in the U.S., up 19% at $44.1 billion; in the Middle East and Africa, up 58% at $12.5 billion; and in India, up 22% at $10.2 billion.

Investment in solar has achieved the highest growth in 2015 among RES. Solar saw a 12% increase to $161 billion and wind a 4% boost to $109.6 billion. Biomass and waste-to-energy suffered a 42% fall to $6 billion; small hydro projects of less than 50 MW a 29% decline to $3.9 billion; biofuels a 35% drop to $3.1 billion; geothermal a 23% setback to $2 billion; and marine (wave and tidal) a 42% slip to just $215 million. Figure 3b shows the sector split for global investment.

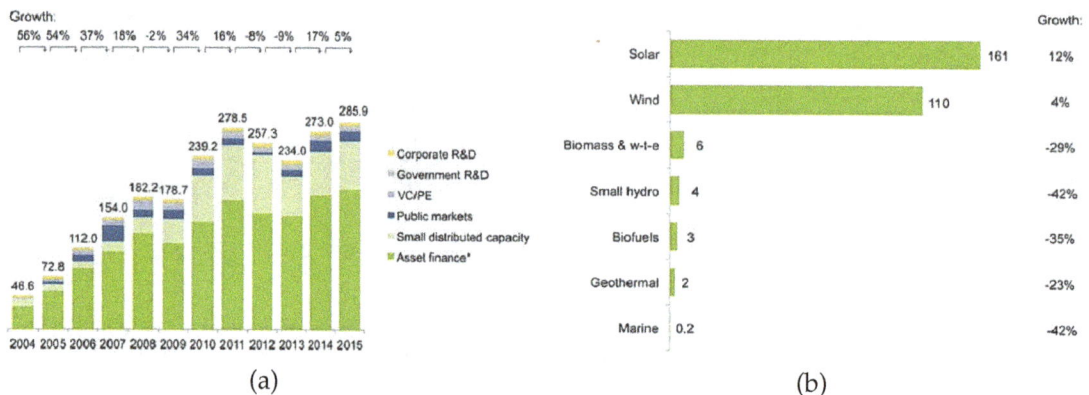

Figure 3. Real MS-stability regions: (**a**) new global investment in RE by asset class; (**b**) new global investment in RE by sector, 2015, and growth in 2014; $BN [36].

In order to pave the way for a wider spread of investing in renewable energy, a number of public support schemes have been considered. These schemes can be divided into two major groups:

- Regulatory price-based mechanisms (a payment for kWh of energy produced)
- Regulatory quantity-based mechanisms (the government sets a desired level of RES, and "green" generators receive tradable certificates according to their production)

Fagiani et al. [7] point out that a dilemma arises here: market risk provides an incentive to make efficient use of resources, thus limiting the cost to society, but it simultaneously deters investors, thus potentially resulting in less RE and higher prices (as they include a higher risk premium). Regarding the support schemes, the literature has argued, especially in recent times, that a key driver of RE investment is keeping investors' risks within reasonable limits. Three particular risk factors can stem from the policy instruments themselves:

- The type of instrument (e.g., feed-in tariffs, tradable green certificates)
- Constantly changing support schemes
- The design details of the particular instrument

Policy characteristics strongly affect the price risk and the quantity risk faced by an investor. However, their scope in mitigating other sources of technical risk [37,38] and financial risk [39] is more limited. Uncertain returns on these investments are generally considered a major cause for concern for developers and investors alike.

It was stated earlier that, when valuing renewable energy projects, there is uncertainty stemming from the long-term, costly, dependency on a feed-in tariff system and random behavior of prices associated with the energy source itself. When considering solar as an RES, one of the sources of uncertainty is future solar power, as well as future electricity prices. When a storage system is considered, the uncertainty remains the same; although the storage system is in place to make the energy source more predictable, there is still uncertainty in how much solar power we will see at a given hour, as well as uncertainty in the price of electricity at a given hour. We can use real options in this setting to determine the optimal time of launching the project and assess the value of real options.

3. Renewable Energy Sector in Egypt

In this section, an overview of the RE sector in Egypt is provided. Furthermore, we show the support schemes, which are introduced by the government to increase investment opportunities in generating electricity using Solar power. In 2008, the government announced the strategic plan to reach 20% of the total electricity generated from RE by 2020 vs. 9.1% in 2013. The country enjoys a total annual global solar irradiance of up to 2.6 TWh/m² and a total annual sunshine duration of up to 4000 h yearly [40]. The World Bank acknowledged Egypt's solar power advantage. It explains that there are many best areas for solar energy. Figure 4 shows the potential of solar energy in Egypt.

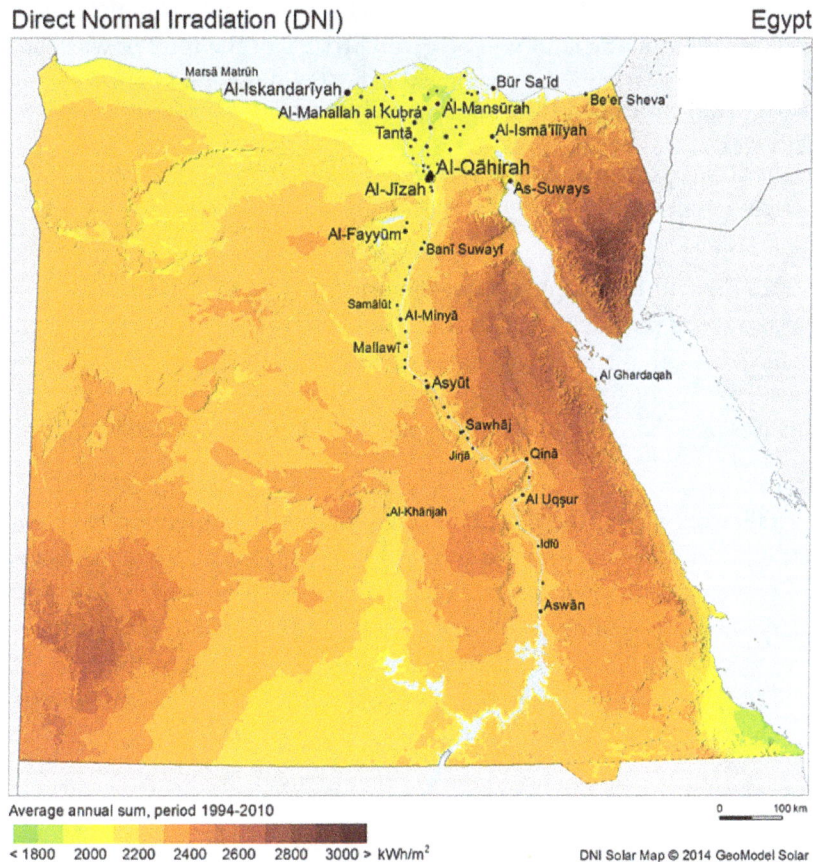

Figure 4. Potential of solar and wind energy in Egypt (source: Solar GIS http://solargis.info).

Fossil fuels have shared 91% of the electricity generation, in addition to 9% from RES. Of the 9% RE generation, there is 7.7% hydro-power, 1.2% wind and 0.1% solar [41]. In recent years, the Egyptian Electricity Holding Company (EEHC) has faced a gap on the power supply side, which caused recurring power cuts from 2012 to 2015. This gap will increase if the lack of investment in energy generation, both conventional and renewable, continues.

The Egyptian government has adopted an ambitious plan to reach 20% of the total electricity generated from RE by 2020, including 2% solar. The target is expected to be met by reaching the solar energy target of 3500 MW installed capacities up to 2027 [41] vs. the total capacity of 140 MW in 2014 [42]. The Egyptian government has introduced the following policies to foster the increasing of the RE energy contribution:

1. Public competitive bidding:
    Issuing tenders internationally requesting the private sector to supply power from RE projects.
2. Third party access (TPA):
    Investors are allowed to build and operate RE power plants to satisfy their electricity needs or to sell electricity to other consumers though the national grid.
3. Feed-in tariff (FIT):
    In September 2014, the government passed the key Feed-In Tariff Law (the feed-in tariff enacted by decree 1947/2014 [43,44]), triggering wide interest from international developers and investors. The main parameters of the feed-in tariffs are:

    *   Solar power stations: The value of the tariff is divided into five scales according to the production capacity of the station, and the value of the tariff will be fixed during the contract period, which reaches 25 years.
    *   Land allocation: Through the use of the craft scheme for a period of time equal to the contract period. Furthermore, the land will be given just 2% of the total power generated revenue from the plant. In addition, the customs will be 2% of the total items cost.
    *   Electricity: That produced through renewable energy stations has priority access to the electricity grid.
    *   Government support and guarantee: For power stations that exceed 500 kW, include low-interest credit facilities.

4. Net metering:
    In January 2013, EgyptERA adopted a net-metering policy that allows small-scale renewable energy projects to feed electricity to the grid. Generated surplus electricity will be discounted from the balance through the net-metering process.
5. Quota system:
    Heavy industries will be obliged to use a percentage of their electricity consumption from RE sources.

One of the challenges facing the Egyptian government to implement the RE strategy is that solar power plant investment is irreversible and uncertain. The solar energy projects are long-term, costly and depend on a feed-in tariff system. The real option framework, which takes into account investment irreversibility, uncertainty and flexibility in RE investment, was addressed in [11,16,45]. In the following, we derive new classes of numerical methods, the L3CM methods for SDEs. We discuss the applicability of the L3CM methods to approximate a stochastic process arising from real options analysis for the underlying asset in assessing the uncertainty investment.

4. The Lobatto3C-Milstein Method for SDEs

Numerical methods are needed for real option valuation in cases where analytic solutions are either unavailable or not easily compatible. In this work, we construct L3CM as a new numerical method, which can be used to approximate the stochastic process for the underlying asset in real option valuation. We consider the Itô SDEs of the form:

$$dy(t) = f(y(t))dt + g(y(t))dW(t) \quad y(t_0) = y_0 \quad t \in [t_0, T] \tag{1}$$

where $f(y(t))$ is the drift coefficient, $g(y(t))$ is the diffusion coefficient and Wiener process $W(t)$ is defined on a given probability space $(\Omega, \mathcal{F}, P)$ with a filtration $\{\mathcal{F}_t\}_{t \geq 0}$ satisfying the usual conditions, whose increment $\Delta W(t) = W(t + \Delta t) - W(t)$ is a Gaussian random variable $N(0, \Delta t)$.

Recently, there have been several attempts to construct numerical methods based on split-step techniques, to improve the fundamental analysis containing the convergence and stability of numerical solutions for SDEs. It is well known that there are many A-stable split-step numerical methods with a convergence order of 0.5 for scalar linear SDEs, such as the SSBE and SSθ methods. The split-step numerical methods with a convergence order of 1.0 are constructed for SDEs, for example the DSSBM method and SSAMM method. Unfortunately, we can see that the MS stability conditions of these methods for linear SDEs have some restrictions for the parameters and step-size h. Furthermore, Figure 5 shows that these methods are not A-stable.

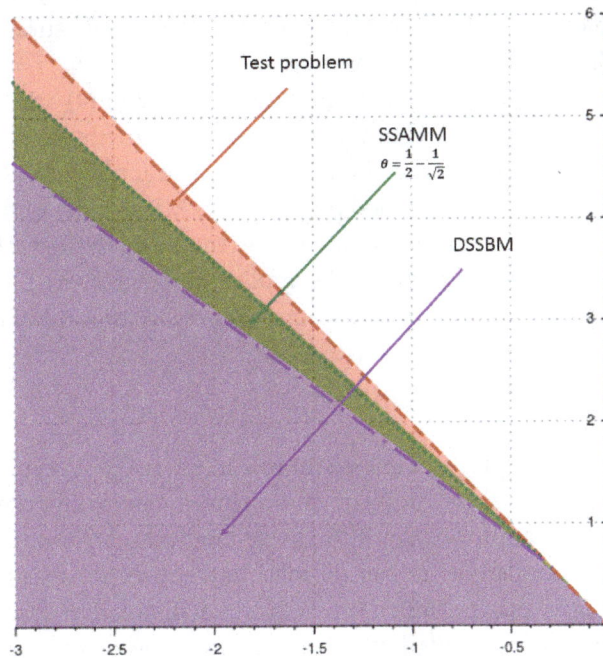

Figure 5. Real MS-stability regions of drifting split-step backward Milstein (DSSBM), split-step Adams–Moulton–Milstein (SSAMM) and the test problem.

In the following, in order to improve the numerical stability properties, the Lobatto3C-Milstein (L3CM) methods are derived for SDEs (1). The Lobatto3C note that, the Lobatto3C methods are L-stable (strong stability) and have been used successfully in solving stiff initial value ODE systems [35]) methods have the following form (the basic information about the Lobatto3C methods is presented in Appendix A):

$$Y_{ni} = y_n + h \sum_{j=1}^{s} a_{ij} f(t_n + c_j h, Y_{nj}) \quad i = 1, 2, ..., s \tag{2}$$

$$y_{n+1} = y_n + h \sum_{j=1}^{s} b_j f(t_n + c_j h, Y_{nj}) \tag{3}$$

Now, using the Lobatto3C Formula (2) and (3) in collusion with the Milstein method, we derive the L3CM methods for SDEs (1) as follows:

$$\tilde{Y}_{ni} \quad = \quad y_n + h \sum_{j=1}^{s} a_{ij} f(\tilde{Y}_{nj}), \quad i = 1, 2, ..., s \tag{4}$$

$$\hat{y}_n \quad = \quad y_n + h \sum_{j=1}^{s} b_j f(\tilde{Y}_{nj}) \tag{5}$$

$$y_{n+1} \quad = \quad \hat{y}_n + g(\hat{y}_n)\Delta W_n + \frac{1}{2} g'(\hat{y}_n) g(\hat{y}_n) \left[(\Delta W_n)^2 - h \right] \tag{6}$$

where y_n is an approximation to $X(t_n)$, $s \geq 2$ is the stage value, the coefficients a_{ij}, b_j characterize the L3CM methods, with increments $\Delta W_n := W(t_{n+1}) - W(t_n)$ being independent $N(0, h)$-distributed Gaussian random variables and $y(0) = y_0$. Moreover, y_n is $\{\mathcal{F}_{t_n}\}$-measurable at the mesh-point t_n. It is well known that the two implicit equations need to be solved for $\tilde{Y}_{ni}, i = 1, ..., s$ and $\hat{y}_n$.

Eissa et al. [46] provided that the L3C2Mmethod (i.e., s= 2 in (4) and (5)) converges strongly with order 1.0 under the Lipschitz condition and the linear growth condition. Furthermore, the mean-square (MS) stability of the L3C2M method is investigated for SDEs with both real and complex parameters. It is shown that the L3C2M method preserves the MS-stability of the exact solutions under no restriction on the step-size in the mean-square sense. In addition, the method is A-stable. In the following, the new L3C2M method (4) to (6) will be applied in the geometric Brownian motion (GBM), which is used to model ROV (note that this geometric Brownian motion (GBM) is a special case of SDEs (1)).

5. The Real Option Framework

The valuation of real option plays an important role in the real option planning. The framework for the real option provides a special viewpoint in valuing investment with uncertainty. There are many different methods that can be applied to an ROV; these methods can be categorized into analytical and numerical methods. They can be further divided into subsections, as represented in Figure 6. Schulmerich [47] gave an overview, in-depth discussion and mathematical descriptions of some specific methods. The ROV process can be divided into five steps as follows [48]:

1. Finding uncertainty investment opportunity.
2. The probability distribution of the uncertainties is approximated.
3. Know and analyze available real options.
4. Real option valuation.
5. Develop real options mind-set: by comparing the value of the options and the cost to obtain options, a set of strategies and decisions can be reached. Meanwhile, the mind-set regarding flexibility that is available and different is established.

In this section, we consider the solar plant power as an uncertainty investment opportunity. We develop a model to assess the value of a deferred option. At any stage of the project, the model can inform a strategic option to defer the project. Based on the particular characteristics of the real option in RE investment, a deferred option of the solar power plant is considered, where the cash flows are uncertain. We assume that the revenues will start the operation time of the solar power plant (i.e., we consider the cost of delay (the time lag between the decision to establish and the actual production)).

In this real option framework, we distinguish the numerical methods of the ROV. We examine the applicability of the L3C2M method of ROV and compare the results with that of the FDM and MC methods. The L3C2M method is integrated with option theory and the four economic elements, cost, value, risk and flexibility, to value the real option. The real option framework considers the volatility in solar energy price during the project lifetime, and the development lags between launch of the project and start if the production. The decision maker is facing an uncertain utility stream for investment. The valuation of real options helps the decision maker to evaluate the investment opportunity.

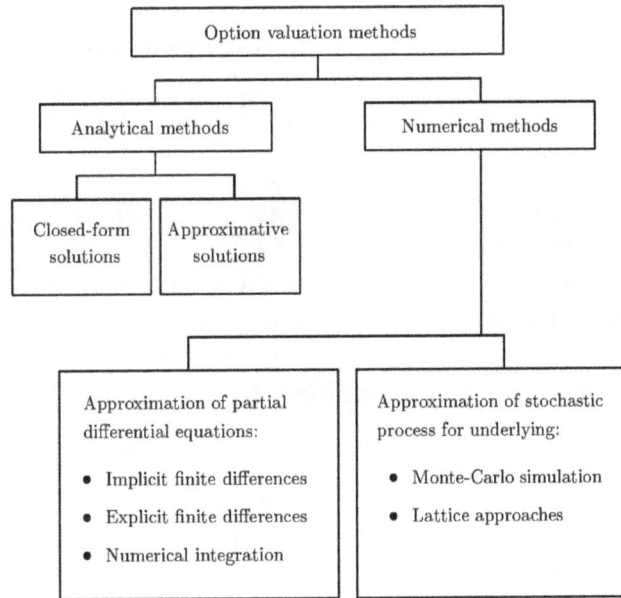

Figure 6. Classification of real option valuation (ROV) methods.

5.1. Framework Application

With the solar energy investment, the solar energy is the underlying asset. The value of the asset is based on two variables, the estimation of the installed capacity (MW) of the solar energy power plant and the pricing system. To value a solar energy investment as a real option, we need to make assumptions about a number of variables as follows

1. Availability of the solar energy source:
    At the outset, since this is not known with certainty, the availability of renewables has to be estimated. The investor can estimate the installed capacity (MW) of the solar energy plant and produced energy (kWh) by environmental assessment studies.
2. Estimated cost of establishing the solar energy plant:
    The estimated development cost is the exercise price of the option. The cost of establishing the solar energy plant can be estimated by feasibility studies for the projects.
3. Time to expiration of the option:
    The life of an RE option can be defined as a contract period; that period will be the lifetime of the option. For example, the contract in the sector of RE is a long-term contract of approximately 20 to 25 years.
4. Variance in the value of the cash flows:
    The variance in the value of the cash flows is determined by two factors, variability in the pricing system of the RE and variability in the estimate of the availability of the RE. In the more realistic case where the average of the RE resources and the RE price can change over time, the option becomes more difficult to value.
5. Cost of delay:
    Since the net production revenue cannot be started instantaneously, a time lag has to be allowed between the decision to establish the solar energy plant, and the actual production is the cost of delay (If the cash flows are evenly distributed over time and the exclusive rights last n years (20 years), the annual cost of delay can be written as: $\frac{1}{n} = \frac{1}{20} = 5\%$ a year. Though, this cost of delay rises each year, to $\frac{1}{19}$ in Year 2, $\frac{1}{18}$ in Year 3, and so on, making the cost of delaying the exercise larger over time.).

5.2. Stochastic Model

In this model, the L3C2M numerical method is examined. GBM is used to model the ROV. Suppose that we are seeking a valuation of a project with a finite lifetime $t \in [0, T]$. The cash flows S from the investment are stochastic with a standard deviation σ and risk-free interest rate r. Hence, the evolution of cash flows over time is described as:

$$dS(t) = rS(t)dt + \sigma S(t)dW(t) \quad t \in [0, T] \tag{7}$$

In the following, we derive a valuation for the investment case study problem (7) using the L3C2M numerical method. The SDEs (7) describes the paths of cash flows for the lifetime of solar power plant $t \in [0, T]$. The path values of $S(t)$ can be calculated iteratively by the L3C2M method, which is introduced in the previous section. The future steps depend on the type of real option.

5.2.1. The Deferred Option

If we assume that a project requires an initial up-front investment of I (initial cost) and that the present value of expected cash inflows computed right at time T is $S(T)$, the value of the defer option at time T is denoted by $V(S, T)$ as follows:

$$V(S, T) = e^{-rT}E[\max(S(T) - I, 0)] \tag{8}$$

The value of the real option can be determined by calculating the expected value in (8) for a given n paths, as an approximation to the expected value. The value of $S(t)$ can be determined using the L3C2M method for each path. Finally, we compare the value of real options (8) with the value of real options, which are computed by the FDM and MC method to show the efficiency of our method L3C2M.

6. A Case Study: 140-MW Solar Power Plant in Kuraymat, Egypt

In this section, we present numerical solutions for an actual case study of the solar power plant project in Egypt and analyze the numerical results. We test the evaluation model for the deferred option using the L3C2M numerical method. We demonstrate the efficiency of numerical method on the real options framework by comparing with FDM and MC methods.

Our data below are for the solar combined cycle power plant in Kuraymat, Egypt, the estimates of key parameters. They are relevant for computing revenues and initial cost over the useful lifetime. Through detailed information published in the annual report of New and Renewable Energy Authority (NREA) 2012/2013 [42], we could get and estimate the following information about the project:

- The installed capacity is $C = 140$ MW, including the solar share of 20 MW (think of the total area of the integrated solar field being about 644,000 m^2 and the total solar collectors is about 1920 solar collectors containing 53,760 mirrors) (NREA annual report 2012/2013 [42]).
- The total cost is about $I = 340$ \$ million, and the development lag is four years (NREA annual report 2012/2013 [42]).
- The lifetime of the project $T = 25$ years (the feed-in tariff enacted by decree 1947/2014 [43,44]).
- The tariff applied to the electricity generated from solar was $P = 0.1434$ \$/kWh (the feed-in tariff enacted by decree 1947/2014 [43,44]).
- The risk-free interest rate considered is $r = 8.75\%$, which corresponds to the 10-year Egypt government debt in September 2014 (source: Egypt Central Bank [49]).

In the following, we will estimate the discount cash-flow and the variance of the purchase price of electricity from solar power plants.

6.1. Estimate Discount Cash Flows

The feed-in tariff is generally claimed to be the most effective method for promoting RE. Let P denote the fixed tariff applied to the electricity generated from the solar power plant. According to [43,44], the feed-in tariff was enacted in October 2014 and provides for a sophisticated pricing system, differentiating between solar projects, as well as project installed capacity. The keys of the pricing system are indicated; those that are relevant to international investors are:

- 500 kW up to 20 MW: $0.136
- 20 MW up to 50 MW: $0.1434

The capacity of the project is $C = 140$ MW, including the solar share of 20 MW. Therefore, the feed-in tariff is considered to be $P = 0.1434$ $/kWh. Using the total produced energy (GWh) in a given year in Table 1 [42,50], the average of producing energy (kWh/year) is estimated to be $S_y = 305 \times 10^6$ kWh/year.

Table 1. The total produced energy (GWh) per year.

2010/2011	2011/2012	2012/2013	Average
206	479	230	305

The discount cash flow CF, in U.S. million dollars, of the investment under this scheme, which considers development lag, is [51]:

$$CF = S_0 = \frac{0.1434 \times 305 \times 10^6 \times 25}{(1.9)^2} = 302.8878$$

6.2. Estimate the Volatility

In July 2014, the Egyptian government issued its decree 1257/2014, which determines the increase of the electricity future price gradually over five years from 2014 to 2019 [52]. This decision was made within the Egyptian government plan to reduce the energy support. In October 2014, the Egyptian government issued the feed-in tariff enacted by decree 1947/2014 [43,44], which determines the purchase price of electrical energy supplied to the Egyptian company to transport electricity, from the plants producing the electricity from RES. Furthermore, we reconsider this price after two years from the date of publication of the decree, commensurate with the change in the selling price of electricity for the user.

Using electricity selling prices stated in the decree 1257/2014 [52] and following [15], we can derive the regression model whose residuals allows us to compute the volatility:

$$\sigma = 0.1045$$

In the following Table 2, we summarized all of the data sources for the case study.

Table 2. Parameters used in the investment option case. NREA, New and Renewable Energy Authority.

Parameter	Symbol	Value	Unit	Source
Current CF from investment	S_0	302.8878	$US million	Section 6.1
Fixed investment cost	I	340	$US million	NREA annual report 2012/2013 [42]
Time to invest	T	25	Years	Feed-in tariff decree 1947/2014 [43,44]
S.d. of cash flows	σ	0.1045		Section 6.2
Risk-free discount rate	r	0.0875		Egypt Central Bank [49]

6.3. Valuation of the Deferred Option

We consider the inputs in Table 2 to discuss the deferred option model as follows. We use the closed-form solution to benchmark the numerical results. A close resemblance to the pricing of a European call option (In finance, a European option can be exercised only at the expiration time of the option, while an American option can be exercised at any point of time during the option lifetime. Given the price of underlying security P and the strike price S, the payoff for a call option is defined as $\max(P - S, 0)$ and for a put option as $\max(S - P, 0)$.) with the Black–Scholes equation [27]. Plugging the given parameters into the closed-form Black–Scholes equation yields

$$V_{exact} = 264.7410$$

6.3.1. L3C2M Method

We derive a numerical solution with the L3C2M method for the investment option. In addition to the parameters listed in the Table 2, we have additional parameter $\theta = 0.8$, $h = 0.145$, such that the sample size is $N = T/h$, and we compute 5000 different discredited Brownian paths over the lifetime ($M = 5000$). We get from (8):

$$V_{L3C2M} = 264.7611$$

If we compare the value of the method with the exact solution, we find that the value of the method is very close to the exact solution. Moreover, note that the investment is valuated naturally in the whole domain with both methods. Comparing the option values, we note that the error in both methods is approximately the same and decreases rapidly with the length of the time steps. Figure 7 shows the mean-square error at time T versus the step-size h analyzed in the log-log diagram.

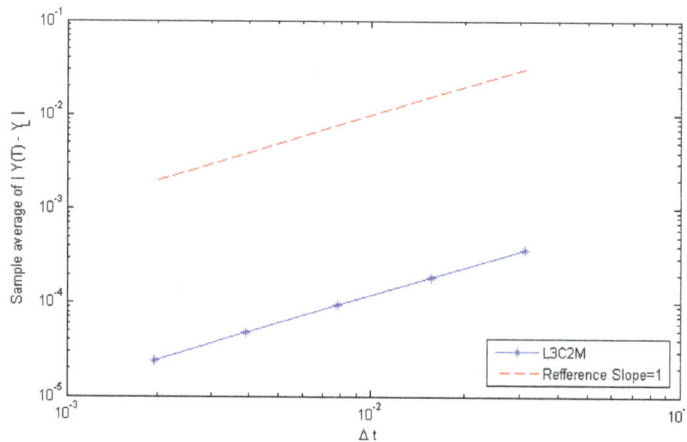

Figure 7. The MS error for the L3C2Mmethod.

6.3.2. Monte Carlo Simulation

Following [9], we run the MC simulation with the parameters given in Table 2. Using a sample size of $n_{max} = 1.5 \times 10^6$ and the 95% confidence level, the simulation yields the value of the investment option:

$$V_{MC} = 264.8050 \pm 0.3161$$

We note that the value is reasonably close to the exact value. To investigate the convergence properties, we run the simulation with smaller sample sizes, descending evenly to $n_{min} = 5 \times 10^4$. The results of the simulation are presented in Figure 8 along with the 95% confidence level.

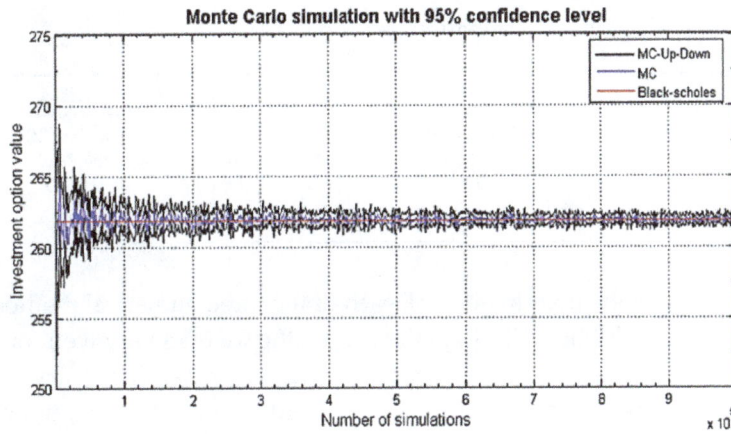

Figure 8. The value of the investment option (blue) and the 95% confidence level with an MC simulation in comparison to the analytical solution (red).

6.3.3. Finite Difference Method

Finally, following [9], we solve the investment option case with FDM. We derive a numerical solution with the explicit and implicit interpolation scheme. In addition to the parameters listed in the Table 2, we have to set additional parameters for the grid. Limiting the domain to $X = 900$ with $N = 250$ nodes and using $M = 10^5$ time steps, we obtain:

$$V_{FDM,exp} = 264.7458$$
$$V_{FDM,imp} = 264.7362$$

Comparing the values to the exact solution, we note that the values are very close to the exact solution with both methods. Moreover, we note that the investment is valuated naturally in the whole domain with FDM, which is not possible for example with the MC method due to path independence. The corresponding error plot of the values in log-log scale is given in Figure 9.

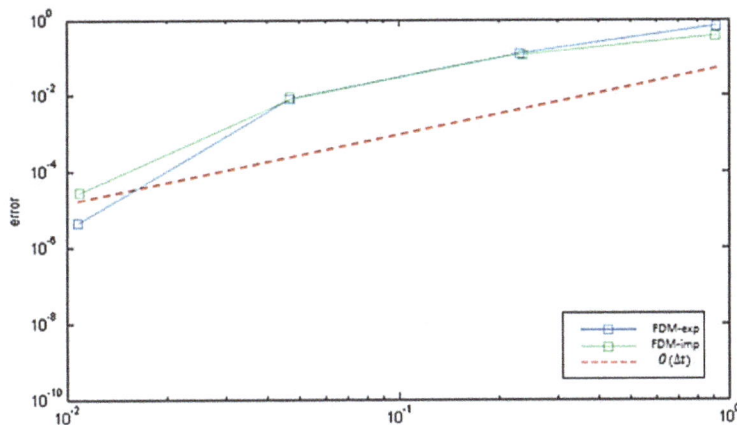

Figure 9. Absolute error for the explicit and implicit finite difference methods (FDM).

6.4. Discussion of the Results

Using the accuracy of the numerical solution as the only metric is problematic since increasing the number of iterations by one step does not equal increasing the grid size by one node. The comparison of the numerical methods for the investment case with respect to fixed absolute error and wall-clock time in seconds is presented in Table 3.

Table 3. Comparison of the numerical methods.

	MC	FDM-Exp	FDM-Imp	L3C2M
Inputs	100×10^6	(80, 9000)	(80, 9000)	(5000, 172)
Value (V)	264.8050	264.7458	264.7362	264.7611
Clock time	48.2573	0.7747	0.6002	0.0695
Error $\left(\frac{V_{Num.} - V_{exact}}{V_{exact}} \right)$	0.00024165	0.00001804	0.00001804	0.0000132

From Table 3, we conclude that: In general, each of the three numerical methods has values that are very close to the exact solution. Although the MC method works very well for pricing European options, approximates every arbitrary exotic option, is flexible in handling varying and even higher dimensional financial problems, the convergence of the MC method is very slow and takes a long run-time compared to other methods. FDM converges faster than the MC method and is more accurate; they are fairly robust and good for pricing options where there are the possibilities of early exercise, but FDM has become uncommonly used today, particularly amongst practitioners, due to the required mathematical sophistication; these too cannot readily be used for high-dimensional problems and also are very complicated in implementation. Finally, we can see that the L3C2M method outperforms all of the other methods in efficiency, converges faster than other methods and is considered simple in implementation compared to other methods of the case study with the given parameters.

7. Conclusions

In this paper, we address the real option valuation of an uncertainty investment in a solar power plant project and the optimal time to invest under the support program of Egypt: a feed-in tariff, electricity price and transitory subsidy. Three sources of uncertainty are considered: the electricity price, the level of solar generation and feed-in tariff. We construct a new general numerical method, the Lobatto3C-Milstein (L3CM) method, to use in the stochastic process of real option valuation, when the analytic solutions are lacking. Our real option framework differs from the previous work since; the new numerical L3CM method is integrated with option theory and the four economic elements, cost, value, risk and flexibility, to value a real option. We examine the L3CM method with two commonly-used methods, finite difference methods (FDM) and the Monte Carlo (MC) method, in an option valuation for investment with uncertainty in a case study.

Acknowledgments: The authors thanks the anonymous referees for careful reading and many helpful suggestions to improve the presentation of this paper. This research was partly financed by NSFC Grants 71350005 and 91646106. Additionally, this work is supported by the Higher Education Commission of Egypt.

Author Contributions: Mahmoud A. Eissa carried out the numerical method for SDEs, the numerical solution of the valuation problems and was relatively more involved in MATLAB. Boping Tian dealt more with the conception, theoretical issues and analysis of the results. All authors read and approved the final manuscript.

Conflicts of Interest: The authors declare that they have no competing interests.

Appendix A. Background of Lobatto3C Methods

The fundamental analysis containing the convergence and stability for numerical methods for differential equations is provided in [35,53–55]. The families of RKmethods based on Lobatto quadrature formulas are one of several classes of fully-implicit RK methods possessing good stability properties for ODEs. The number 3 is usually found in the literature associated with Lobatto methods. Ehle [56] introduced the Lobatto 3A, 3B and 3C classes. The general definition of the Lobatto3C methods are due to [57,58]. For more information about the fundamental properties of Lobatto methods, we recommend [35,53].

The classes of s-stage Lobatto methods are given in [35]:

$$Y_{ni} = y_n + h \sum_{j=1}^{s} a_{ij} f(t_n + c_j h, Y_{nj}) \quad i = 1, 2, ..., s \tag{A1}$$

$$y_{n+1} = y_n + h \sum_{j=1}^{s} b_j f(t_n + c_j h, Y_{nj}) \tag{A2}$$

where the stage value s satisfies $s \geq 2$ and the coefficients a_{ij}, b_j and c_j characterize the Lobatto methods. The s intermediate values Y_{nj} for $j = 1, ..., s$ are called the internal stages and can be considered as approximations to the solution at $t_n + c_j h$. The main numerical approximation at $t_{n+1} = t_n + h$ is given by y_{n+1}. Lobatto methods are characterized by $c_1 = 0$ and $c_s = 1$. For a fixed value of s, the various families of Lobatto methods share the same coefficients b_j and c_j. In addition, the coefficients a_{ij} vary depending on the classes of Lobatto methods. For the Lobatto3C class, the a_{ij} is defined as:

$$a_{i1} = b_1 \quad i = 1, ..., s \tag{A3}$$

and determined the remaining a_{ij} by $C(s-1)$. The coefficients of the Lobatto3C methods can be displayed by the Butcher tableau in Figure A1.

$$
\begin{array}{c|cc}
0 & \frac{1}{2} & -\frac{1}{2} \\
1 & \frac{1}{2} & \frac{1}{2} \\
\hline
 & \frac{1}{2} & \frac{1}{2}
\end{array}
\qquad
\begin{array}{c|ccc}
0 & \frac{1}{6} & -\frac{1}{3} & \frac{1}{6} \\
\frac{1}{2} & \frac{1}{6} & \frac{5}{12} & -\frac{1}{12} \\
1 & \frac{1}{2} & \frac{2}{3} & \frac{1}{2} \\
\hline
 & \frac{1}{2} & \frac{2}{3} & \frac{1}{2}
\end{array}
$$

Figure A1. The Lobatto3C methods of order two (**left**) and order four (**right**).

The stability properties of the numerical methods for deterministic ODEs are reported in [35]. In the following, we present the well-known results for Lobatto methods in a way that helps to motivate the SDEs analysis.

Proposition A1. *(See [35]) The s-stage Lobatto3C methods (A1) and (A2) applied to the scalar test equation:*

$$dX(t) = \lambda X(t)dt \quad t > 0 \quad X(0) = X_0 \neq 0 \tag{A4}$$

where $\lambda \in \mathbb{C}$ is a constant, yields:

$$y_{n+1} = R(\lambda, h) y_n \tag{A5}$$

with:

$$R(Z) = 1 + Z b^T (I - ZA)^{-1} \mathbf{1} \tag{A6}$$

where $b^T = (b_1, ..., b_s)$ $A = (a_{ij})_{i,j=1}^{s}$ $\mathbf{1} = (1, ..., 1)^T$ and I is the identity matrix. $R(Z)$ is called the stability function of the numerical method, which can be written for implicit methods as a rational function with numerator and denominator of degree $\leq s$ as follows:

$$R(Z) = \frac{P(Z)}{Q(Z)} \quad degP = k \quad degQ = j \tag{A7}$$

Let S_L be the stability domain for the Lobatto3C methods (A1) and (A2), then the method with stability function (A7) is A-stable if and only if $|R(iy)| \leq 1$ for all real y, and $R(Z)$ is analytic for $ReZ < 0$. In addition, using the definition of the method coefficients (A3) and (Proposition 3.8 in [35]), we find that the method also is L-stable. Furthermore, the Lobatto3C methods are characterized by non-stiff order $(2s - 2)$, being not symmetric, algebraically stable and B-stable, and the stability function $R(z)$ is given by $(s-2, s)$ Padé approximation to e^z. Therefore, the the Lobatto3C methods (A1) and (A2) are described as excellent methods for stiff problems.

References

1. World Wildlife Fund (WWF). *The Energy Report—100% Renewable Energy by 2050*; WWF: Gland, Switzerland, 2011.

2. International Energy Outlook (IEO). 2016. Avilable online: http://www.eia.gov/forecasts/ieo/ (accessed on 5 August 2016).

3. Andersson, B.A.; Azar, C.; Holmberg, J.; Karlsson, S. Material constraints for thin-film solar cells. *Energy* **1998**, *23*, 407–411.

4. Parida, B.; Iniyan, S.; Goic, R. A review of solar photovoltaic technologies. *Renew. Sustain. Energy Rev.* **2011**, *15*, 1625–1636.

5. EIA (2016) Updated Capital Cost Estimates for Electricity Generation Plants. Available online: www.eia.gov/forecasts/ieo/pdf/0484(2016).pdf (accessed on 5 August 2016).

6. Wiser, R.; Bolinger, M.; Barbose, G. Using the federal production tax credit to build a durable market for wind power in the United States. *Electr. J.* **2007**, *20*, 77–88.

7. Fagiani, R.; Barquín, J.; Hakvoort, R. Risk-based assessment of the cost-efficiency and the effectivity of renewable energy support schemes: Certificate markets versus feed-in tariffs. *Energy Policy* **2013**, *55*, 648–661.

8. Daim, T.U.; Amer, M.; Brenden, R. Technology roadmapping for wind energy: Case of the Pacific Northwest. *J. Clean. Prod.* **2012**, *20*, 27–37.

9. Sorsimo, A. Numerical Methods in Real Option Analysis. Master's Thesis, Aalto University, Helsinki, Finland, 2015.

10. Toke, D. Renewable financial support systems and cost-effectiveness. *J. Clean. Prod.* **2007**, *15*, 280–287.

11. Dixit, A.K.; Pindyck, R.S. *Investment under Uncertainty*; Princeton University Press: Princeton, NJ, USA, 1994.

12. Graham, J.R.; Harvey, C.R. The theory and practice of corporate finance: Evidence from the field. *J. Financ. Econ.* **2001**, *60*, 187–243.

13. Graham, J.; Harvey, C. How do CFOs make capital budgeting and capital structure decisions? *J. Appl. Corp. Financ.* **2002**, *15*, 8–23.

14. Fernandes, B.; Cunha, J.; Ferreira, P. The use of real options approach in energy sector investments. *Renew. Sustain. Energy Rev.* **2011**, *15*, 4491–4497.

15. Abadie, L.M.; Chamorro, J.M. Valuation of wind energy projects: A real options approach. *Energies* **2014**, *7*, 3218–3255.

16. Gazheli, A.; Di Corato, L. Land-use change and solar energy production: A real option approach. *Agric. Financ. Rev.* **2013**, *73*, 507–525.

17. Kloeden, P.E; Platen, E. *Numerical Solution of Stochastic Differential Equations.* Springer: Berlin, Germany, 1999.

18. Kloeden, P.E.; Platen, E. Higher-order implicit strong numerical schemes for stochastic differential equations. *J. Stat. Phys.* **1992**, *66*, 283–314.

19. Higham, D.J.; Mao, X.; Stuart, A.M. Strong convergence of Euler-type methods for nonlinear stochastic differential equations. *SIAM J. Numer. Anal.* **2002**, *40*, 1041–1063.

20. Ding, X.; Ma, Q.; Zhang, L. Convergence and stability of the split-step θ-method for stochastic differential equations. *Comput. Math. Appl.* **2010**, *60*, 1310–1321.

21. Huang, C. Exponential mean square stability of numerical methods for systems of stochastic differential equations. *J. Comput. Appl. Math.* **2012**, *236*, 4016–4026.

22. Wang, P.; Liu, Z. Split-step backward balanced Milstein methods for stiff stochastic systems. *Appl. Numer. Math.* **2009**, *59*, 1198–1213.

23. Guo, Q.; Li, H.; Zhu, Y. The improved split-step θ methods for stochastic differential equation. *Math. Methods Appl. Sci.* **2014**, *37*, 2245–2256.

24. Voss, D.A.; Khaliq, A.Q. Split-step Adams-Moulton Milstein methods for systems of stiff stochastic differential equations. *Int. J. Comput. Math.* **2015**, *92*, 995–1011.

25. Tian, B.; Eissa, M.A; Zhang, S. Two families of theta milstein methods in a real options framework. In Proceedings of the 5th Annual International Conference on Computational Mathematics, Computational Geometry and Statistics (CMCGS 2016), Singapore, 18–19 January 2016.

26. Copeland, T.E.; Weston, J.F.; Shastri, K. *Financial Theory and Corporate Policy*, 4th ed.; Pearson Addison Wesley: New York, NY, USA, 2005.

27. Black, F.; Scholes, M. The pricing of options and corporate liabilities. *J. Political Econ.* **1973**, 81, 637–654.

28. Merton, R.C. Theory of rational option pricing. *Bell J. Econ. Manag. Sci.* **1973**, 4, 141–183.

29. Brennan, M.J.; Schwartz, E.S. Finite difference methods and jump processes arising in the pricing of contingent claims: A synthesis. *J. Financ. Quant. Anal.* **1978**, 13, 461–474.

30. Boyle, P.P. Options: A monte carlo approach. *J. Financ. Econ.* **1977**, 4, 323–338.

31. Fadugba, S.; Nwozo, C.; Babalola, T. The comparative study of finite difference method and Monte Carlo method for pricing European option. *Math. Theory Model.* **2012**, 2, 60–66.

32. Pringles, R.; Olsina, F.; Garcés, F. Real option valuation of power transmission investments by stochastic simulation. *Energy Econ.* **2015**, 47, 215–226.

33. Thompson, M.; Barr, D. Cut-off grade: A real options analysis. *Resour. Policy* **2014**, 42, 83–92.

34. Sauer, T. *Numerical Analysis*; Pearson: Boston, MA, USA, 2005.

35. Hairer, E.; Wanner, G. *Solving Ordinary Differential Equations II: Stiff and Differential-Algebraic Problems*; Springer: Berlin, Germany, 1996.

36. Global Trends in Renewable Energy Investment 2016, Frankfurt School UNEP Collaborating Centre for Climate and Sustainable Energy Finance. 2016. Avilable online: http://fs-unep-centre.org/publications/global-trends-renewable-energy-investment-2016 (accessed on 5 August 2016).

37. Hirth, L. The market value of variable renewables: The effect of solar wind power variability on their relative price. *Energy Econ.* **2013**, 38, 218–236.

38. Scorah, H.; Sopinka, A.; van Kooten, G.C. The economics of storage, transmission and drought: Integrating variable wind power into spatially separated electricity grids. *Energy Econ.* **2012**, 34, 536–541.

39. Gross, R.; Blyth, W.; Heptonstall, P. Risks, revenues and investment in electricity generation: Why policy needs to look beyond costs. *Energy Econ.* **2010**, 32, 796–804.

40. Egypt in Transition: Infrastructure and Development, British Expertise. Available online: http://www.britishexpertise.org/bx/upload/Events/Egyptintransition_Spring2015_LR.pdf (accessed on 5 August 2016).

41. Egyptian Electricity Holding Company (EEHC). *Annual Report for FY 2012/2013*; EEHC: Cairo, Egypt, 2013.

42. New and Renewable Energy Authority (NREA). Annual Report 2012/2013. Available online: http://nrea.gov.eg/annual%20report/Annual_Report_2012_2013_eng.pdf (accessded on 5 August 2016).

43. Amereller, Investing in Renewable Energy in Egypt, January 2015. Available online: http://amereller.de/fileadmin/PDFs/ALC_Investing-in-Renewable-Energies-Jan-2015.pdf (accessed on 5 August 2016).

44. New and Renewable Energy Authority (NREA). Available online: http://nrea.gov.eg/beta/Investors/Legislation (accessed on 5 August 2016).

45. Min, K.J.; Lou, C.; Wang, C.H. An exit and entry study of renewable power producers: A real options approach. *Eng. Econ.* **2012**, 57, 55–75.

46. Eissa, M.A; Xiao, Y.; Yang, Z.Y. Lobatto3C-Milstein methods for stiff stochastic differential equations. *Appl. Numer. Math.* **2016**, under review.

47. Schulmerich, M. *Real Options Valuation: The Importance of Interest Rate Modelling in Theory and Practice*; Springer Science & Business Media: Berlin, Germany, 2010.

48. Wang, T. Analysis of Real Options in Hydropower Construction Projects—A Case Study in China. Ph.D. Thesis, Massachusetts Institute of Technology, Cambridge, MA, USA, 2003.

49. The Central Bank of Egypt. Available online: http://www.tradingeconomics.com/egypt/interest-rate (accessed on 5 August 2016).

50. New and Renewable Energy Authority (NREA), Annual Report 2009/2010. Available online: http://nrea.gov.eg/annual%20report/annual2010En.pdf (accessed on 5 August 2016).

51. Stern, N. *Real Option*; New York University: New York, NY, USA, 2007.

52. Egypt's Information Portal. Available online: http://www.eip.gov.eg/Documents/StudiesDetails.aspx?id=2142 (accessed on 5 August 2016).

53. Hairer, E.; Nrsett, S.P.; Wanner, G. *Solving Ordinary Differential Equations I: Nonstiff Problems*; Springer: Berlin, Germany, 1993.

54. Ramadan, M.; ElDanaf, T.S.; Eissa, M.A. System of ordinary differential equations solving using cellular neural networks. *J. Adv. Math. Appl.* **2014**, 3, 182–194.

55. Ramadan, M.; ElDanaf, T.S.; Eissa, M.A. Approximate solutions of partial differential equations using cellular neural networks. *Int. J. Sci. Eng. Investig.* **2015**, *4*, 14–21.
56. Ehle, B.L. On Pade Approximations to the Exponential Function and a-Stable Methods for the Numerical Solution of Initial Value Problems. Ph.D. Thesis, University of Waterloo, Waterloo, ON, Canada, 1969.
57. Axelsson, O. A note on a class of strongly A-stable methods. *BIT Numer. Math.* **1972**, *12*, 1–4.
58. Chipman, F. A-stable Runge-Kutta processes. *BIT Numer. Math.* **1971**, *11*, 384–388.

8

Thermo-Economic Comparison and Parametric Optimizations among Two Compressed Air Energy Storage System Based on Kalina Cycle and ORC

Ruixiong Li, Huanran Wang *, Erren Yao and Shuyu Zhang

School of Energy and Power Engineering, Xi'an Jiaotong University, Xi'an 710049, Shaanxi, China;
happy09english@stu.xjtu.edu.cn (R.L.); yao.erren@stu.xjtu.edu.cn (E.Y.); zhangshuyu@stu.xjtu.edu.cn (S.Z.)
* Correspondence: huanran@xjtu.edu.cn

Academic Editor: Antonio Calvo Hernández

Abstract: The compressed air energy storage (CAES) system, considered as one method for peaking shaving and load-levelling of the electricity system, has excellent characteristics of energy storage and utilization. However, due to the waste heat existing in compressed air during the charge stage and exhaust gas during the discharge stage, the efficient operation of the conventional CAES system has been greatly restricted. The Kalina cycle (KC) and organic Rankine cycle (ORC) have been proven to be two worthwhile technologies to fulfill the different residual heat recovery for energy systems. To capture and reuse the waste heat from the CAES system, two systems (the CAES system combined with KC and ORC, respectively) are proposed in this paper. The sensitivity analysis shows the effect of the compression ratio and the temperature of the exhaust on the system performance: the KC-CAES system can achieve more efficient operation than the ORC-CAES system under the same temperature of exhaust gas; meanwhile, the larger compression ratio can lead to the higher efficiency for the KC-CAES system than that of ORC-CAES with the constant temperature of the exhaust gas. In addition, the evolutionary multi-objective algorithm is conducted between the thermodynamic and economic performances to find the optimal parameters of the two systems. The optimum results indicate that the solutions with an exergy efficiency of around 59.74% and 53.56% are promising for KC-CAES and ORC-CAES system practical designs, respectively.

Keywords: integrated energy storage system; CAES; Kalina cycle; ORC; thermo-economic

1. Introduction

Due to the increasing depletion of fossil fuels and the deterioration of global environmental pollution, integrating renewable energy sources into the power system has developed rapidly during the last few decades. However, the inherent nature of renewable energy, i.e., randomness and intermittence, has greatly restricted the large-scale utilization of renewable energies, mostly in the power grid [1,2]. Energy storage technology can deal with the encountered situation by storing the excess renewable energies and releasing the energy to balance the difference between energy demand and supply [3].

At present, pumped hydroelectric energy storage (PHES) and compressed air energy storage (CAES) are two energy storage technologies suitable for large-scale energy storage applications [4,5]. Compared with PHES, CAES has the advantages of low investment costs, fast construction time and high economic feasibility. The system may contribute to creating a flexible energy system with a better utilization of fluctuating renewable energy sources [3,6]. During the charge period at the off-peak time, electricity produced by renewable energy sources converts into internal energy of compressed air, which is stored in an under-ground cavern; during the discharge period at peak demand, the compressed air is heated up first and then expanded in the turbine to produce electricity [4,7].

In recent years, many works regarding the CAES system have been conducted in order to improve the applicability of the system [8]. Grazzini and Milazzo [9] presented a thermodynamic analysis for an adiabatic compressed air energy storage system (ACAES) and proposed the variable configuration system with a variable compression ratio to address the un-steady operation of the components. Wolf and Budt [1] introduced a low-temperature adiabatic compressed air energy storage (LTA-CAES) plant, which can avoid all of the technical challenges of adiabatic CAES designs. Arabkoohsar et al. [10–12] proposed a CAES system equipped with an ancillary solar heating system for a large-scale PV farm in Brazil, and the thermodynamic and economic analyses are conducted to select the best operation strategy of the power plant. Abbaspour et al. [13] and Zhao et al. [14] conducted a preliminary dynamic behavior analysis for CAES integrated with wind power to find the optimization and improvement methods of the system. Fu et al. [15] described a new gas turbine power generation system coupled with conventional CAES technology to improve the thermal efficiency at least 5% over that of the existing system. Bouman [16] and Gulagi [17] discussed the environmental impacts and importance associated with a CAES or ACAES system as a means of balancing the electricity output.

Although the CAES system has had remarkable attention paid to it, the large-scale CAES system needs a suitable underground geology to store the compressed air, which leads to the great limitation of the commercial application of CAES. Thus, developing a small-scale CAES system, which could use an artificial air vessel instead of a natural cavern to store compressed air, has become the main trend in recent years [5]. Because of the advantages of the gas engine in the small-scale industrial system, integrating the gas engine with CAES could not only boost the system output power, but also improve the efficiency of energy utilization and operation flexibility in the system [5]. Ibrahim et al. [18] presented a hybrid wind-diesel CAES to heighten diesel power output, increase engine lifetime and reduce the fuel consumption and greenhouse gas emissions. Yao et al. [5] proposed a novel combined cooling, heating and power system (CCHP) combined with CAES and a gas engine; the system increases the output power of the CCHP system and realizes energy cascade utilization. Nielsen et al. [19] and Basbous et al. [20] exposed new CAES systems based on diesel engines, which can be applied to remote areas and ships.

The mentioned systems, due to the waste heat from the engine or compressor exhaust gas not being utilized efficiently and flexibility, cause much destruction of the available energy. The organic Rankine cycle (ORC) and Kalina cycle (KC) have been proven to be two worthwhile technologies to fulfill the residual heat recovery for energy systems [21,22]. Therefore, the two cycles (i.e., ORC and KC) can be employed in the CAES system to utilize the waste heat and improve the efficiency of the system. To the best of the authors' knowledge, there is only one paper related to the CAES combined with KC at present. Zhao et al. [8] proposed an integrated energy system based on CAES and the Kalina cycle, which can recover the waste heat of exhaust from a low pressure turbine during the discharge process. However, the heat produced by the compressor is provided to users directly, which will lead to the irrational utilization of high-grade energy when the exhaust temperature of the compressor is high. Therefore, we consider the possibility of recovering heat from the exhaust gas of the compressor and the gas engine by means of KC and ORC in this paper to improve the operating efficiency of the system greatly and expand the scope of application of the energy storage system.

This paper is organized into the following sections: Section 2 describes the schematic diagram of two novel CAES systems. Then, the mathematical modelling of each component, including thermodynamics and exergy model, is presented in Section 3. The sensitivity analysis and optimization of the system are shown in Section 4. Finally, the conclusions are summarized in Section 5. The main originality of this paper is summarized as follows:

(a)   In this paper, two novel combined CAES systems which have never been presented are proposed.

(b)   Compared with ACAES, the thermal storage vessel is not employed in the proposed system. In addition, the output power of the system can increase with the increasing power demand of the electric network.

(c)   The heat produced during the charging and discharging stage is used to generate electricity, which can improve the operation efficiency.

2. System Description

The schematic of the compressed air energy storage (CAES) system is shown in Figure 1. The working process of the system can be divided into two periods: charge period and discharge period. During the charge period, the compressor is driven by renewable energies to compress the ambient air to high-pressure compressed air (from Streams 1 to 5). The compressed air enters the evaporator for after-cooling (Streams 2 and 3) and then is stored in the air storage vessel (Stream 5). The compression heat produced by Evaporators 1 and 2 (EP1 and EP2) will be provided to the heat recovery cycle, i.e., the Kalina cycle (KC) or organic Rankine cycle (ORC).

During the discharge period, the high-pressure air passes through the throttle valve (THV), which is used to adjust the air flow rate (Stream 6) and then enters the regenerator (Reg) for preheating (Stream 6). After that, the compression air expands in the air turbine to drive the generator to produce electricity (Stream 7). After the expansion, the mixture of the exhaust air from the air turbine and the fuel enters the gas engine (GE) (Stream 8). The energy produced by the GE includes three parts: the power energy, the heat of exhaust gas and the heat of jacket water (i.e., used to cool the GE to prevent the inside temperature of the GE from becoming too high). The jacket water heat will be absorbed by EP4 (Stream 12) for the heat recovery cycle. The exhaust gas heat will be used to preheat the compressed air (Stream 9) coming from the air storage vessel first, and then, the residual heat of the exhaust gas will be employed for the heat recovery cycle in EP3 (Stream 10).

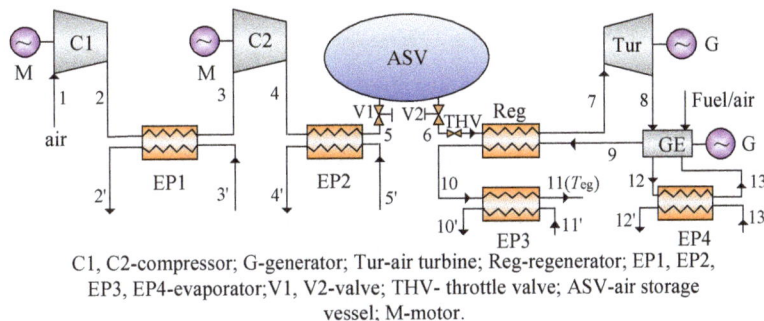

C1, C2-compressor; G-generator; Tur-air turbine; Reg-regenerator; EP1, EP2, EP3, EP4-evaporator; V1, V2-valve; THV- throttle valve; ASV-air storage vessel; M-motor.

Figure 1. Schematic diagram of the CAES system.

Figure 2 illustrates the schematic diagram of ORC combined with CAES (ORC-CAES). During the charge or discharge stage, the working fluid absorbs the heat energy from the exhaust gas in the evaporator to produce superheated vapor. After that, the vapor is expanded in the turbine (Tur$_O$) to the output power (stream n'). The exhaust gas of the turbine is cooled in the regenerator (Reg$_O$) (Stream 14'), condensed by cooling water (Con$_O$) (Stream 15'), compressed (Pum$_O$) (Stream 16') and cooled by the regenerator (Reg$_O$) (Stream 17'). Finally, the working fluid enters the evaporator (stream m') to carry out the next cycle. For the working fluid in the ORC, R245fa is selected as the working fluids, due to its good thermodynamic performance and small negative impact on the environment [23,24].

The schematic of the combination of the KC and CAES (KC-CAES) is shown in Figure 3. The KC could provide a better match to a heat source than the constant-temperature evaporation of a pure substance (water/steam) [21]. This cycle employs an ammonia-water mixture as its working fluid. During the working process, the mixture splits into four streams (basic solution, working solution, rich solution and poor solution) with different concentrations to provide the flexibility of optimizing the heat recovery system [21,25]. During the charge or discharge stage, the ammonia-water mixture absorbs the heat energy from the exhaust gas with a high temperature and pressure in the evaporator to produce superheated vapor. Then, the ammonia-water vapor expands in the ammonia turbine (Tur$_K$)

(stream n′); meanwhile, power is generated. After that, the ammonia turbine exhaust is cooled in the regenerator (HReg$_K$) (Stream 18′) and diluted with ammonia-poor liquid in a mixture (M1) to become basic solution (Stream 20′). In the condenser (LCon$_K$), the basic solution is condensed to saturated liquid (Stream 21′), which will be compressed to an intermediate pressure (Stream 22′) in low pressure pump (LPum$_K$) after leaving LCon$_K$ by cooling water. The diverter (D1) divides working solution into two parts (Stream 22′). One of them is heated by the regenerator (LReg$_K$, HReg$_K$) (Streams 23′ to 24′) and then enters the separator (Stream 25′) that separates the medium into the ammonia-poor solution and the ammonia-rich vapor; the other part (Stream 26′) mixes directly with the rich ammonia solution (Stream 28′) in the mixer (M2) to become the working solution (Stream 26′). Finally, the working solution will be cooled, condensed by cooling water (HCon$_K$), compressed (HPum$_K$) and sent to the evaporator (Streams 29′ to m′) [26].

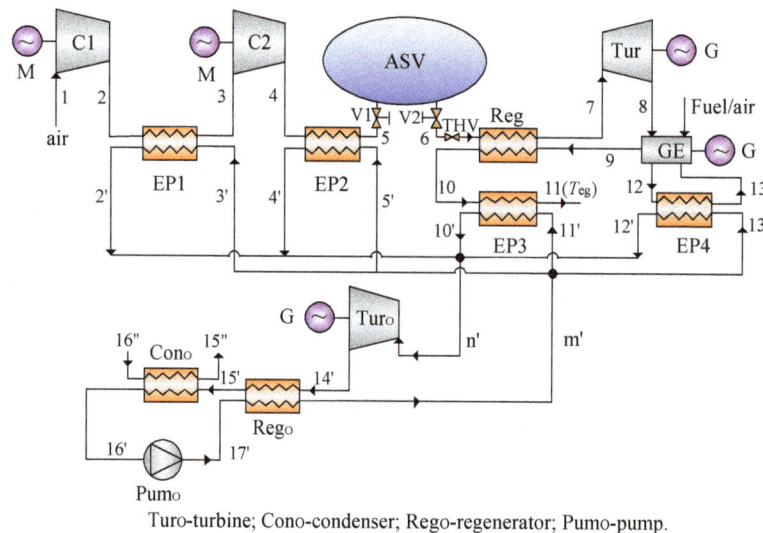

Turo-turbine; Cono-condenser; Rego-regenerator; Pumo-pump.

Figure 2. Schematic diagram of the ORC-CAES system.

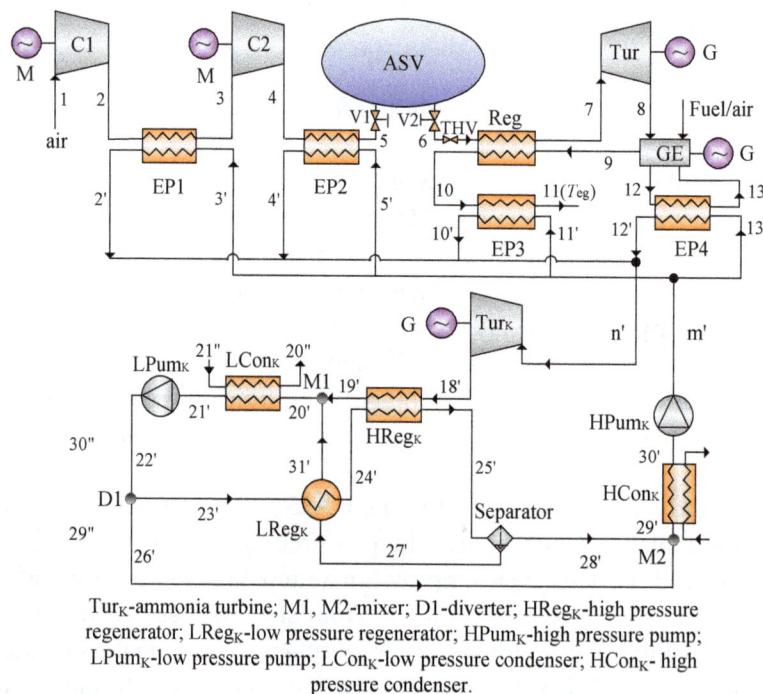

Tur$_K$-ammonia turbine; M1, M2-mixer; D1-diverter; HReg$_K$-high pressure regenerator; LReg$_K$-low pressure regenerator; HPum$_K$-high pressure pump; LPum$_K$-low pressure pump; LCon$_K$-low pressure condenser; HCon$_K$- high pressure condenser.

Figure 3. Schematic diagram of the Kalina cycle (KC)-CAES system.

3. Mathematical Modeling

The mathematical models of the novel energy storage system include the CAES, the KC and the ORC. The following assumptions are made [8,9,26]:

(1) The air is treated as an ideal gas in the CAES system;
(2) Ambient air is composed of 78.12% nitrogen, 20.96% oxygen and 0.92% argon;
(3) The temperature and pressure of ambient atmospheric environment are 298.15 K and 101.325 kPa, respectively;
(4) The fuel of the gas engine is composed of 100% methane;
(5) In the KC and ORC, the working fluid in the condenser is cooled by water coming from the atmospheric environment.

3.1. CAES Mathematical Model

3.1.1. Turbine and Compressor

For the compression process, the power consumption of the compressor can be obtained:

$$\dot{W}_{Comp} = \frac{k R_g T_{in} \dot{m}_{Comp} \left(\pi_{Comp}^{(k-1)/k} - 1 \right)}{(k-1)\eta_{Comp}}, \tag{1}$$

where $\dot{m}_{Comp}$ is the air mass flow rate of the compressor, T_{in} stands for the inlet temperature of the compressor, k is the isentropic exponent and π_{Comp} and η_{Comp} represent the pressure ratio and isentropic efficiency of the compressor, respectively.

The outlet temperature of the compressor can be given as:

$$T_{out} = T_{in} \left[\frac{\left(\pi_{Comp}^{(k-1)/k} - 1 \right)}{\eta_{Comp}} + 1 \right], \tag{2}$$

Then, the isentropic efficiency of the compressor can be expressed as:

$$\eta_{Comp} = \frac{h_{out,s} - h_{in}}{h_{out} - h_{in}}, \tag{3}$$

In the expansion process, the power output by the turbine can be calculated by the following equations:

$$\dot{W}_{Tur} = \frac{\eta_{Tur} k R_g T_6 \dot{m}_{Tur} \left(1 - \pi_{Tur}^{-(k-1)/k} \right)}{(k-1)}, \tag{4}$$

where $\dot{m}_{Tur}$ is the air mass flow rate of the turbine, T_6 is the inlet temperature of the turbine and π_{Tur} and η_{Tur} represent the pressure ratio and isentropic efficiency of the turbine, respectively.

The outlet temperature and isentropic efficiency of the turbine are shown below:

$$T_7 = T_6 \left[1 - \eta_{Tur} \left(1 - \pi_{Tur}^{-(k-1)/k} \right) \right], \tag{5}$$

$$\eta_{Tur} = \frac{h_6 - h_7}{h_6 - h_{7,s}}. \tag{6}$$

3.1.2. Gas Engine

The power output of the gas engine is the product of the power efficiency η_{GE} and fuel consumption rate $\dot{Q}_{GE}$:

$$\dot{W}_{GE} = \dot{Q}_{GE} \cdot \eta_{GE}, \tag{7}$$

The fuel consumption rate by the gas engine can be written as:

$$\dot{Q}_{GE} = \dot{m}_{fuel} \cdot LHV \tag{8}$$

where LHV denotes the low calorific value of fuel and $\dot{m}_{fuel}$ is the mass flow rate of the fuel.

3.1.3. Air Storage Vessel

According to the conservation law of mass, the following equation can be obtained:

$$\sum m_{in,ASV} = \sum m_{out,ASV} \tag{9}$$

where $\sum m_{in,ASV}$ and $\sum m_{out,ASV}$ are the total input and output mass of air, respectively.

Due to the heat dissipation existing in the air storage vessel (ASV), the outlet temperature of compressed air from the ASV can be considered as ambient temperature:

$$T_6 = T_0 \tag{10}$$

where T_0 is the ambient temperature.

3.2. ORC and KC Mathematical Model

3.2.1. Turbine and Pump

The power output of the turbine can be expressed as:

$$\dot{W}_{Tur,HRC} = \dot{m}_{Tur,HRC} \cdot \left(h^{in}_{Tur,HRC} - h^{out}_{Tur,HRC} \right), \tag{11}$$

where $\dot{m}_{Tur,HRC}$ denotes the mass flow rate of the working fluid in the heat recovery cycle and $h^{in}_{Tur,HRC}$ and $h^{out}_{Tur,HRC}$ are the inlet and outlet enthalpy of turbine, respectively. The subscript HRC represents the heat recovery cycle: ORC or KC.

The isentropic efficiency of the turbine is:

$$\eta_{Tur,HRC} = \frac{h^{in}_{Tur,HRC} - h^{out}_{Tur,HRC}}{h^{in}_{Tur,HRC} - h^{out}_{Tur,HRC,s}}. \tag{12}$$

The power consumption and isentropic efficiency of the pump can be written as:

$$\dot{W}_{Pum,HRC} = \dot{m}_{Pum,HRC} \left(h^{out}_{Pum,HRC} - h^{in}_{Pum,HRC} \right), \tag{13}$$

$$\eta_{Pum,HRC} = \frac{h^{out}_{Pum,HRC,s} - h^{in}_{Pum,HRC}}{h^{out}_{Pum,HRC} - h^{in}_{Pum,HRC}}, \tag{14}$$

where $h^{in}_{Pum,HRC}$, $h^{out}_{Pum,HRC}$ are the inlet and outlet enthalpy of the pump.

3.2.2. Mixer, Separator and Diverter

Assuming that there is no energy loss in the mixer, separator and diverter, the following model about the mixer, separator and diverter can be used:

$$\begin{cases} \sum \dot{m}^{in}_{mixer} = \sum \dot{m}^{out}_{mixer} \\ \sum \dot{m}^{in}_{diverter} = \sum \dot{m}^{out}_{diverter} \\ \sum \dot{m}^{in}_{seperator} = \sum \dot{m}^{out}_{seperator} \end{cases}, \tag{15}$$

$$\begin{cases} \sum \dot{m}_{\text{mixer}}^{\text{in}} x_{\text{mixer}}^{\text{in}} = \sum \dot{m}_{\text{mixer}}^{\text{out}} x_{\text{mixer}}^{\text{out}} \\ \sum \dot{m}_{\text{diverter}}^{\text{in}} x_{\text{diverter}}^{\text{in}} = \sum \dot{m}_{\text{diverter}}^{\text{out}} x_{\text{diverter}}^{\text{out}} \\ \sum \dot{m}_{\text{seperator}}^{\text{in}} x_{\text{seperator}}^{\text{in}} = \sum \dot{m}_{\text{seperator}}^{\text{out}} x_{\text{seperator}}^{\text{out}} \end{cases} , \tag{16}$$

$$\begin{cases} \sum \dot{m}_{\text{mixer}}^{\text{in}} h_{\text{mixer}}^{\text{in}} = \sum \dot{m}_{\text{mixer}}^{\text{out}} h_{\text{mixer}}^{\text{out}} \\ \sum \dot{m}_{\text{diverter}}^{\text{in}} h_{\text{diverter}}^{\text{in}} = \sum \dot{m}_{\text{diverter}}^{\text{out}} h_{\text{diverter}}^{\text{out}} \\ \sum \dot{m}_{\text{seperator}}^{\text{in}} h_{\text{seperator}}^{\text{in}} = \sum \dot{m}_{\text{seperator}}^{\text{out}} h_{\text{seperator}}^{\text{out}} \end{cases} , \tag{17}$$

where x represents the concentration of ammonia-water.

3.3. Heat Exchanger Model

In the heat exchanger, the working fluid may work in different thermodynamic states, namely superheated state, two-phase state and sub-cooled state, which causes the heat transfer process in the heat exchanger to be divided into many regions. In the condenser, the working fluid is liquefied from one state (such as the superheated state or saturated liquid state) to another state (liquid state); while the working fluid may work in the sub-cooled region, two-phase region and superheated region in the evaporator [27]. However, the superheated region and the sub-cooled region can be considered as a single-phase region compared to a two-phase region [27].

The plate heat exchanger is employed in this paper due to the high heat transfer coefficient, good acclimatization, compact size and fewer materials' consumption [28,29]. According to the law of energy conservation, the heat released by hot fluid is equal to the value of heat absorbed by cold fluid. For the evaporator and regenerator, the following equations can be employed:

$$(h_{\text{hot,in}} - h_{\text{hot,in}}) \dot{m}_{\text{hot}} = (h_{\text{cold,in}} - h_{\text{cold,in}}) \dot{m}_{\text{cold}} \tag{18}$$

(A) Single-phase flow:

The heat rate of the heat exchanger can be given as [30,31]:

$$\dot{Q} = UA\Delta t_m, \tag{19}$$

where U represents the overall heat transfer coefficient of the heat exchanger and Δt_m is the logarithmic mean temperature difference (LMTD) that can be calculated as:

$$\Delta t_m = \frac{\Delta t_{\max} - \Delta t_{\min}}{\ln(\Delta t_{\min}/\Delta t_{\max})} \tag{20}$$

where $\Delta t_{\max}$ and $\Delta t_{\min}$ are the maximum and minimum temperature difference between cold and hot fluid, respectively.

According to the law of energy conservation, the overall heat transfer coefficient of the heat exchanger is:

$$\frac{1}{U} = \frac{1}{h_{\text{hot}}} + \frac{\delta}{\lambda} + \frac{1}{h_{\text{cold}}}, \tag{21}$$

where h_{hot} and h_{cold} are the heat transfer coefficient of hot fluid and cold fluid, respectively. δ is the thickness of the plate, and λ is the thermal conductivity.

The equivalent diameter of flow channel has been presented by Wang [28] and can be defined as:

$$D_{\text{h}} = \frac{4A}{C} = \frac{4Lb}{2(L+b)}, \tag{22}$$

where L, b are the channel width and channel spacing, respectively [28,31]. C is the wetted perimeter of the cross-section. A is the cross-section area of the channel.

The mass velocity of fluid:

$$G = \frac{\dot{m}_f}{NLb},$$

(23)

where N is the number of the channels [28,31].

The Reynolds number (Re) is a dimensionless quantity that is used to predict similar flow patterns in different fluid flow situations, which can be obtained by Equation (24) [5]:

$$Re = \frac{GD_h}{\mu},$$

(24)

where μ is the viscosity of fluid.

The Prandtl number (Pr) is the ratio of momentum diffusivity to thermal diffusivity [31]:

$$Pr = \frac{c_p \mu}{\lambda},$$

(25)

where λ and c_p are the thermal conductivity and specific heat of the fluid, respectively.

The Nusselt number (Nu) is the ratio of convective to conductive heat transfer across the boundary, which is calculated by Equation (26) for the single-phase region [32].

$$Nu = 0.724 \left(\frac{6\beta}{\pi} \right)^{0.646} Re^{0.583} Pr^{1/3},$$

(26)

where β stands for the chevron angle of the plates, being expressed by [32].

Therefore, the heat transfer coefficient HTC for single-phase flow can be expressed as:

$$HTC = \frac{\lambda \cdot Nu}{D_h}.$$

(27)

(B) Two-phase flow:

The properties of fluid in the two-phase region are not constant, which will lead to inaccurate calculation results in the single-phase flow model. Thus, the two-phase region is divided into a number of small regions where the properties of the fluid keep almost constant [28,33,34]. The LMTD and overall heat transfer coefficient for each region can be obtained by Equations (20) and (21).

In the condenser, the heat transfer coefficient on the hot side for each region can be given by [28,33]:

$$Nu_{\text{Con},i} = 4.118 Re_i^{0.4} Pr_i^{1/3};$$

(28)

while in the evaporator, the heat transfer coefficient on the cold side for each region can be expressed as [28,33]:

$$Nu_{\text{Eva},i} = 1.926 \left[(1 - x_{m,i}) + x_{m,i} \left(\frac{\rho_l}{\rho_v} \right)^{0.5} \right] Re_i^{0.5} Pr_i^{1/3} Bo_i^{0.5},$$

(29)

where Bo_i is the boiling number and $x_{m,i}$ is the dryness of the fluid [28,33]. Re_i, Pr_i can be obtained using Equations (24) and (25), respectively.

3.4. Exergy Model of Systems

Exergy analysis reveals the location, the magnitude and the sources of thermodynamic inefficiencies within the energy system [5]. In addition, the exergy analysis can be used to improve the system efficiency by determining the sources and magnitude of irreversibility.

The total exergy of a material stream can be divided into two parts: chemical exergy (Ex_{ix}^{ch}) and physical exergy (Ex_{ix}^{ph}) [35]:

$$Ex_{ix} = Ex_{ix}^{\text{ph}} + Ex_{ix}^{\text{ch}},$$

(30)

$$Ex_{ix}^{ph} = m[h_{ix} - h_0 - T_0(s_{ix} - s_0)], \tag{31}$$

$$Ex_{ix}^{ch} = m\left(\sum_{jx} x_{jx}\varepsilon^0 + RT_0\sum_{jx} x_{jx}\ln x_{jx}\right), \tag{32}$$

where x_{jx} is the mole fraction and ε^0 is the standard chemical exergy of the involved chemical component.

3.5. Performance Evaluation Criteria

In order to evaluate the thermodynamic and economic performance of the proposed system, the following system indicators are introduced [36–38]:

The round-trip efficiency η of the system is the ratio of the total energy generated $\sum \dot{W}_{out}$ (i.e., the power produced by the sum of ammonia turbine, air turbine and gas engine) and the total energy consumed by the system (i.e., the power consumed by compressor and pump $\sum \dot{W}_{in}$ and the fuel consumption of the gas engine $\dot{Q}_{GE}$).

$$\eta = \frac{\sum \dot{W}_{out}}{\dot{Q}_{GE} + \sum \dot{W}_{in}}, \tag{33}$$

Exergy efficiency used to characterize the utilization of the total effective energy of the system is defined as the ratio between exergy product and exergy fuel [6,39,40]. The exergy product and the exergy fuel are defined by considering the desired result produced by the component and the resources expended to generate this result [21,35].

$$\varepsilon = \frac{\dot{Ex}_{P,tot}}{\dot{Ex}_{F,tot}}, \tag{34}$$

where $\dot{Ex}_{P,tot}$ is the exergy product of the whole system and $\dot{Ex}_{F,tot}$ is the sum of input power and the exergy of input fuel.

In order to evaluate the relationship between fuel consumption and output power of the system, the heat rate (HR) is introduced [1,41,42]:

$$HR = \frac{\dot{Q}_{GE}}{\sum \dot{W}_{out}}, \tag{35}$$

The total investment cost per total output power ($ICPP$) as a widely-used economic indicator is used to calculate the approximate total investment cost of a given system for different scales [5,28]:

$$ICPP = \frac{C_{total}}{\sum \dot{W}_{in}}, \tag{36}$$

where C_{total} is the total investment of the novel systems. The cost functions of the involved components in the systems are listed in Appendix A.

4. Results and Discussions

4.1. Sensitivity Analysis

The main thermodynamic parameters of the components in the systems are listed in Table 1. In this paper, the concentration of ammonia-water is 0.8 [43,44]. From the previous studies [25,44], the value of the temperature of exhaust gas (T_{eg}) reveals the ability of waste recovery by the heat recovery cycle. In this paper, the compression ratio and temperature of the exhaust gas (T_{eg}) are the key parameters to find the different requirements for the two proposed systems to recover the heat efficiently.

Table 1. The simulation condition of the proposed CAES system.

Term	Unit	Value
Ambient temperature	K	298.15
Ambient pressure	MPa	0.10
Pinch temperature difference	K	8
Turbine isentropic efficiency	-	0.9
Compressor isentropic efficiency	-	0.9
Pump isentropic efficiency	-	0.7
Work solution concentration	-	0.55
Rich solution concentration	-	0.80
Basic solution concentration	-	0.25
Gas engine power efficiency	-	0.42
Rated air flow rate of compression	kg/s	98.80
Volume of air storage vessel	m^3	30,000.00

Figure 4 shows the effect of pressure ratio of the compressor (π_{Comp}) on the exergy efficiency ($\varepsilon_{ORC\text{-}CAES}$), round-trip efficiency ($\eta_{ORC\text{-}CAES}$) and heat rate ($HR_{ORC\text{-}CAES}$) of the ORC-CAES system. With the increasing π_{Comp}, both $\varepsilon_{ORC\text{-}CAES}$ and $\eta_{ORC\text{-}CAES}$ decrease slowly at first; after the π_{Comp} reaches 4.3, the system performance indicators decrease quickly. However, as the π_{Comp} increases, the $HR_{ORC\text{-}CAES}$ increases from the beginning to the end. This is because the increasing π_{Comp} causes the increasing irreversible loss of the compressor, which would lead to the decrement of exergy efficiency and the round-trip efficiency. In addition, the final pressure of the air storage vessel increases with the increment of π_{Comp}, which will cause the increasing of the fuel consumption, resulting in the increase of the heat rate finally. Figure 5 shows the behaviors of the heat rate ($HR_{KC\text{-}CAES}$), round-trip efficiency ($\eta_{KC\text{-}CAES}$) and exergy efficiency ($\varepsilon_{KC\text{-}CAES}$) of the KC-CAES systems. It can be seen that with increasing π_{Comp}, $\varepsilon_{KC\text{-}CAES}$ and $\eta_{KC\text{-}CAES}$ of the KC-CAES system decrease, whereas $HR_{KC\text{-}CAES}$ increases. The reason for these trends is similar to the ORC-CAES system. Moreover, comparing Figure 4 with Figure 5, the exergy efficiency and the round-trip efficiency of KC-CAES are higher than that of ORC-CAES. This is because the boiling point of the ammonia-water can be adjusted to suit the heat input temperature of the heat source, resulting in a good temperature match between the heat source and the working fluid of KC.

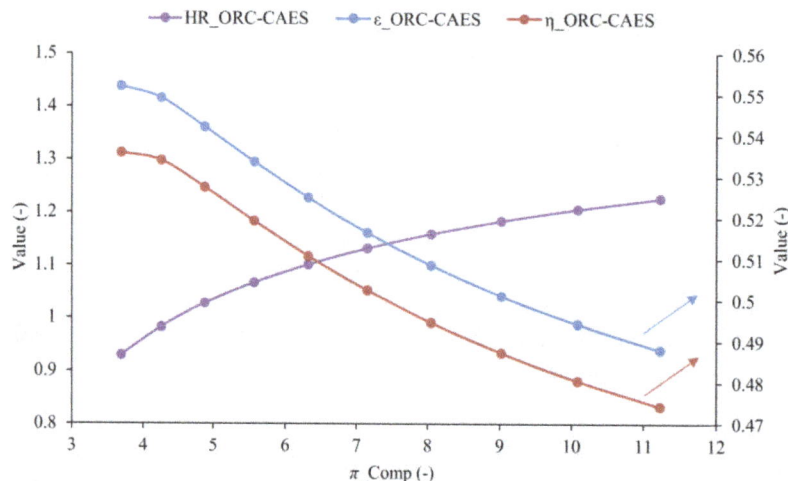

Figure 4. Effect of π_{Comp} on the performance parameters of ORC-CAES.

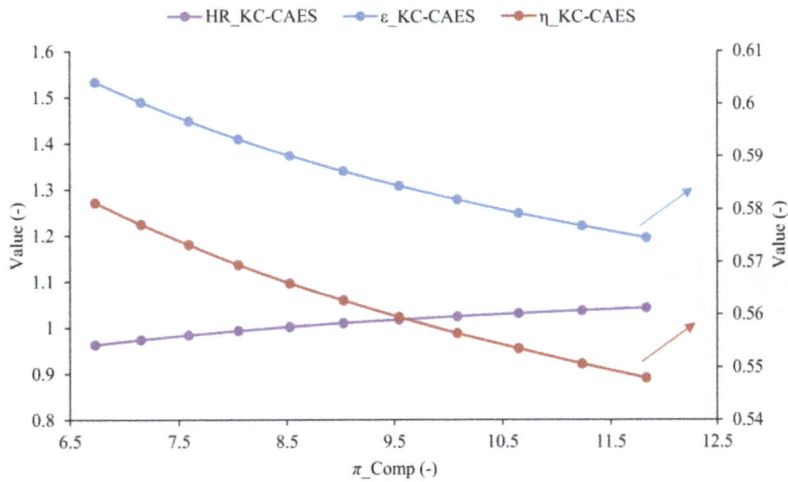

Figure 5. Effect of π_{Comp} on performance parameters of KC-CAES.

The effect of the outlet temperature of the exhaust gas (T_{eg}) on exergy efficiency ($\varepsilon_{ORC\text{-}CAES}$), round-trip efficiency ($\eta_{ORC\text{-}CAES}$) and the heat rate ($HR_{ORC\text{-}CAES}$) of the ORC-CAES system are given in Figure 6. With the increasing T_{eg} from 300 K to 325 K, $\varepsilon_{ORC\text{-}CAES}$ and $\eta_{ORC\text{-}CAES}$ of the system decrease slowly, whereas $HR_{ORC\text{-}CAES}$ increases slowly. This is because the exergy destruction of the compressor, gas engine and air turbine keep constant due to the fixed operation parameters; however, the increment in the outlet temperature of the exhaust gas leads to the decrement of the temperature difference between the inlet and outlet of the hot fluid of the evaporator, resulting in a decreasing heat recovery in the evaporator. After T_{eg} reaches 325 K, with the increasing T_{eg}, both the $\varepsilon_{ORC\text{-}CAES}$ and $\eta_{ORC\text{-}CAES}$ have remarkable plunges and then keep almost constant with the increase of temperature, whereas the $HR_{ORC\text{-}CAES}$ rises at first and then keeps constant. The reasons are the change of the thermodynamics property of the exhaust gas and that there is a breaking of the performance parameters of ORC-CAES systems when the T_{eg} is in the range of 325 to 330 K. Figure 7 illustrates the effect of the outlet temperature of the turbine on the enthalpy and entropy of the exhaust gas. There is a sudden increase of the enthalpy and entropy as the outlet temperature increases, which is mainly caused by the changing of the compression factor of nitrogen.

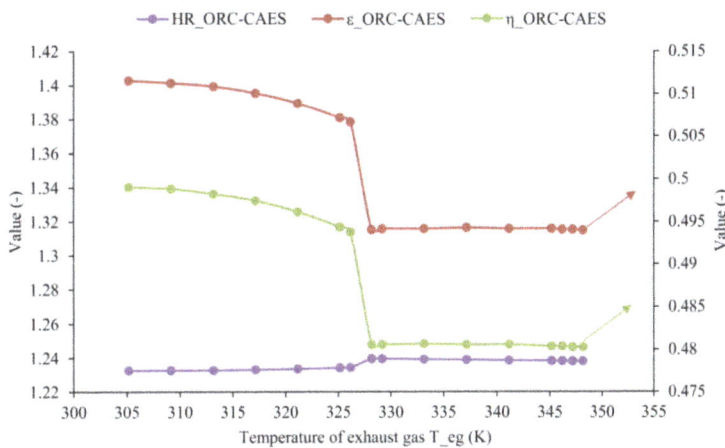

Figure 6. Effect of the outlet temperature of the exhaust gas on the performance parameters of the ORC-CAES.

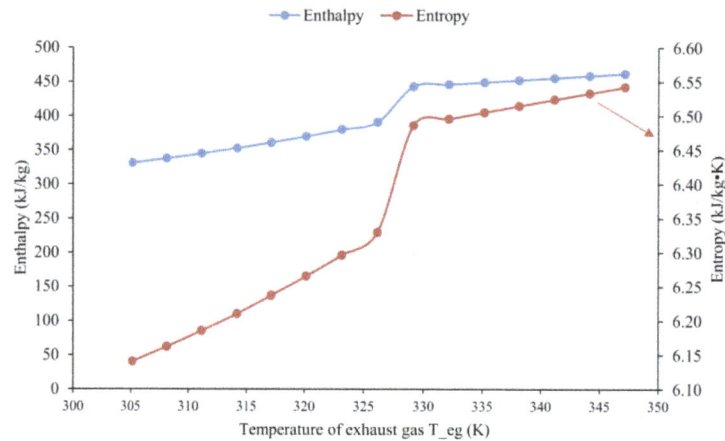

Figure 7. Effect of the outlet temperature of the exhaust gas on the enthalpy and entropy.

Figure 8 exhibits the effect of the temperature of the exhaust gas of the KC-CAES system (T_{eg}) on the exergy efficiency ($\varepsilon_{KC-CAES}$), round-trip efficiency ($\eta_{KC-CAES}$) and heat rate ($HR_{KC-CAES}$). Similar to the effect of T_{eg} on the exergy efficiency and round-trip efficiency of ORC-CAES, as T_{eg} increases from 300 K to 320 K, the performance the parameters of the system decreases. However, there is little difference of the performance parameters between ORC-CAES and KC-CAES as T_{eg} is higher than 320 K. This is because the thermodynamics property (entropy, density, etc.) of ammonia-water changes greatly compared to R245fa as the temperature increases, resulting in the power produced by the heat recovery cycle changing greatly.

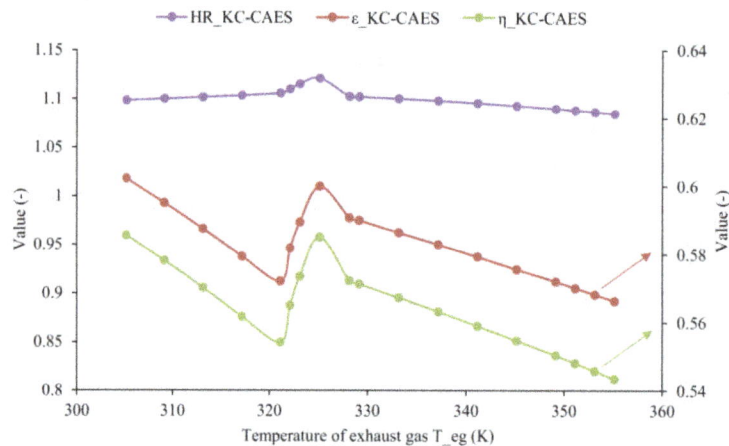

Figure 8. Effect of the outlet temperature of the exhaust gas on the performance parameters of the KC-CAES.

4.2. Parameter Optimization of Systems Based on the Genetic Algorithm

The sensitivity analysis shows the effect of a single system parameter on the system performance. In this section, the multi-objective optimization is carried out to find the trade-off between the thermodynamic performance and the economic performance of the system. The genetic algorithm (GA) is a method for solving both constrained and unconstrained optimization problems based on a natural selection process that mimics biological evolution. The algorithm repeatedly modifies a population of individual solutions. At each step, the GA selects individuals randomly from the current population and uses them as parents to produce the children for the next generation. Over successive generations, the population "evolves" toward an optimal solution [45,46].

To operate the system in the optimal state, the NSGA-Π is employed with two decision variables: the temperature of the exhaust gas (ORC-CAES: 305–347.5 K, KC-CAES: 305–355 K) and the pressure ratio of the compressor (ORC-CAES: 3.7 to 11.2, KC-CAES: 6.5 to 11.8); setting the population size, generations and function tolerance to 100, 200 and 1×10^{-4} for the algorithm, to find the maximize exergy efficiency (ε) and minimize the total investment cost per total output power (*ICPP*).

Figure 9 shows the Pareto optimal solutions for the ORC-CAES system between exergy efficiency and *ICPP*. In the multi-objective optimization, each point located on the Pareto frontier is a potential optimum solution. It can be seen that the maximum exergy efficiency can reach 55.2% as the *ICPP* achieves the maximum value of 0.83 k\$/kW (Point E), whereas the minimum exergy efficiency of 50.1% is obtained, and the *ICPP* also attains the minimum value of 0.57 k\$/kW (Point F). Since it is impossible to make each objective at its optimum value simultaneously, i.e., to achieve the greatest value of exergy efficiency and the lowest value of *ICPP* (Point B), the process of decision-making is mostly performed based on engineering experiences [27,29]. While in this paper, the design point in the Pareto front, which has the shortest distance from the hypothetical Point B, is selected as the final optimal design point (Point A) [29]. In addition, the Pareto front is flat when the exergy efficiency is below approximately 53.56%. Afterward, the *ICPP* increases substantially in the course of the increase in exergy efficiency. With the help of the above decision-making process, Point A with the exergy efficiency value of 53.56% and the *ICPP* of 0.67 k\$/kW is selected as the final optimal solution that is promising for engineering design.

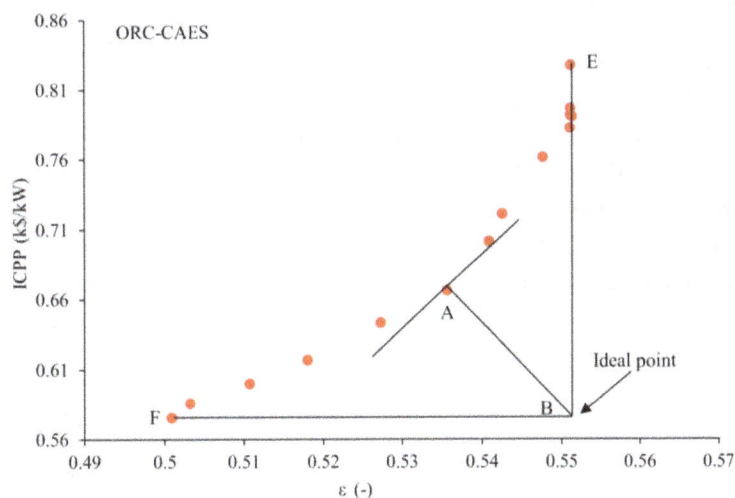

Figure 9. Pareto frontier of the ORC-CAES system for *ICPP* versus ε using multi-objective GA.

Figure 10 gives the Pareto optimal solutions for the KC-CAES system between exergy efficiency and *ICPP*. The minimum exergy efficiency achieves 57.3% with the minimum *ICPP* (Point H), whereas as the investment cost obtains the maximum, the maximum exergy efficiency is obtained (Point G). Point D, which is a hypothetical point, is the ideal point with the minimum investment cost and maximum exergy efficiency. Point C possesses the shortest distance from the hypothetical point and is selected as the final optimal solution that is promising for engineering design.

Thus, the final optimal parameters of the two systems are shown in Table 2. Comparing with the ORC-CAES system, the KC-CAES system can achieve a lower temperature of the exhaust gas under the optimum conditions, resulting in less waste of the exhaust heat and higher exergy efficiency; besides, the KC-CAES system can operate at lower capital cost at the optimum conditions.

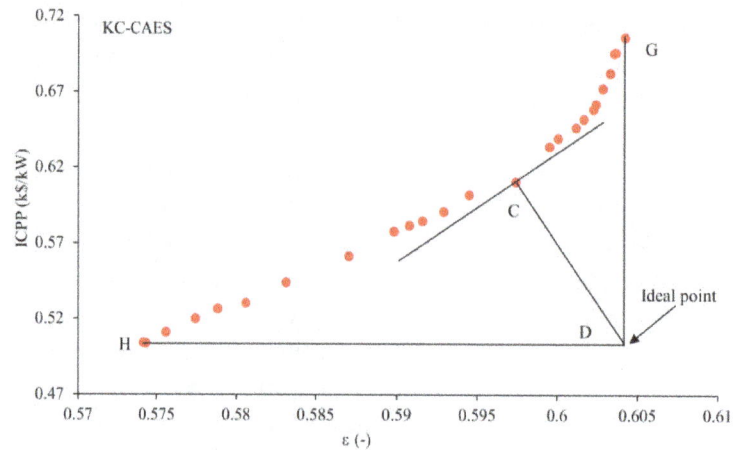

Figure 10. Pareto frontier of KC-CAES for *ICPP* versus *ε* using multi-objective GA.

Table 2. Optimal parameters of the two systems.

System	π_{Comp}	T_{eg} (K)	Exergy Efficiency (%)	*ICPP* (k$/kW)
KC-CAES	7.33	323	59.74	0.61
ORC-CAES	6.42	344	53.56	0.67

The relative efficiency of the two proposed systems, including relative round-trip efficiency $((\eta_{HRC\text{-}CAES} - \eta_{CAES})/\eta_{HRC\text{-}CAES})$ and relative exergy efficiency $((\varepsilon_{HRC\text{-}CAES} - \varepsilon_{CAES})/\varepsilon_{HRC\text{-}CAES})$, based on the single CAES system are given in Table 3. It can be seen that the second law efficiency of the proposed KC-CAES and ORC-CAES system improves 27.32% and 16.83% compared to that of the single CAES system, respectively. The round-trip efficiency of the combined two systems improves 27.35% (KC-CAES) and 16.89% (ORC-CAES). Moreover, the KC-CAES system achieves higher exergy and round-trip efficiency than those of the ORC-CAES system. The reason is that KC could provide a better match to a heat source and could achieve higher efficiency (20.5%) operation than that of ORC (11.7%).

Table 3. Comparison of the efficiency between KC-CAES and ORC-CAES at the optimal point (based on the single CAES).

System	Relative Exergy Efficiency (%)	Relative Round-Trip Efficiency (%)	Heat Recovery Cycle Efficiency (%)
KC-CAES	27.32	27.35	20.5
ORC-CAES	16.83	16.89	11.7

5. Conclusions

In this paper, two novel compressed air energy storage systems (CAES), i.e., CAES combined with the organic Rankine cycle (ORC-CAES) and CAES combined with the Kalina cycle (KC-CAES), are proposed to utilize renewable energy sources efficiently by recovering the residual heat recovery. The dependence of the system's thermodynamic performance on the compression ratio and temperature of the exhaust is investigated by a sensitivity analysis. Besides, the multi-objective optimization is conducted to find the optimal parameters of the two systems. The main conclusions are summarized as follows:

(1) Firstly, increasing the pressure ratio of the compressor, the exergy efficiency and round-trip efficiency of the two energy systems decrease, resulting in the operation performance of the systems become worse. Secondly, the effect of the temperature of exhaust gas on KC-CAES is

more obvious than that of the ORC-CAES; the KC-CAES system possesses higher operation efficiency and a lower heat rate, which illustrates that the Kalina cycle is more suitable as a heat recovery cycle of the CAES system.

(2) For ORC-CAES, due to the changes in the total investment cost per total output power (*ICPP*) being relatively slow when the efficiency is less than 53.56%, the *ICPP* increases significantly when the efficiency is more than 53.56%; the optimal solutions around exergy efficiency of 53.56% are suggested for industrial applications. Moreover, the exergy efficiency value of 59.74% has been selected for the KC-CAES system as the final optimal solution, which is based on the design point in the Pareto front having the shortest distance from the hypothetical point.

Acknowledgments: The authors are grateful to the support given by the National Natural Science Foundation of China (51676151).

Author Contributions: Ruixiong Li dealt with the detailed calculations, the simulation and was responsible for drafting and revising the whole paper. Erren Yao revised the grammar and the structure of the whole paper. Shuyu Zhang and Huanran Wang gave some useful suggestions on the system modelling.

Conflicts of Interest: The authors declare no conflict of interest.

Abbreviations

The following abbreviations are used in this manuscript:

A	area of components (m^2)
b	channel spacing (m)
Bo	boiling number
C	capital cost
c_p	specific heat capacity at constant pressure (J/kg·K)
Ex	exergy (kJ)
G	mass velocity
h	enthalpy (kJ/kg)
HTC	heat transfer coefficient (W/m²·K)
k	isentropic exponent
K	constant
L	average distance between the channel (m)
LHV	lower heating value (J/kg)
$\dot{m}$	flow rate (kg/s)
N	number of channel
Nu	Nusselt number
p	pressure (pa)
Pr	Prandtl number
Re	Reynolds number
R_g	universal gas constant (J/kg·K)
T	temperature (K)
U	overall heat transfer coefficient (W/m²·K)
V	volume of components (m^3)
$\dot{W}$	power (kW)
x	mass fraction of ammonia

Greek letters

β	chevron angle
δ	thickness of the plate (m)
ε	exergy efficiency
η	round-trip efficiency
λ	thermal conductivity of fluid (W/m·K)
μ	dynamic viscosity (N·s/m²)
π	pressure ratio
ρ	density (kg/m^3)

Subscripts and superscripts

ch	chemical
cold	cold fluid
Comp	compressor
Con	condenser
Div	diverter
eg	exhaust gas

Eva	evaporator
GE	gas engine
Hhex	high pressure regenerator
hot	hot fluid
HRC	heat recovery cycle
in	input of the system
ix	material stream
jx	chemical components
K	Kalina cycle
Lhex	low pressure regenerator
O	ORC
out	output of the system
P	product
ph	physical
PS	proposed system
Pum	pump
s	isentropic
tot	total
Tur	turbine
x	percentage of components

Appendix A

The large-scale application of the energy system can be determined by economic operation, which means lower-cost and higher profits. Therefore, an economic analysis of a novel compressed air energy storage (CAES) system is conducted in order to optimize the business-economic profits on the market. The centrifugal pump is employed in the two proposed systems.

The total investment of the novel systems:

$$C_{Total,PS} = C_{Total,CAES} + C_{Total,HRC}, \tag{A1}$$

where the subscript PS represents the proposed system, ORC-CAES (CAES combined with ORC) or KC-CAES (CAES combined with Kalina cycle), and the subscript HRC denotes the heat recovery cycle: KC (Kalina cycle) or ORC (organic Rankine cycle); $C_{Total,CAES}$ is the overall capital cost of the CAES, and $C_{Total,HRC}$ is the overall capital cost of the heat recovery cycle.

(1) Economic model of KC and ORC:

The overall capital cost of the system is determined by the summation of individual components:

$$C_{Total,HRC} = \sum C_{Ci,HRC} + \sum C_{Cj,HRC} \tag{A2}$$

where $C_{Ci,HRC}$ and $C_{Cj,HRC}$ represent the capital cost of components in the heat recovery cycle.

The capital cost of components (including turbine, separator, mixer or diverter) in the cycle is expressed as [29,47]:

$$C_{Ci,HRC} = \frac{584.6}{397} F_{Ci,MP} F_S C_{Ci,HRC}^0, \tag{A3}$$

where the subscript Ci represents components in the cycle, such as the turbine, separator, mixer or diverter. $F_{Ci,MP}$ is the material and pressure factor of the components (stainless steel). F_S is the additional factor. $C_{Ci,HRC}^0$ is the basic cost of the components made from carbon steel and can be given by [29,47]:

$$\log C_{Ci,HRC}^0 = K_{1,Ci,HRC} + K_{2,Ci,HRC} \log \dot{W}_{Ci,HRC} + K_{3,Ci,HRC} \left(\log \dot{W}_{Ci,HRC} \right)^2, \tag{A4}$$

where $\dot{W}_{Ci,HRC}$ is the power output for the turbine, whereas $\dot{W}_{Ci,HRC}$ represents the volume for the separator, mixer or diverter. $K_{1,Ci,HRC}$, $K_{2,Ci,HRC}$ and $K_{3,Ci,HRC}$ are respectively constants for the components' type.

For other components (including the pump and heat exchanger), the capital cost is given by [4,29,47]:

$$C_{Cj,HRC} = \frac{584.6}{397}\left(B_{1,Cj} + B_{2,Cj}F_{Cj,M}F_{Cj,P}\right)F_S C^0_{Cj,HRC} \tag{A5}$$

where Cj represents the pump or heat exchanger. $B_{1,Cj}$ and $B_{2,Cj}$ are constants for the component type. $F_{Cj,M}$ and $F_{Cj,P}$ are the material factor and pressure factor of components (stainless steel). $C^0_{Cj,HRC}$ is the basic cost of components made from carbon steel.

$$\log C^0_{Cj,HRC} = K_{1,Cj,HRC} + K_{2,Cj,HRC}\log \dot{W}_{Cj,HRC} + K_{3,Cj,HRC}\left(\log \dot{W}_{Cj,HRC}\right)^2, \tag{A6}$$

where $K_{1,Cj,HRC}$, $K_{2,Cj,HRC}$ and $K_{3,Cj,HRC}$ are constants for the components. $\dot{W}_{Cj,HRC}$ is the consumption power for the pump.

$$\log F_{Cj,P} = C_{1,Cj} + C_{2,Cj}\log P_{Cj} + C_{3,Cj}\left(\log P_{Cj}\right)^2 \tag{A7}$$

Note that $C_{1,Cj}$, $C_{2,Cj}$ and $C_{3,Cj}$ are constants for the component type. P_{Cj} is the design pressure of the corresponding components [4,29,47].

(2) Economic model of CAES:

The economic models proposed by Couper et al. [48–50] are employed in calculating the capital cost of CAES.

$$C_{Total,CAES} = C_{Tur} + C_{Comp} + C_{Reg} + C_{GE} + C_{vessel}, \tag{A8}$$

where C_{Tur}, C_{Comp}, C_{GE} and C_{vessel} denote the air turbine (Tur), compressor (C1 and C2), gas engine (GE) and air storage vessel.

Equations for evaluating the capital cost of the CAES components are shown in Table A1.

Table A1. Capital cost model of each component of the CAES.

Components	Investment Model (k$)
Air turbine, Tur	$C_{Tur} = 0.110\left(\dot{W}_{Tur}\right)^{0.81}$ [49,50]
Compressor, C1 or C2	$C_{Comp} = 0.790\left(\dot{W}_{Comp}\right)^{0.62}$ [49,50]
Regenerator, Reg	$C_{Reg} = 1.218 f_d f_m f_p \exp\left[8.821 - 0.30863(\ln A_{Reg}) + 0.0681(\ln A_{Reg})^2\right]10^{-4}$ [49]
Gas engine, GE	$C_{GE} = 0.65\dot{W}_{GE}\left(16137\dot{W}_{GE}^{-0.3799}\right)10^{-4}$ [48]
Air storage vessel	$C_{vessel} = 10^{-4}\times 1.218 \exp\left[2.631 + 1.3673(\ln V_{vessel}) - 0.06309(\ln V_{vessel})^2\right]$ [49]

f_d: type correction factor; f_m: material correction factor; f_p: pressure correction factor. $\dot{W}_{Tur}$, $\dot{W}_{Comp}$ and $\dot{W}_{GE}$ are respectively the output power of the turbine, the power consumption of the compressor and the output power by the gas engine. A_{Reg} and V_{vessel} present the area of the regenerator and the volume of the air storage vessel.

References

1. Wolf, D.; Budt, M. LTA-CAES—A low-temperature approach to Adiabatic Compressed Air Energy Storage. *Appl. Energy* **2014**, *125*, 158–164. [CrossRef]
2. Dudiak, J.; Kolcun, M. Integration of renewable energy sources to the power system. In Proceedings of the 14th International Conference on Environment and Electrical Engineering (EEEIC), Krakow, Poland, 10–12 May 2014.
3. De Bosio, F.; Verda, V. Thermoeconomic analysis of a Compressed Air Energy Storage (CAES) system integrated with a wind power plant in the framework of the IPEX Market. *Appl. Energy* **2015**, *152*, 173–182. [CrossRef]
4. Yao, E.; Wang, H.; Liu, L.; Xi, G. A Novel Constant-Pressure Pumped Hydro Combined with Compressed Air Energy Storage System. *Energies* **2015**, *8*, 154–171. [CrossRef]
5. Yao, E.; Wang, H.; Wang, L.; Xi, G.; Maréchal, F. Thermo-economic optimization of a combined cooling, heating and power system based on small-scale compressed air energy storage. *Energy Convers. Manag.* **2016**, *118*, 377–386. [CrossRef]

6.   Tessier, M.J.; Floros, M.C.; Bouzidi, L.; Narine, S.S. Exergy analysis of an adiabatic compressed air energy storage system using a cascade of phase change materials. *Energy* **2016**, *106*, 528–534. [CrossRef]

7.   Zhang, Y.; Yang, K.; Li, X.; Xu, J. The thermodynamic effect of thermal energy storage on compressed air energy storage system. *Renew. Energy* **2013**, *50*, 227–235. [CrossRef]

8.   Zhao, P.; Wang, J.; Dai, Y. Thermodynamic analysis of an integrated energy system based on compressed air energy storage (CAES) system and Kalina cycle. *Energy Convers. Manag.* **2015**, *98*, 161–172. [CrossRef]

9.   Grazzini, G.; Milazzo, A. Thermodynamic analysis of CAES/TES systems for renewable energy plants. *Renew. Energy* **2008**, *33*, 1998–2006. [CrossRef]

10.   Arabkoohsar, A.; Machado, L.; Farzaneh-Gord, M.; Koury, R.N.N. Thermo-economic analysis and sizing of a PV plant equipped with acompressed air energy storage system. *Renew. Energy* **2015**, *83*, 491–509. [CrossRef]

11.   Arabkoohsar, A.; Machado, L.; Farzaneh-Gord, M.; Koury, R.N.N. The first and second law analysis of a grid connected photovoltaic plant equipped with a compressed air energy storage unit. *Energy* **2015**, *87*, 212–214. [CrossRef]

12.   Simpore, S.; Garde, F.; David, M.; Marc, O.; Castaing-Lasvignottes, J. Design and Dynamic Simulation of a Compressed Air Energy Storage System (CAES) Coupled with a Building, an Electric Grid and a Photovoltaic Power Plant. *Clima* **2016**, *4*, 1–11.

13.   Abbaspour, M.; Satkin, M.; Mohammadi-Ivatloo, B.; Hoseinzadeh Lotfi, F.; Noorollahi, Y. Optimal operation scheduling of wind power integrated with compressed air energy storage (CAES). *Renew. Energy* **2013**, *51*, 53–59. [CrossRef]

14.   Zhao, P.; Wang, M.; Wang, J.; Dai, Y. A preliminary dynamic behaviors analysis of a hybrid energy storage system based on adiabatic compressed air energy storage and flywheel energy storage system for wind power application. *Energy* **2015**, *84*, 825–839. [CrossRef]

15.   Fu, Z.; Lu, K.; Zhu, Y. Thermal System Analysis and Optimization of Large-Scale Compressed Air Energy Storage (CAES). *Energies* **2015**, *8*, 8873–8886. [CrossRef]

16.   Bouman, E.A.; Øberg, M.M.; Hertwich, E.G. Environmental impacts of balancing offshore wind power with compressed air energy storage (CAES). *Energy* **2016**, *95*, 91–98. [CrossRef]

17.   Gulagi, A.; Aghahosseini, A.; Bogdanov, D.; Breyer, C. Comparison of the Potential Role of Adiabatic Compressed Air Energy Storage (A-CAES) for a Fully Sustainable Energy System in a Region of Significant and Low Seasonal Variations. In Proceedings of the International Renewable Energy Storage Conference, Düsseldorf, Germany, 15–17 March 2016.

18.   Ibrahim, H.; Younès, R.; Ilinca, A.; Dimitrova, M.; Perron, J. Study and design of a hybrid wind–diesel-compressed air energy storage system for remote areas. *Appl. Energy* **2010**, *87*, 1749–1762. [CrossRef]

19.   Nielsen, R.F.; Haglind, F.; Larsen, U. Design and modeling of an advanced marine machinery system including waste heat recovery and removal of sulphur oxides. *Energy Convers. Manag.* **2014**, *85*, 687–693. [CrossRef]

20.   Basbous, T.; Younes, R.; Ilinca, A.; Perron, J. Optimal management of compressed air energy storage in a hybrid wind-pneumatic-diesel system for remote area's power generation. *Energy* **2015**, *84*, 267–278. [CrossRef]

21.   Bombarda, P.; Invernizzi, C.M.; Pietra, C. Heat recovery from Diesel engines: A thermodynamic comparison between Kalina and ORC cycles. *Appl. Therm. Eng.* **2010**, *30*, 212–219. [CrossRef]

22.   Lecompte, S.; Huisseune, H.; van den Broek, M.; Vanslambrouck, B.; De Paepe, M. Review of organic Rankine cycle (ORC) architectures for waste heat recovery. *Renew. Sustain. Energy Rev.* **2015**, *47*, 448–461. [CrossRef]

23.   Dai, Y.; Wang, J.; Gao, L. Parametric optimization and comparative study of Organic Rankine Cycle (ORC) for low grade waste heat recovery. *Energy Convers. Manag.* **2008**, *50*, 576–582. [CrossRef]

24.   Yang, K.; Zhang, H.; Wang, Z.; Zhang, J.; Yang, F.; Wang, E.; Yao, B. Study of zeotropic mixtures of ORC (organic Rankine cycle) under engine various operating conditions. *Energy* **2013**, *58*, 494–510. [CrossRef]

25.   Yari, M.; Mehr, A.S.; Zare, V.; Mahmoudi, S.M.S.; Rosen, M.A. Exergoeconomic comparison of TLC (trilateral Rankine cycle), ORC (organic Rankine cycle) and Kalina cycle using a low grade heat source. *Energy* **2015**, *83*, 712–722. [CrossRef]

26.   Kim, Y.M.; Shin, D.G.; Favrat, D. Operating characteristics of constant-pressure compressed air energy storage (CAES) system combined with pumped hydro storage based on energy and exergy analysis. *Energy* **2011**, *36*, 6220–6233. [CrossRef]

27.   Wang, M.; Wang, J.; Zhao, P.; Dai, Y. Multi-objective optimization of a combined cooling, heating and power system driven by solar energy. *Energy Convers. Manag.* **2015**, *89*, 289–297. [CrossRef]

28. Wang, J.; Yan, Z.; Wang, M.; Ma, S.; Dai, Y. Thermodynamic analysis and optimization of an (organic Rankine cycle) ORC using low grade heat source. *Energy* **2013**, *49*, 356–365. [CrossRef]

29. Wang, J.; Yan, Z.; Wang, M.; Li, M.; Dai, Y. Multi-objective optimization of an organic Rankine cycle (ORC) for low grade waste heat recovery using evolutionary algorithm. *Energy Convers. Manag.* **2013**, *71*, 146–158. [CrossRef]

30. Guo, Z.Y.; Liu, X.B.; Tao, W.Q.; Shah, R.K. Effectiveness–thermal resistance method for heat exchanger design and analysis. *Int. J. Heat Mass Transf.* **2010**, *53*, 2877–2884. [CrossRef]

31. Guo, J.; Xu, M.; Cheng, L. The application of field synergy number in shell-and-tube heat exchanger optimization design. *Appl. Energy* **2009**, *86*, 2079–2087. [CrossRef]

32. García-Cascales, J.R.; Vera-García, F.; Corberán-Salvador, J.M.; Gonzálvez-Maciá, J. Assessment of boiling and condensation heat transfer correlations in the modelling of plate heat exchangers. *Int. J. Refrig.* **2007**, *30*, 1029–1041. [CrossRef]

33. Coletti, F. *Heat Exchanger Design Handbook*; Begell House: Danbury, CT, USA, 2016.

34. Biematht, R.W.; Soane, D.S. Department of Chemical Engineering. *Am. Chem. Soc.* **2011**, *115*, 1394–1402.

35. Zhang, X.; He, M.; Zhang, Y. A review of research on the Kalina cycle. *Renew. Sustain. Energy Rev.* **2012**, *16*, 5309–5318. [CrossRef]

36. Yin, J.L.; Wang, D.Z.; Kim, Y.; Lee, Y. A hybrid energy storage system using pump compressed air and micro-hydro turbine. *Renew. Energy* **2014**, *65*, 117–122. [CrossRef]

37. Oldenburg, C.M.; Pan, L. Porous Media Compressed-Air Energy Storage (PM-CAES): Theory and Simulation of the Coupled Wellbore–Reservoir System. *Transp. Porous Med.* **2013**, *97*, 201–221. [CrossRef]

38. Pei, P.; Korom, S.F.; Ling, K.; He, J.; Gil, A. Thermodynamic impact of aquifer permeability on the performance of a compressed air energy storage plant. *Energy Convers. Manag.* **2015**, *97*, 340–350. [CrossRef]

39. Vongmanee, V. The renewable energy applications for uninterruptible power supply based on compressed air energy storage system. In Proceedings of the IEEE Symposium on Industrial Electronics and Applications, Kuala Lumpur, Malaysia, 4–6 October 2009; pp. 827–830.

40. Kemble, S.; Manfrida, G.; Milazzo, A.; Buffa, F.; Kemble, S.; Buffa, F. Thermoeconomics of a ground-based CAES plant for peak-load energy production system. *J. Nanosci. Nanotechnol.* **2016**, *32*, 1–16.

41. Zavattoni, S.A.; Barbato, M.C.; Pedretti, A.; Zanganeh, G.; Steinfeld, A. High Temperature Rock-bed TES System Suitable for Industrial-scale CSP Plant—CFD Analysis Under Charge/Discharge Cyclic Conditions. *Energy Proced.* **2014**, *46*, 124–133. [CrossRef]

42. Kirn, Y.M.; Favrat, D. Energy and exergy analysis of a micro-compressed air energy storage and air cycle heating and cooling system. *Energy* **2010**, *35*, 213–220.

43. Singh, O.K.; Kaushik, S.C. Energy and exergy analysis and optimization of Kalina cycle coupled with a coal fired steam power plant. *Appl. Therm. Eng.* **2013**, *51*, 787–800. [CrossRef]

44. Fu, W.; Zhu, J.; Zhang, W.; Lu, Z. Performance evaluation of Kalina cycle subsystem on geothermal power generation in the oilfield. *Appl. Therm. Eng.* **2013**, *54*, 497–506. [CrossRef]

45. Fwa, T.F.; Chan, W.T.; Tan, C.Y. Genetic-Algorithm Programming of Road Maintenance and Rehabilitation. *J. Transp. Eng.* **1996**, *122*, 246–253. [CrossRef]

46. Yang, M.D.; Chen, Y.P.; Lin, Y.H.; Ho, Y.F.; Lin, J.Y. Multiobjective optimization using nondominated sorting genetic algorithm-II for allocation of energy conservation and renewable energy facilities in a campus. *Energy Build.* **2016**, *122*, 120–130. [CrossRef]

47. Atrens, A.D.; Gurgenci, H.; Rudolph, V. Economic Optimization of a CO2—Based EGS Power Plant. *Energy Fuel* **2011**, *25*, 3765–3775. [CrossRef]

48. Li, H. Environomic Modeling and Multi-Objective Optimisation of Integrated Energy Systems for Power and Cogeneration. Ph.D Thesis, École Polytechnique Fédérale de Lausanne, Zurich, Switzerland, 2006.

49. Couper, J.R.; Penney, W.R.; Fair, J.R.; Walas, S.M. *Chemical Process Equipment*, 2nd ed.; Gulf Professional Publishing: Houston, TX, USA, 2009.

50. Ulrich, G.D. *A Guide to Chemical Engineering Process Design and Economics*; Wiley: New York, NY, USA, 1984.

9

Thermal Analysis of a Solar Powered Absorption Cooling System with Fully Mixed Thermal Storage at Startup

Camelia Stanciu, Dorin Stanciu * and Adina-Teodora Gheorghian

Department of Engineering Thermodynamics, University Politehnica of Bucharest, Splaiul Independentei 313, 060042 Bucharest, Romania; camelia.stanciu@upb.ro (C.S.); adina.gheorghian@upb.ro (A.-T.G.)
* Correspondence: dorin.stanciu@upb.ro

Academic Editor: Antonio Calvo Hernández

Abstract: A simple effect one stage ammonia-water absorption cooling system fueled by solar energy is analyzed. The considered system is composed by a parabolic trough collector concentrating solar energy into a tubular receiver for heating water. This is stored in a fully mixed thermal storage tank and used in the vapor generator of the absorption cooling system. Time dependent cooling load is considered for the air conditioning of a residential two-storey house. A parametric study is performed to analyze the operation stability of the cooling system with respect to solar collector and storage tank dimensions. The results emphasized that there is a specific storage tank dimension associated to a specific solar collector dimension that could ensure the longest continuous startup operation of the cooling system when constant mass flow rates inside the system are assumed.

Keywords: solar energy; ammonia-water absorption cooling; parabolic trough collector; fully mixed thermal storage

1. Introduction

Absorption systems are widely studied as they are an eco-friendly alternative to conventional compression chillers. The energy input is waste heat or a renewable heat source, such as non-conventional solar or geothermal heat. Another benefit is that absorption units operate with environmental friendly working fluids. By combining the two mentioned advantages over mechanical compression cooling systems, one can achieve a reduction of the negative impact on the environment.

A detailed state of the art review of solar absorption refrigeration systems was published by Kalogirou [1]. Different analyses and numerical simulations have been performed by researchers in the field, leading to increased interest. Koroneos et al. [2] emphasized in their study that among all installed worldwide solar thermal assisted cooling systems, 69% are absorption cycle-based. Most of the published works on solar cooling systems are concentrated on absorption cycle systems operating with LiBr-H_2O solution and flat plate solar collectors. As Duffie and Beckman [3] emphasized, the temperature limitations of flat plate collectors imposed the use of LiBr-H_2O based systems. Ammonia-water based systems require higher temperature heat sources and thus are less used with flat plate collectors.

The potential of the ammonia–water absorption refrigeration system in Dhahran, Saudi Arabia, was evaluated by Khan et al. [4] for a cooling capacity of 10 kW driven by a 116 m^2 of evacuated tube solar collector. The system was coupled with dual storages of ice and chilled water used alternatively function on solar energy availability and in accordance with the cooling demands of a 132 m^3 room.

A case study about converting an existing conventional ice-cream factory located in Isparta, Turkey to a solar energy based one is presented by Kizilkan et al. [5]. The authors proposed a system which involves the use of a parabolic trough solar collector instead of an existing electrically heated boiler. Also, instead of actual vapor compression refrigeration system, a H_2O-LiBr absorption refrigeration system is proposed for cooling the ice-cream mixtures. The authors found that the daily energy savings which can be achieved using the parabolic trough solar collector system are about 98.56%.

Lu and Wang [6] presented an experimental performance investigation and economic analysis of three solar cooling systems. The first system consisted of an evacuated tube U pipe solar collector driving a silica gel–water adsorption chiller; the second one was composed of a high efficiency compound parabolic concentrating solar collector connected to single-effect H_2O-LiBr absorption chillers; and the last one was made up of a medium temperature parabolic trough collector coupled to a double effect H_2O-LiBr absorption chiller. The results showed that the highest solar coefficient of performance (COP) was attained by the third system. The parabolic trough collector can drive the cooling chiller from 14:30 to 17:00 in an environment where the temperature is about 35 °C.

Ghaddar et al. [7] presented modelling and simulation of a H_2O-LiBr solar absorption system for Beirut. The results showed that the minimum collector area should be 23.3 m^2 per ton of refrigeration and the optimum water storage capacity should be 1000 to 1500 L in order to operate seven hours daily only on solar energy.

A comparison of three novel single-stage combined absorption cycles (NH_3/H_2O, $NH_3/NaSCN$ and $NH_3/LiNO_3$) to the Goswami cycle was performed by Lopez-Villada et al. [8]. The studied cycles were driven by an evacuated tube collector, a linear Fresnel collector and a parabolic trough collector. The authors simulated the systems for a whole year in Sevilla, Spain, using TRNSYS software 2004 (Transient System Simulation Tool, developed by Solar Energy Laboratory, University of Wisconsin, Madison, WI, USA). They concluded that an evacuated tube collector is a more suitable solar technology for such systems.

Li et al. [9] investigated the experimental performance of a single-effect H_2O-LiBr absorption refrigeration system (of 23 kW refrigeration capacity) driven by a parabolic trough collector of aperture area 56 m^2 for air conditioning of a 102 m^2 meeting room located in Kunming, China, and analyzed appropriate methods for improving the cooling performance.

Another H_2O-LiBr solar driven system was presented by Mazloumi et al. [10] for a 120 m^2 room located in Iran, whose peak cooling load is 17.5 kW. The authors proposed a minimum parabolic trough collector area of 57.6 m^2 and associated hot water storage tank volume of 1.26 m^3. The system operates between 6.49 h and 18.82 h (about 6 a.m. to 7 p.m.).

As one may notice, the study of H_2O-LiBr absorption refrigeration systems is widespread in the technical literature. Nevertheless, there are papers presenting comparisons between the operation of absorption systems using different working fluids. Among them, a comparison between NH_3-H_2O, H_2O-LiBr and other four mixtures is presented by Flores et al. [11]. When computing system performances, the authors found that H_2O-LiBr system has a small range of vapor temperature operation due to crystallization problems. Their study reports an operation generator temperature of about 75 °C–95 °C for H_2O-LiBr system, while for NH_3-H_2O system working under the same conditions, the range is higher, namely 78 °C–120 °C. The chosen working conditions were set to 40 °C condensation temperature, 10 °C vaporization temperature and 35 °C absorber one, while the cooling load was 1 kW. Key highlights of the above literature review are presented in Table 1.

Concluding the above literature study, better performances are reported for H_2O-LiBr systems in comparison to NH_3-H_2O ones in air conditioning applications, but they operate in a lower and narrower range of vapor generator temperatures, due to the fluid's risk of crystallization, thus NH_3-H_2O mixtures might still be good candidates for solar absorption cooling.

Complementary to the above published results, the present paper presents a thermodynamic analysis of a system composed of a parabolic trough collector, a solar tubular receiver, a fully mixed storage tank and a simple effect one stage NH_3-H_2O absorption cooling system. Time dependent cooling load is considered for a residential building occupied by four persons, two adults and two students [12]. The minimum necessary dimensions of the parabolic trough collector (PTC) and storage tank are determined in order to cover the whole day cooling load. Under these conditions, the objective of the work is to simulate daily operation for the whole system in July, in Bucharest (44.25° N latitude) and to emphasize the sensitivity of the operation stability to storage tank and PTC dimensions. Variation of the storage tank water temperature is represented along the entire day, under the considered variable cooling load, putting into evidence the turn on and turn off timings and thus the possible operation interval of the system.

Table 1. Literature review data.

ACS Type	Cooling Load	Solar Collector	Storage Tank	Reference
NH_3-H_2O	10 kW (for a 132 m^3 room), between 8:45 and 15:00	42 m^2 ETC at 25° tilt, Saudi Arabia, clear sky in March, peak solar radiation 1000 W/m^2, water, 900 L/h, 3 bar (80 °C)–6 bar(>100 °C), peak generator temperature 120 °C 32% collector efficiency	No hot thermal storage between collector and ACS, ice cold storage	[4]
H_2O-LiBr	40 kW, ice-cream factory, between 12:00 and 15:00	143.63 m^2 PTC (42.12 m × 3.5 m), 55.19 kW, Turkey, 500 W/m^2 constant solar radiation, therminol-VP1 oil (25–212.7 °C at 3.9 bar), 87 °C constant generator temperature	No hot thermal storage between collector and ACS	[5]
H_2O-LiBr simple effect	16 °C chilled water, Between 10:00 and 16:00 (6 h in sunny days)	CPC, 4 m^2/kW cooling, China, Sunny day water, 0.02 kg/(sm^2) 80–125 °C outlet water temperature 50% collector efficiency(125 °C)	Hot water tank (no data)	[6]
H_2O-LiBr double effect	16 °C chilled water, Between 14:30 and 17:00 (2.5 h in sunny days)	80 m^2 PTC (2 m^2/kW cooling), 40 concentrating ratio, China, sunny day, tracking, 125–150 °C outlet water or oil temperature 44% collector efficiency (150 °C)	Hot water tank (no data)	[6]
H_2O-LiBr	10.5 kW (for a 150 m^2 house), 7 h/day	23.3 m^2/TR [1] FPC at 15° tilt (80.6 m^2 optimum), Beirut, water, 7 m^3/h optimum, 50–85 °C generator temperature	Hot water tank 1300 L optimum (13–19 L/m^2 collector)	[7]
NH_3-H_2O (compared to others)	175 kW peak, chilled water from 12 to 7 °C	600 m^2 ETC (3.05 m^2 aperture) and Fresnel (3.39 m^2 aperture) at 20° tilt, PTC (5.39 m^2 aperture) at 0° tilt, Spain, optimum temperature 90–120 °C (ETC is the best choice for NH_3-H_2O cycle)	Hot water tank 36 m^3	[8]
H_2O-LiBr	23 kW (for a 102 m^2 room)	56 m^2 PTC (26 m × 2.5 m), China Water, 0.602 kg/s 35%–45% collector efficiency (clear sky)	Hot water tank 1 m^3	[9]
H_2O-LiBr	17.5 kW peak (120 m^2 room), between 9 and 19	57.6 m^2 PTC (2 m wide), Iran Water preheated at 71 °C, 1800 kg/h max 92 °C storage tank water temperature	Hot water tank 1.26 m^3	[10]
H_2O-LiBr	1 kW	75–95 °C desorber temperature	-	[11]
NH_3-H_2O		78–120 °C desorber temperature	-	

[1] TR = tone of refrigeration, 3.51 kW.

2. Considered Cooling Load

The proposed system is designed to cover the cooling load of a residential building located in Bucharest (Romania). The house is composed by two storeys, having a living surface of 73.65 m^2 on the ground-floor and 59.05 m^2 on the first-floor. The walls are made of autoclaved aerated concrete brickwork, insulated with 10 cm polystyrene at the exterior side. Thermo-insulated and double glazed windows are considered. The global heat transfer coefficient U was computed for each building element (wall, door, floor, ceiling, window, etc.) considering conduction through the element structure and either interior and exterior convection for exterior elements, or twice interior convection for interior elements. Thickness and thermal conductivity for the layers of the wall structure are detailed in paper [12], interior convection heat transfer coefficient was considered 8 W·m^{-2}·K^{-1} for walls and 5.8 W·m^{-2}·K^{-1} for ceilings, while the exterior one was 17.5 W·m^{-2}·K^{-1}. These values were chosen in accordance with Romanian norms [13]. The corresponding computed values for the global heat transfer coefficient are presented in Table 2.

Table 2. Overall heat transfer coefficient values for building elements.

Building Element	U (W/(m²·K))
Exterior wall	0.298
Interior wall	0.887
Exterior and interior doors	2.32
Exterior windows	2.564
Ground floor	3.000
Ceiling above the ground floor	2.182
Mansard ceiling	0.371

The occupants are a family of four, two adults and two students, performing ordinary daily activities. The cooling load was computed summing up all external and internal loads, namely all heat rates exchanged between the building and ambient, all sensible and latent heat rates corresponding to perspiration and exhalation of occupants, humidity sources, electronic equipment, appliances and artificial lightening. Thermal inertia of the walls was considered. Also, the external heat rate was computed taking into account each wall orientation with respect to the Sun, and accordingly the time-dependent solar radiation reaching each vertical wall under clear sky conditions. More details about these calculations are given in [12].

The time variation of this load on 15 July for an interior desired temperature of 22 °C, is presented in Figure 1. Note that this temperature value belongs to the lower part of the acceptable range of operative temperatures recommended by ASHRAE Standard [14] for residential buildings.

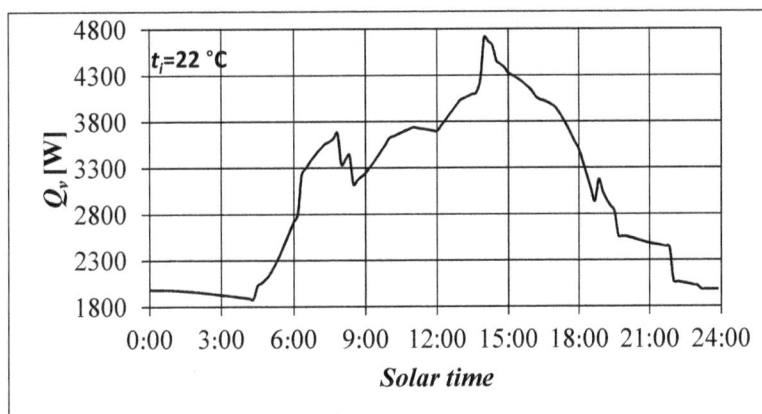

Figure 1. Required cooling load, Qv.

As one can notice, three peaks are apparent for the cooling load, around 8:00, 14:00 and 19:00, respectively. They are caused by the internal load contribution corresponding to occupants' activity hours inside the building. The step downwards in the evening, around 22:00, corresponds to the moment when occupants are suspending their activity. The cooling load records a daily minimum around 5:00 and a daily maximum of 4709 W around 14:00.

In the present work, a cooling period between 9:00 and 18:00 will be considered. When the cooling system is considered off, the temperature inside the room is computed from the energy balance equations considering incoming solar energy when available, occupants' activity and losses to or gains from the ambient. Obviously it differs from the set temperature of 22 °C. Thus, when the cooling system is turned on, the initial temperatures for building elements are those computed at that time. Consequently, in an indirect manner, the extra cooling loads not covered by the cooling system are considered by means of increased initial temperatures and building thermal inertia.

3. Description of the System

The considered solar driven system is shown in Figure 2. A parabolic trough collector (PTC) with a tubular receiver is used to catch solar radiation for heating the desorber of an absorption cooling system. Typical dimensions are between 1 to 3 m for aperture width and 2 to 10 m for its length, as reported by Fernandez-Garcia et al. [15] for solar-driven cooling applications. A single trough of 2.9 m by 10 m dimensions is expected to be used in the studied case. The collector is considered oriented fixed on an East-West direction, facing South and tilted at a fixed angle of 30° all day. These fixed collector constraints have the advantage of avoiding moving parts inside the system allowing lower acquisition and operating costs for this small-scale application. Also, for the considered latitude, the East-West orientation of the PTC provides an energy availability 6% lower with respect to a North-South orientation, as reported by Sharma et al. [16]. Nevertheless, it is here preferred since it offers a higher mechanical stability of the structure during windy days. Regarding the fixed tilt, its value was chosen so that maximum beam radiation is intercepted by the aperture plane around noon [17].

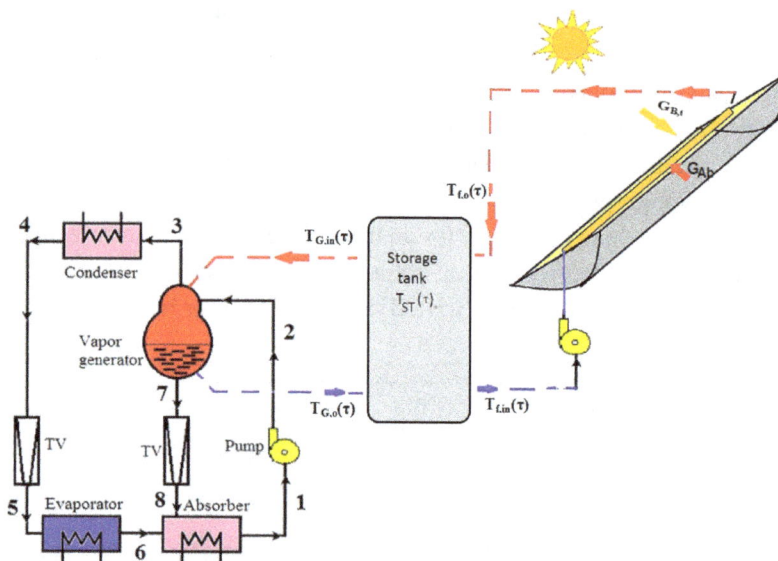

Figure 2. Simple effect one stage solar powered absorption cooling system.

Inside the tubular receiver, liquid water is used as a heat career. The trend is to reduce operating costs. In this regard, one of the options is to employ a fluid that can be used for both heat collection and as a thermal storage medium [18]. According to technical literature regarding the operation of PTC, water can be used as working fluid for temperatures up to 220 °C; in this case the reported operating pressure is 10 kgf/cm^2 (9.8 bar); a pressurized expansion tank is used to maintain the pressure of circulating water in the closed system, allowing water to expand with rising temperature [19]. Nitrogen is used to regulate pressure variations. The disadvantage of using water as a working fluid is that a high pressure hot water storage tank must be employed and special safety precautions should be considered. A regularly monitor pressure should be mounted and safety relief valves should be set only by trained personal. Also the system should be enhanced by sensors to automatically defocus the trough from its position in case the water temperature exceeds the maximum allowable limit of 220 °C [19]. A heavy-duty quality steel tank should be used. Commercial ones are made of austenitic stainless steel 304, 316, 316 L or 316 Ti, as reported by manufacturers [20].

The water enters the receiver tube at temperature T_{fi} and exits at a superior temperature T_{fo}, as effect of the absorbed solar radiation. When exiting the receiver, the hot water enters a storage tank (ST).

After fully mixing with the existing water, a certain mass flow rate of ST water leaves the tank and heats the ammonia-water solution inside the vapor generator (at state 2 of temperature $T_{G,in}$, in Figure 2) of a classical one stage absorption cooling system (ACS). By "classical" one means a basic configuration for the system to operate: absorber, desorber (vapor generator), condenser, evaporator, throttling valves, liquid pump and all necessary connection devices. As a result of the heat exchange process in the desorber, first ammonia vapors leave the vapor generator at state 3 and fuel the refrigerating part of the absorption system, creating the cooling effect in the evaporator. The remaining solution, lean in ammonia, leaves the vapor generator at state 7, passes through the throttling valve and enters the absorber where it mixes again with the ammonia vapors leaving the evaporator (at state 6). The mass flow rate of water leaving the vapor generator and returning to the storage tank has now a lower temperature $T_{G,o}$.

4. Thermodynamic Modeling of the System

The thermodynamic model consists in applying the First Law of Thermodynamics for the whole system and system components. Conduction, convection and radiation heat exchange laws complete the system of equations. The mathematical model is presented for each computing stage, namely:

1.   the parabolic trough collector (PTC);
2.   the fully mixed storage tank (ST);
3.   the absorption cooling system (ACS).

The following general assumptions are made for the present study:

i.   clear sky conditions are assumed for the ambient and solar radiation data;
ii.   time dependent cooling load (see Figure 1) is applied;
iii.   thermal inertia of ACS and PTC is negligible with respect to that of the storage tank. As a result, unsteady model is considered only for the storage system. All other components of the system are modeled in steady state conditions;
iv.   a fully mixed storage tank is considered. As a result, at each time $T_{G,in} = T_{f,in} = T_{ST}$.

4.1. Solar Radiation and Ambient Data

Time averaged measured data between 1991 and 2010 for solar radiation and between 2000 and 2009 for ambient temperature are generated with Meteonorm V7.1.8.29631 software [21] for the Otopeni meteorological station (close to Bucharest). The data were extracted for 15 July with a time step of 10 min for clear sky conditions. The time variations of beam radiation on the PTC tilt surface, G_{Bt}, and ambient temperature, T_a, are shown in Figure 3. In the model, a wind speed w of 0.2 m/s was considered constant during the entire operating day.

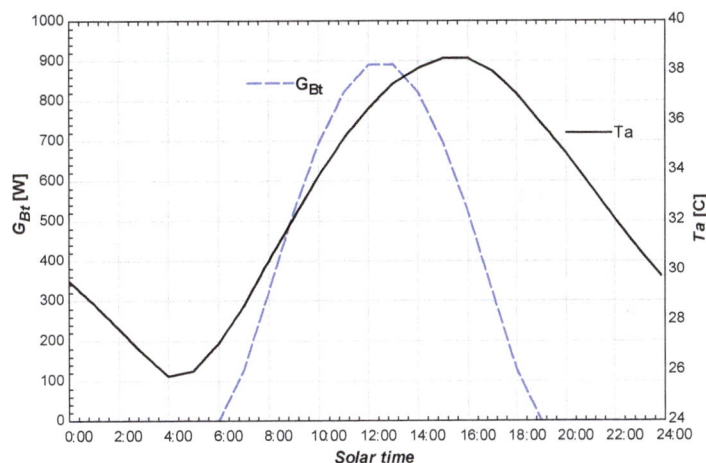

Figure 3. Beam (direct) solar radiation G_{Bt} received by the PTC facing South and tilted at $\beta = 30°$ and ambient temperature T_a in Bucharest, on 15 July; Data generated with Meteonorm V7.1.8.29631 [21].

4.2. Parabolic Trough Collector and Tubular Receiver Model

The parabolic trough collector is characterized by an opening of width H_{PTC} and a length L_{PTC}.

Its effective optical efficiency is computed as $\eta_{env} = \eta_{opt}K_\theta$. The term η_{opt} takes into account the optical losses for normal solar incident irradiance. These losses are due to receiver shadowing, tracking and geometry errors, dirt on the collector mirror and receiver, mirror reflectance, etc. In this paper a common value of 0.80 was adopted [22]. The incident angle modifier K_θ counts for the losses when solar irradiance is not normal to collector aperture. It takes into account the incidence losses, end shadowing of the through, reflection and refraction losses, etc. The incidence angle modifier was computed according to [23]:

$$\dot{Q}_{in,R} = \eta_{env}H_{PTC}L_{PTC}G_B = \eta_{opt}H_{PTC}L_{PTC}G_{Bt} \tag{1}$$

This heat rate is partially transmitted through the glass, absorbed by the pipe and so used to heat the fluid inside the receiver tube; the rest is lost by glass absorption, convection, conduction and radiation heat rates to ambient, as shown in Figure 4. In order to compute all heat rates, the geometry and characteristics of the tubular receiver should be firstly defined.

The tubular receiver is composed of a stainless steel pipe covered by a tubular Pyrex glass cover and has the following dimensions: pipe interior diameter $D_{pi} = 0.051$ m, pipe wall thickness of 0.001 m, tubular glass cover interior diameter $D_{gls,i} = 0.075$ m with a thickness of 0.001 m. The mass flow rate of fluid inside the tube, $\dot{m}_R$, is fixed to a value of 0.1 kg/s in order to maintain a laminar flow. The following glass properties are used: absorptivity $\alpha_{gls} = 0.02$, emissivity $\varepsilon_{gls} = 0.86$, transmittance $\tau_{gls} = 0.935$. For the stainless steel, the absorptivity is $\alpha_p = 0.92$ and the emissivity is dependent on mean inside wall pipe temperature T_{pi} as: $\varepsilon_p = 0.000327 \cdot T_{pi} - 0.065971$ [22].

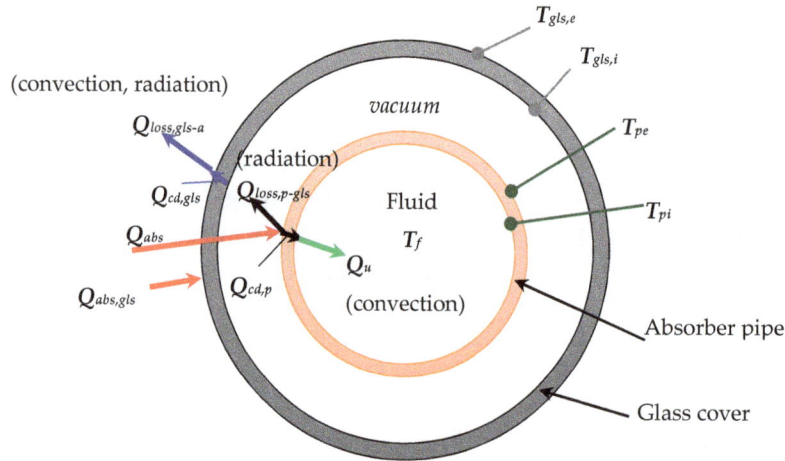

Figure 4. Heat rates associated to the tubular receiver.

The mathematical model of PTC relies on the energy balance equations written on the inner and outer surfaces of absorber pipe and glass cover respectively. According to assumption iii these equations are written in steady state.

The absorbed heat rate by the absorber pipe is computed as:

$$\dot{Q}_{abs,p} = \tau_{gls}\alpha_p\dot{Q}_{in,R} \tag{2}$$

The useful part of $\dot{Q}_{abs,p}$ is transmitted through the pipe by conduction and then by convection to the fluid. Thus, the energy balance equation written on the inner pipe surface looks like:

$$\dot{Q}_u = \dot{Q}_{cd,p} \tag{3}$$

where the conduction heat rate through the tubular pipe is:

$$\dot{Q}_{cd,p} = 2\pi L_{PTC}\lambda_p \frac{T_{pe} - T_{pi}}{\ln \frac{D_{pe}}{D_{pi}}} \tag{4}$$

The fluid receives this heat rate by thermal convection:

$$\dot{Q}_u = \pi D_{pi}L_{PTC}h_f\left(T_{pi} - T_f\right) \tag{5}$$

where T_f is the mean temperature of the water flowing inside the absorber pipe. Since the heat transfer is developing under constant heat flux boundary condition, one may consider that T_f is the arithmetic mean temperature between inlet and outlet temperatures T_{fi} and T_{fo}.

The convection heat transfer coefficient h_f is computed from Nusselt number considered 4.36 as the flow inside the tube is kept laminar:

$$h_f = \frac{Nu_f\lambda_f}{D_{pi}} \tag{6}$$

On the other hand, the useful heat rate received by the fluid can also be written as:

$$\dot{Q}_u = \dot{m}_f c_{p,f}\left(T_{fo} - T_{fi}\right) \tag{7}$$

At each time step, the inlet fluid temperature T_{fi} is equal to the storage tank fluid temperature, T_{ST}, as the fluid is recirculated from the storage tank through the PTC.

The energy balance equation on the pipe exterior surface is:

$$\dot{Q}_{abs,p} = \dot{Q}_u + \dot{Q}_{loss,p-gls} \tag{8}$$

where $\dot{Q}_{abs,p}$ is computed by Equation (2) and $\dot{Q}_{loss,p-gls}$ represents lost heat rate between the pipe and glass cover. A common assumption is to consider vacuum inside this enclosure so that convection is neglected. Thus, $\dot{Q}_{loss,p-gls}$ is entirely due to radiation losses between pipe and glass cover:

$$\dot{Q}_{loss,p-gls} = \pi D_{pe}L_{PTC}\sigma \frac{T_{pe}^4 - T_{gls,i}^4}{\frac{1}{\varepsilon_p} + \frac{1-\varepsilon_{gls}}{\varepsilon_{gls}}\frac{D_{pe}}{D_{gls,i}}} \tag{9}$$

where σ is the Stefann-Boltzman constant.

Further, the energy balance equation on the inner surface of the glass cover is:

$$\dot{Q}_{loss,p-gls} = \dot{Q}_{cd,gls} \tag{10}$$

where:

$$\dot{Q}_{cd,gls} = 2\pi L_{PTC}\lambda_{gls}\frac{T_{gls,i} - T_{gls,e}}{\ln \frac{D_{gls,e}}{D_{gls,i}}} \tag{11}$$

represents the heat rate passing through the glass cover by conduction. The glass conductivity λ_{gls} corresponds to the mean of the inner and outer surface temperatures of glass cover.

Finally, the energy balance equation on the outer surface of glass cover is:

$$\dot{Q}_{cd,gls} + \dot{Q}_{abs,gls} = \dot{Q}_{loss,gls-a} \tag{12}$$

The heat flux absorbed by the glass cover is expressed by:

$$\dot{Q}_{abs,gls} = \alpha_{gls}\dot{Q}_{in,R} \tag{13}$$

The loss heat rate, $\dot{Q}_{loss,gls-a}$ is transmitted from the outer glass surface to ambient through wind convection and radiation so that:

$$\dot{Q}_{loss,gls-a} = \pi D_{gls,e}L_{PTC}h_w\left(T_{gls,e} - T_a\right) + \pi D_{gls,e}L_{PTC}\varepsilon_{gls}\sigma\left(T_{gls,e}^4 - T_{sky}^4\right) \tag{14}$$

The wind convection heat transfer coefficient is computed according to [24]:

$$h_w = 5.7 + 3.8w \tag{15}$$

The set of equations is now completely defined so that solving of unknown temperatures T_{pi}, T_{pe}, $T_{gls,i}$, $T_{gls,e}$, T_{fo} may proceed. The outlet fluid temperature determined from Equation (7) is now a data input to the storage tank module.

4.3. Fully Mixed Storage Tank Model

A 0.16 m³ storage tank is considered for which a constant heat loss coefficient is assumed, $(UA)_{ST} = 11$ W/K [3]. As pointed out above, the temperature inside the storage tank, T_{ST}, is assumed to be uniformly distributed. By using the mathematical expression of the First Law, one obtains the following ordinary differential equation:

$$(mc_p)_{ST}\frac{dT_{ST}}{d\tau} = \dot{Q}_u - \dot{Q}_G - (UA)_{ST}(T_{ST} - T_a) \tag{16}$$

The right hand side in Equation (16) counts for all heat rates exchanged by the storage tank with the exterior: the useful heat flux, $\dot{Q}_u$, received by the water inside the absorber pipe of PTC is computed by Equation (7), $\dot{Q}_G$ is the heat rate transferred from the storage tank to the vapor generator of the ACS, while the last term counts for heat flux losses to the ambient. Since all the above heat fluxes depend on time, one cannot develop an analytical solution for this equation. Thus, a first order explicit discretization with respect to time is employed, which leads to the discrete relation:

$$T_{ST}^{(n+1)} = T_{ST}^{(n)} + \frac{\dot{Q}_u^{(n)} - \dot{Q}_G^{(n)} - (UA)_{ST}\left(T_{ST}^{(n)} - T_a^{(n)}\right)}{(mc_p)_{ST}}\Delta\tau \tag{17}$$

where superscript $(n + 1)$ denotes the properties values at time $\tau + \Delta\tau$ and subscript (n) identifies the values of properties at current time τ.

4.4. Absorption Cooling System Model

For the operation of the considered NH_3-H_2O absorption cooling system, the vaporization temperature is imposed at $t_v = 10$ °C and the above described cooling load $\dot{Q}_v$ is applied.

The condenser and absorber are cooled with ambient air, so that condensation temperature T_c as well as the absorber temperature T_{Ab} are imposed by T_a, time dependent. The heat source required to feed the desorber is the hot water from the storage tank at T_{ST}. Thus the solution temperature at the end of desorbing process is constrained by this value.

According to assumption iii, steady state operation is assumed. The transient response of the ACS module is negligible in comparison to that of the storage tank. Due to this hypothesis, one expects that the obtained results would be overestimated before the storage tank water temperature reaches its maximum value and underestimated afterwards. Further, neglecting the variation of kinetic and potential energies (which is an appropriate assumption for the studied system), the First Law of Thermodynamics becomes:

$$\dot{Q} - \dot{W} = \sum_{o} (\dot{m}h)_o - \sum_{i} (\dot{m}h)_i \tag{18}$$

which is applied to each component of the system.

The energetic and exergetic analyses are detailed in previous works [25]. The set of equations is summarized in Table 3.

Table 3. Heat rates in the ACS.

Component	Thermal Load	Component	Thermal Load
Condenser	$\dot{Q}_C = \dot{m}_0(h_4 - h_3)$	Vapor Generator	$\dot{Q}_G = \dot{m}(h_7 - h_2) + \dot{m}_0(h_3 - h_7)$
Evaporator	$\dot{Q}_V = \dot{m}_0(h_6 - h_5)$	Absorber	$\dot{Q}_{Ab} = \dot{m}(h_8 - h_1) + \dot{m}_0(h_6 - h_8)$

The overall coefficient of performance and exergetic efficiency are expressed by:

$$COP = \frac{\dot{Q}_V}{\dot{Q}_G + |\dot{W}_P|} \tag{19}$$

$$\eta_{Ex} = \frac{\dot{Q}_V \left(\frac{T_a}{T_V} - 1 \right)}{\dot{Q}_G \left(1 - \frac{T_a}{T_{Gm}} \right) + |\dot{W}_P|} \tag{20}$$

In the above relations, $|\dot{W}_P|$ represents the pump consumed power and T_{Gm} is a mean value for the generator temperature, defined as arithmetic mean between temperature values of the states corresponding to the beginning and ending of the desorbing process.

5. Operating Regimes of the System

Depending on solar radiation availability, the cooling time interval and the correct operation conditions of ACS, the following possible operating regimes of the system can occur:

(i)   During night and early morning, the solar radiation is not available, so that the PTC module is inactive and the water circulation through its pipes is stopped. In this case, only the storage tank module is computed and, due to losses to ambient, the fluid temperature T_{ST} decreases.

(ii)   As far as solar radiation is available, both PTC and ST modules are operating. The storage tank water temperature is increasing as a certain mass flow rate is circulated through the PTC pipes.

(iii) At the targeted ACS starting time (9:00 in the studied case), the ACS module can start only if the storage tank water temperature is sufficiently high to ensure appropriate operating conditions in terms of concentration difference between strong and weak solutions (>0.06). Obviously this difference mainly depends on the ambient temperature too, as well as on the cooling load. If the condition is not satisfied, the ACS module is stopped. In this case, only the PTC and ST modules are working and the storage tank water temperature increases. When it reaches the necessary value to fulfill the above condition, the ACS module is started again.

As conclusion, one may find that the ST module is 24 h operating, the PTC module works as long as the solar beam radiation is available, while the ACS module may function either on its entire targeted period (i.e., 9:00–18:00), or on shorter inside intervals of time, as the appropriate operating conditions are fulfilled or not.

6. Numerical Procedure

The simulation of the entire system operation is worked out in EES programming environment [26], by following the algorithm presented in Figure 5. Firstly, the input data of the system are set. The cooling load demand data (see Figure 1) as well as the direct solar radiation and ambient temperature data (see Figure 3) were stored in lookup tables. All other geometrical parameters of the system were specified at the top of main code.

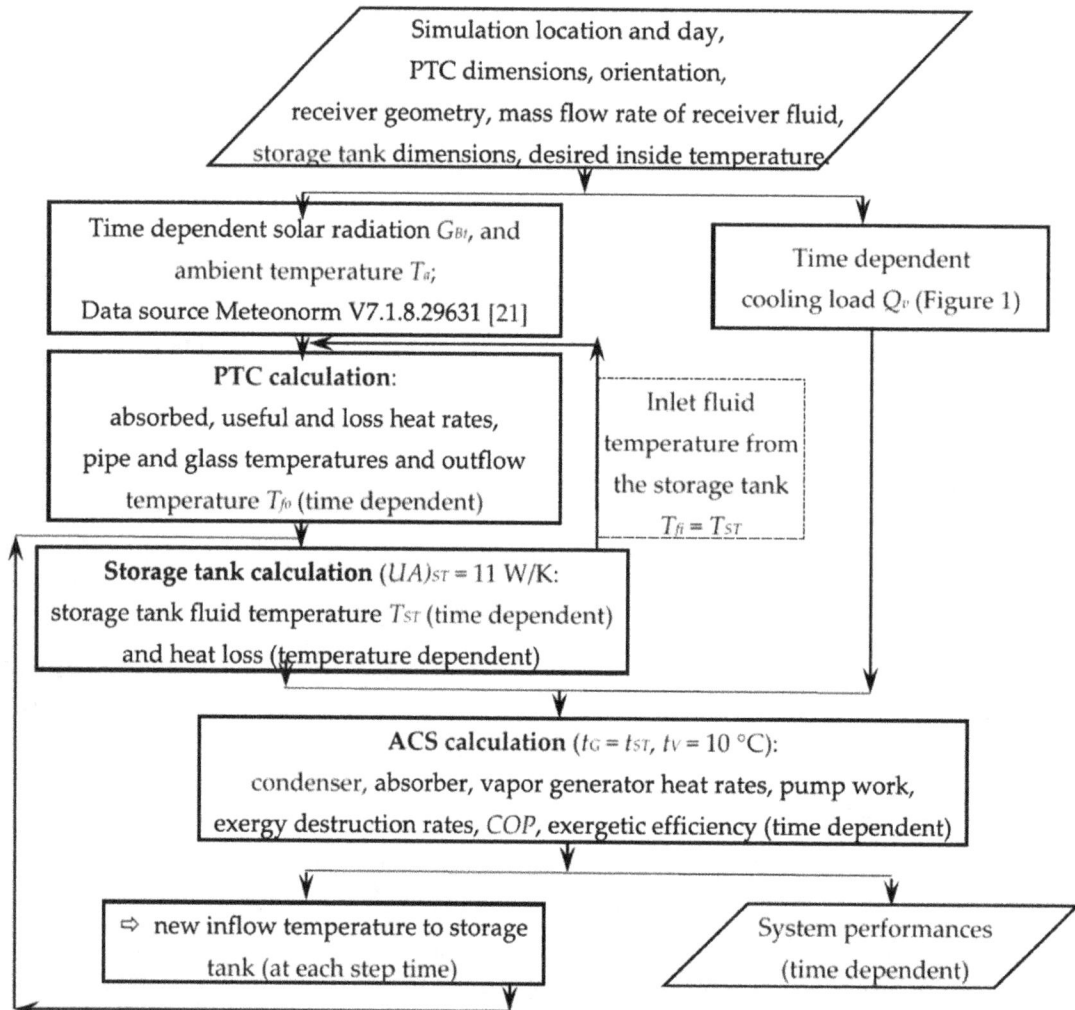

Figure 5. Solving algorithm.

Three modules have been coded for modeling the working regimes of PTC, ACS and ST, respectively. At the current time-step calculation, the PTC module reads G_{Bt} and T_a data from the corresponding lookup tables and receives as input the fluid temperature T_{fi} from the storage tank. Then it solves the Equations (3), (8), (10) and (12). The output of this module is the fluid output temperature T_{fo}, further used as input in the storage tank module.

The ACS module checks the necessary operating conditions and if fulfilled, computes the performances of the cooling system according to Equations (19) and (20). The input data are the temperatures T_a and T_{ST}, as well as the cooling load $\dot{Q}_V$ computed at the current time. The specific properties of the working fluid (ammonia-water pair) are determined by means of a dynamically linked procedure, in which the correlations proposed by Ibrahim and Klein [27] are used. The output in terms of outlet generator fluid temperature T_{Go} is sent to the storage tank module.

The ST module has as inputs the temperatures T_{fo}, T_{Go} and T_{ST} computed at the current time. Losses to the ambient are calculated and Equations (17) is employed to find the storage tank water temperature T_{ST} at the next time, $\tau + \Delta\tau$.

The main program calls these modules according to the operating regimes of the system presented in the previous section.

The computations are started at 00 a.m., 15 July. At this time, the water temperature in the storage tank was considered 10 °C higher than T_a. The time step was set to $\Delta\tau = 10$ min, and was kept constant during the entire day.

7. Results and Discussions

As pointed out before, a PTC of 2.9 m wide and 10 m long is considered. Available solar heat rate $\dot{Q}_S$, input heat rate to the tubular receiver $\dot{Q}_{in,R}$ and useful heat rate transmitted to the fluid $\dot{Q}_u$ are sketched in Figure 6. An important aspect to notice is that solar radiation is available between 6:00 and 19:00.

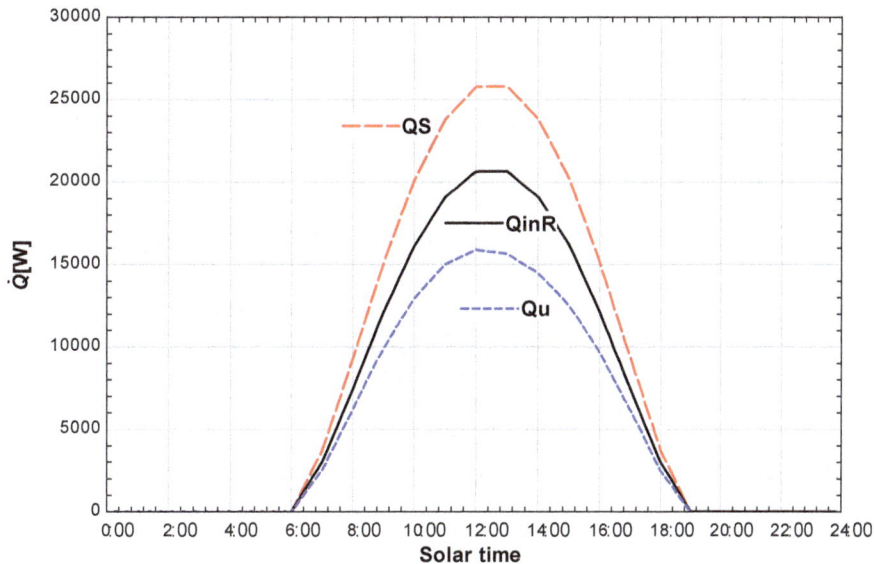

Figure 6. Available solar heat rate $\dot{Q}_S$, input heat rate to the tubular receiver $\dot{Q}_{in,R}$ and useful heat rate transmitted to the fluid $\dot{Q}_u$ for the studied case.

After some initial trials, the storage tank volume was set to 0.16 m³ in order to maintain the water temperature in the operation range of the cooling module as long as possible. For lower tank volumes, the water temperature increased too much in the morning while for higher values, the temperature was too low to cover the considered cooling load.

The starting time of the simulation is 00 a.m. The (i) operating regime described in Section 5 is applied. Till 6:00, the temperature of the fluid in the storage tank is slowly decreasing as heat is lost to the ambient. At 6:00 solar radiation starts reaching the PTC surface, so the water circulation through the PTC is started, corresponding to operation regime (ii). As a consequence, water temperature starts increasing as presented in Figure 7 by the dotted blue line T_{ST}. It raises from 38 °C at 6:00 to 100 °C at 9:00. At 9:00 the absorption cooling system (ACS) is started, so that water is now recirculated also through the desorber of the ACS, heating the NH_3-H_2O solution and assuring its operation for covering the cooling demand. Operating regime (iii) is applied. The storage tank water temperature is still increasing, but with a lower slope. This behavior is due to the following two effects: on one side the solar radiation reaching the PTC surface is increasing in intensity, on the other hand thermal energy of stored water is used in the ACS desorber.

As one may see in Figure 8, the coefficient of performance of the ACS is slowly decreasing and also its exergetic efficiency. A minimum for the exergetic efficiency is met around 14:00. In fact, this value corresponds to the maximum value of the storage tank water temperature, encountered at 13:40. The operation of the ACS is very sensitive to the heat source temperature value, among other parameters. It was previously proved [27] that the exergetic efficiency of such a system has a maximum around a relatively low desorber temperature of about 80–90 °C and then it starts decreasing when increasing the desorber temperature. The same behavior is met here. As the desorber temperature increases, the exergetic efficiency of the ACS decreases. After this peak, the storage tank temperature drastically starts to decrease and when its value arrives to 75 °C the ACS operation is no more possible. This happens at 17:10. As a consequence of temperature decrease, the exergetic efficiency increases on this second working period (Figure 8).

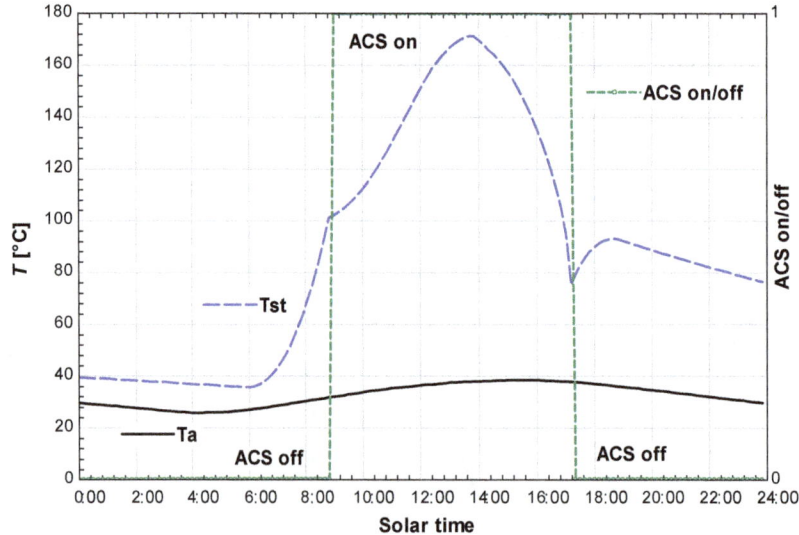

Figure 7. Ambient and storage tank fluid temperatures on working and non-working periods of the ACS; case study 10 m PTC, 0.16 m^3 water storage tank.

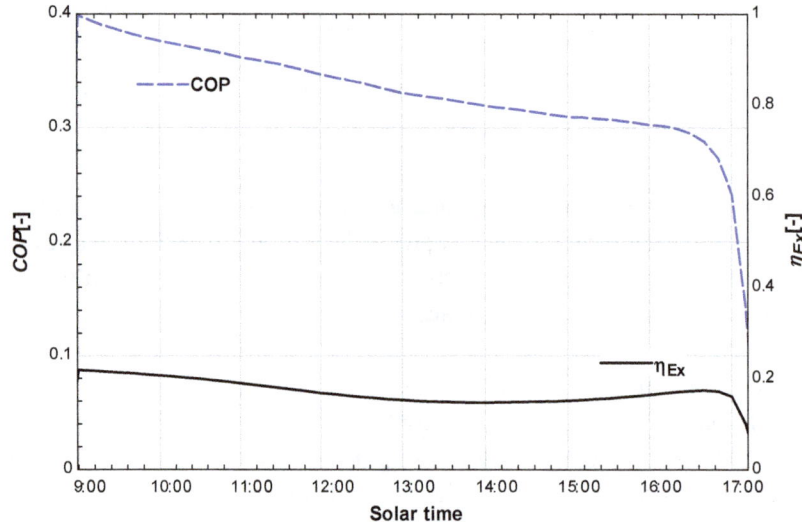

Figure 8. ACS coefficient of performance and exergetic efficiency.

After the ACS is stopped and thus no cooling load can be covered from the available heat source, the storage tank temperature begins to increase for the period of solar radiation availability, i.e., till 19:00. Operating regime (ii) is applied again. After 19:00, (i) operating regime is involved and water temperature slowly decreases due to losses to ambient.

The sensitivity of the storage tank fluid temperature to the variation of the cooling load along the day may be analyzed in detail in Figure 9. One may now observe that the cooling load is covered continuously between 9:00 and 17:10. The storage tank temperature increases continuously up to 13:40. This means that even if the cooling load is increasing and "consuming" more heat from the storage tank (sending the fluid with a lower temperature back to the storage tank), the increase of solar radiation on this period is enough to cover this load. After 13:40 the storage tank temperature slowly decreases to 14:00 when the cooling demand peak is met, and then drastically decreases even if the cooling load is decreasing too. The solar radiation is no more sufficient to maintain T_{ST} above the operation limit so that at 17:10 the ACS is turned off.

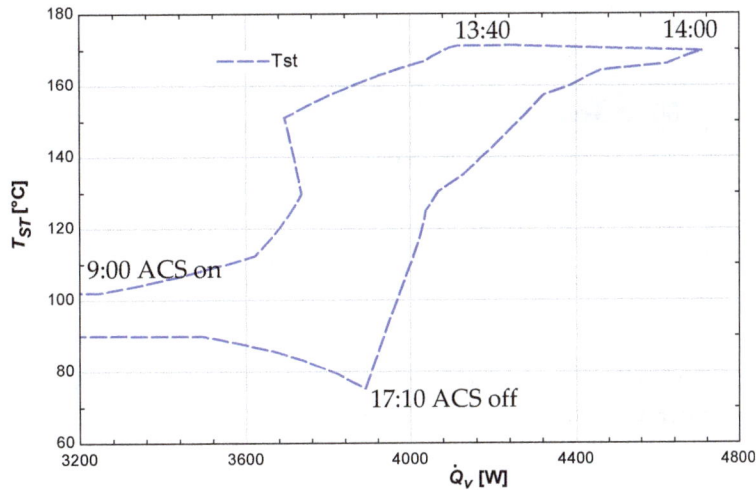

Figure 9. Cooling load influence on the storage tank fluid temperature along the day; case study 10 m PTC, 0.16 m^3 water storage tank.

From the above results, one might think about possibilities for improving the operation of the global system. One of them is to reduce the storage tank temperature so that the exergetic efficiency of the ACS increases. Thus, a simulation was done considering exactly the same parameters except the length of the PTC which was reduced from 10 m to 9.8 m. In this case, the results emphasize a smaller value of the maximum storage tank temperature, but the ACS is turned off automatically earlier, at 16:50, since the desorber temperature has dropped to an insufficient value for the ACS operation (84 °C). As solar radiation is still available, storage tank temperature increases enough to turn on ACS at 17:20, but as it is not sufficiently high, 10 min after, the ACS is turned off again. This is shown in Figures 10 and 11. This case emphasize the instability in ACS operation on the last part of the day. Obviously, such variations should be avoided.

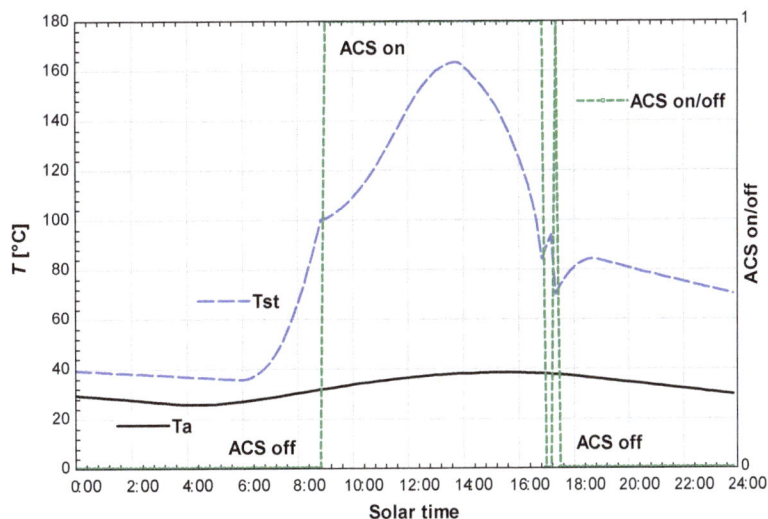

Figure 10. Ambient and storage tank fluid temperatures on working and non-working periods of the ACS; case study 9.8 m PTC, 0.16 m^3 water storage tank.

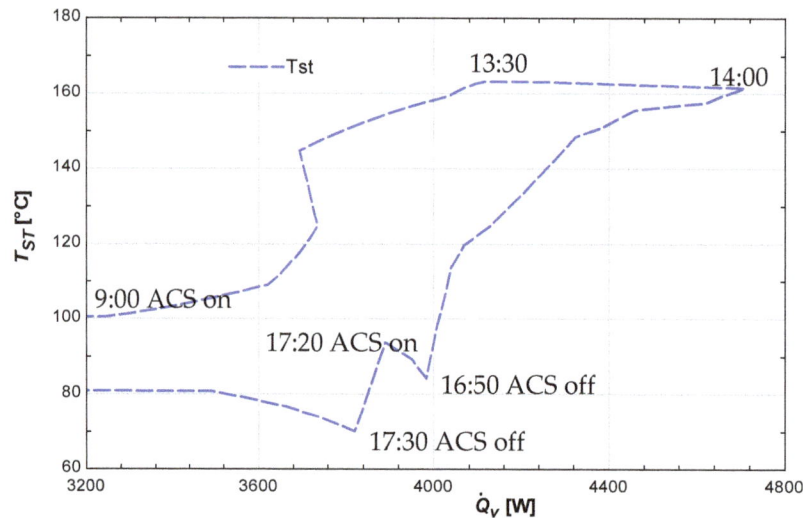

Figure 11. Cooling load influence on the storage tank fluid temperature along the day; case study 9.8 m PTC, 0.16 m^3 water storage tank.

Another improving solution would be to extend the ACS operation period by storing a higher quantity of thermal energy. This could be done by reducing the quantity of storage tank water. When doing a simulation with 0.14 m^3 of water instead of 0.16 m^3, a similar behavior as in the previous case was met. The storage tank temperature reached a higher maximum value (185 °C) in comparison to the previous two cases, but the temperature dropped rapidly so that at 17:10 the ACS was turned off, too. A second attempt of turning on was met at 17:40 but only for 10 min. Results are presented in Figure 12.

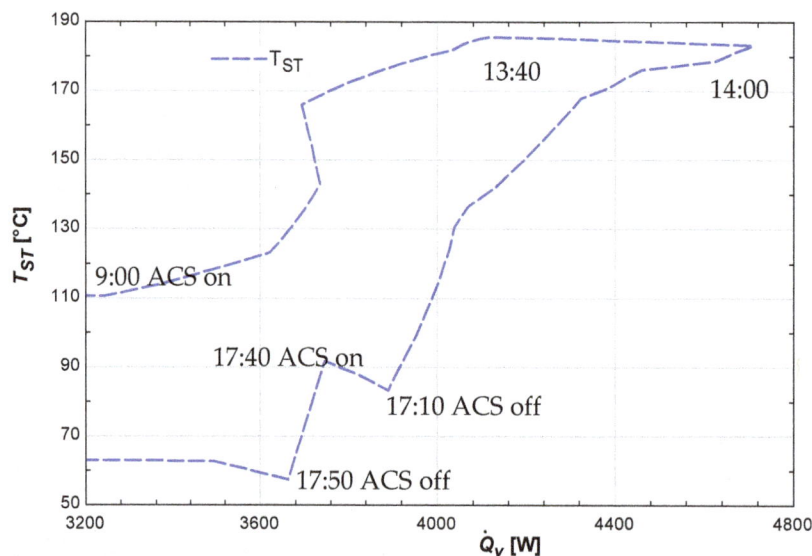

Figure 12. Cooling load influence on the storage tank fluid temperature along the day; case study 10 m PTC, 0.14 m^3 water storage tank.

One may conclude that the ACS operation is very sensitive to storage tank temperature and consequently to design parameters. There should be a narrow range of PTC and storage tank dimensions that fits the ACS operation. In this regard, a sensitivity study with respect to storage tank dimensions for a given PTC length is emphasized in Figure 13 for the ACS exergetic efficiency and in Figure 14 for the dependence of the storage tank temperature on the cooling load.

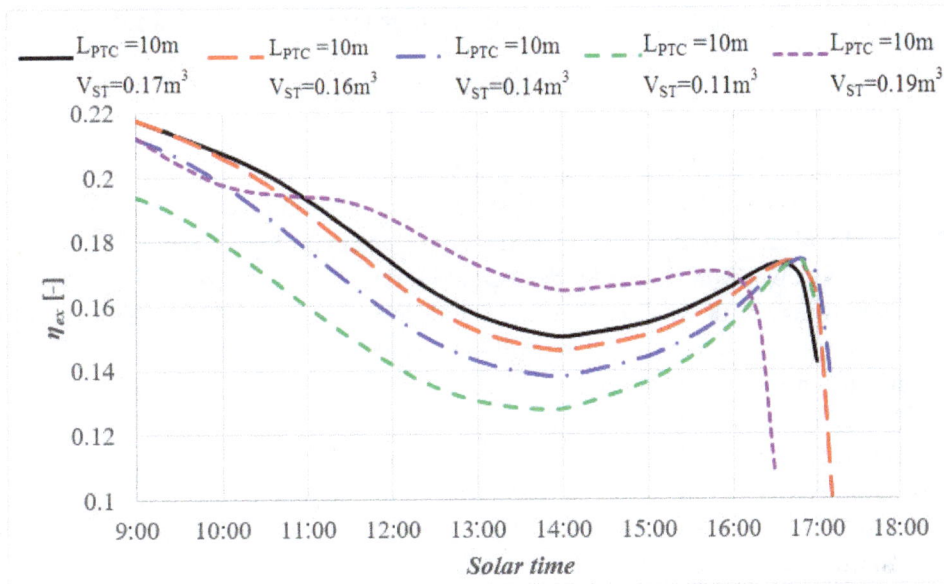

Figure 13. ACS exergetic efficiency—sensitivity with respect to storage tank dimension.

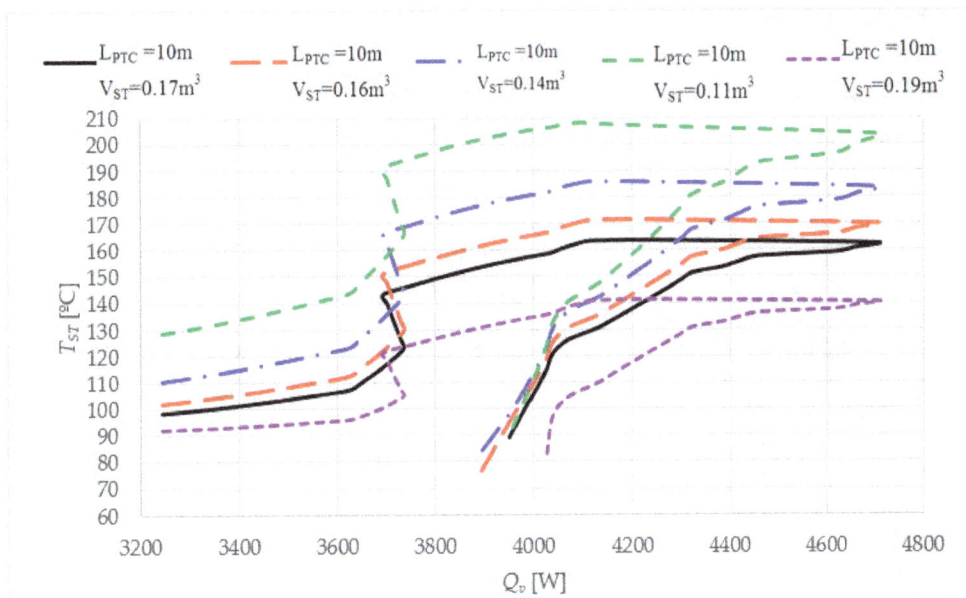

Figure 14. Storage tank temperature—cooling load dependence; sensitivity with respect to storage tank dimension.

One may see that the best ACS exergetic efficiency is obtained when the storage tank temperature is the lowest (the dotted magenta curve in Figures 13 and 14, corresponding to 0.19 m³ storage tank). In this case, the cooling load is not entirely covered on the targeted period 9:00–17:00. The system is turned off at 16:30 as the storage tank water temperature dropped under the desorber operation level. Increasing more the storage tank dimension leads to a lower level of the storage tank water temperature along the day and the ACS operation cannot be covered at all at the required cooling load.

Another important aspect regards the optimum dimensions of the PTC-ST systems for continuous operation. There is a specific storage tank dimension associated to a specific PTC dimension that could ensure the longest possible operation with a good ACS exergetic efficiency (the red dotted curve corresponding to 10 m × 2.9 m PTC and 0.16 m³ water storage tank). The results emphasized the optimum dimensions of the solar collector and storage tank required for a fully solar startup of the system.

8. Conclusions

A solar driven NH_3-H_2O absorption cooling system was analyzed from the point of view of the best sizing of solar-storage part of the global system for the longest possible daily operation in July, at 44.25° N latitude. Measured meteorological data have been employed, generated by Meteonorm software. Cooling load was time dependent computed for a two storeys residential building. A fully mixed hot water storage tank was used to fuel the desorber of the cooling system from the heat collected by a fixed oriented parabolic trough collector. The results emphasized that there is a specific storage tank dimension associated to a specific PTC dimension that could ensure the longest continuous operation of the ACS when constant mass flow rates inside the system are assumed. An initial solar start-up was considered, meaning that the initial temperature of storage tank water was close to the ambient one. The longest continuous operation of the NH_3-H_2O cooling system (from 9 a.m. to 5:10 p.m.) was obtained for a 10 m × 2.9 m PTC aperture dimensions with a 0.16 m^3 storage tank volume. The simulations emphasized that the ACS operation was very sensitive to these values. Any change in PTC or ST dimensions would diminish the operation time of the ACS.

From the best exergetic efficiency point of view, a 0.19 m^3 storage tank capacity is preferred, and as a result, the ACS operation period is reduced by 40 min. Thus, different optima lead to different sets of PTC and ST dimensions.

Further analyses may be thought implementing a control-command unit for variable mass flow rates to the PTC and ACS system, regulating the desorber temperature as needed. A comparative analysis may be envisaged between this studied case and the corresponding fully solar startup with a previous stored thermal energy from an earlier operation of the PTC-ST modules alone.

Acknowledgments: This work was supported by a grant of the Romanian National Authority for Scientific Research and Innovation, CNCS—UEFISCDI, project number PN-II-RU-TE-2014-4-0846.

Author Contributions: Camelia Stanciu contributed to the development of the model and preparation of the manuscript; Dorin Stanciu contributed to the design of the study, analysis and interpretation of the results; Adina-Teodora Gheorghian contributed to the state of the art analysis and preparation of the manuscript.

Conflicts of Interest: The authors declare no conflict of interest. This work was financially supported by a grant of the Romanian National Authority for Scientific Research and Innovation, having no role in the design of the study; in the collection, analyses, or interpretation of data; in the writing of the manuscript, and in the decision to publish the results.

Abbreviations

The following abbreviations are used in this manuscript:

Normal letters

A	area, m^2
D	diameter, m
G	solar irradiation, $W \cdot m^{-2}$
H	height, m
h	specific enthalpy, $J \cdot kg^{-1}$
$K\theta$	incident angle modifier
L	length, m
$\dot{m}$	mass flow rate, $kg \cdot s^{-1}$
$\dot{Q}$	heat rate, W
$\dot{S}$	entropy rate, $W \cdot K^{-1}$
T	temperature, K
U	overall heat transfer coef., $W \cdot m^{-2} \cdot K^{-1}$
w	velocity, $m \cdot s^{-1}$
$\dot{W}$	power, W

Greek letters

α	convection heat transfer coeff., $W \cdot m^{-2} \cdot K^{-1}$
τ	time, s

λ	thermal conductivity, $W \cdot m^{-1} \cdot K^{-1}$
σ	Stefann-Boltzman constant, $5.67 \times 10^{-9}\ W \cdot m^{-2} \cdot K^{-4}$

Subscripts

a	ambient
Ab	absorber/absorbed
Bt	beam (radiation) on tilt surface
C	condenser
cd	conduction heat trasnfer
CV	control volume
cv	convection heat transfer
f	fluid inside the receiver
fi	inlet fluid to receiver
fo	outlet fluid from receiver
gen	generation, creation
Gm	mean value, referring to generator
gls,i	interior glass cover surface
gls,e	exterior glass cover surface
i	inlet
$loss$	losses
o	outlet
P	pump
pe	exterior pipe wall
pe	interior pipe wall
PTC	parabolic through collector
ST, st	storage tank
u	useful
v	vaporization

References

1. Kalogirou, S.A. Solar thermal collectors and applications. *Prog. Energy Combust. Sci.* **2004**, *30*, 231–295. [CrossRef]

2. Koroneos, C.; Nanaki, E.; Xydis, G. Solar air conditioning systems and their applicability—An exergy approach. *Resour. Conserv. Recycl.* **2010**, *55*, 74–82. [CrossRef]

3. Duffie, J.; Beckman, W. *Solar Engineering of Thermal Processes*; John Wiley & Sons Inc.: Hoboken, NJ, USA, 2006.

4. Khan, M.M.A.; Ibrahim, N.I.; Saidur, R.; Mahbubul, I.M.; Al-Sulaiman, F.A. Performance assessment of a solar powered ammonia–water absorption refrigeration system with storage units. *Energy Convers. Manag.* **2016**, *126*, 316–328. [CrossRef]

5. Kizilkan, O.; Kabul, A.; Dincer, I. Development and performance assessment of a parabolic trough solar collector-based integrated system for an ice-cream factory. *Energy* **2016**, *100*, 167–176. [CrossRef]

6. Lu, Z.S.; Wang, R.Z. Experimental performance investigation of small solar air-conditioning systems with different kinds of collectors and chillers. *Sol. Energy* **2014**, *110*, 7–14. [CrossRef]

7. Ghaddar, N.K.; Shihab, M.; Bdeir, F. Modelling and simulation of solar absorption system performance in Beirut. *Renew. Energy* **1997**, *10*, 539–558. [CrossRef]

8. Lopez-Villada, J.; Ayou, D.; Bruno, J.C.; Coronas, A. Modelling, simulation and analysis of solar absorption power-cooling systems. *Int. J. Refrig.* **2014**, *39*, 125–136. [CrossRef]

9. Li, M.; Xu, C.; Hassanien, R.H.E.; Xu, Y.; Zhuang, B. Experimental investigation on the performance of a solar powered lithium bromide-water absorption cooling system. *Int. J. Refrig.* **2016**, *71*, 46–59. [CrossRef]

10. Mazloumi, M.; Naghashzadegan, M.; Javaherdeh, K. Simulation of solar lithium bromide–water absorption cooling system with parabolic trough collector. *Energy Convers. Manag.* **2008**, *49*, 2820–2832. [CrossRef]

11. Flores, V.H.F.; Roman, J.C.; Alpirez, G.M. Performance analysis of different working fluids for an absorption refrigeration cycle. *Am. J. Environ. Eng.* **2014**, *4*, 1–10.

12. Stanciu, C.; Şoriga, I.; Gheorghian, A.; Stanciu, D. Comfort air temperature influence on heating and cooling loads of a residential building. *IOP Conf. Ser. Mater. Sci. Eng.* **2016**, *147*. [CrossRef]

13. Romanian Norm. ORDINUL Nr.1574 din 15.10.2002 Pentru Aprobarea Reglementarii Tehnice "Normativ Pentru Proiectarea la Stabilitate Termica a Elementelor de Inchidere ale Clădirilor", Indicativ C107/702. Norm for Designing Thermal Stability of Building Closing Elements. Available online: http://www.ce-casa.ro/wp-content/uploads/2014/02/c107_7_2002.pdf (accessed on 27 December 2016). (In Romanian)

14. American Society of Heating, Refrigerating, and Air-Conditioning Engineers (ASHRAE). *ANSI/ASHRAE Standard 55-2004, Thermal Environmental Conditions for Human Occupancy*; ASHRAE: Atlanta, GA, USA, 2010.

15. Fernandez-Garcia, A.; Zarza, E.; Valenzuela, L.; Perez, M. Parabolic-trough solar collectors and their applications. *Renew. Sustain. Energy Rev.* **2010**, *14*, 1695–1721. [CrossRef]

16. Sharma, V.M.; Nayak, J.K.; Kedare, S.B. Shading and available energy in a parabolic trough concentrator field. *Sol. Energy* **2013**, *90*, 144–153. [CrossRef]

17. Stanciu, D.; Stanciu, C.; Paraschiv, I. Mathematical links between optimum solar collector tilts in isotropic sky for intercepting maximum solar irradiance. *J. Atmos. Sol.-Terr. Phys.* **2016**, *137*, 58–65. [CrossRef]

18. Blake, D.M.; Moens, L.; Hale, M.J.; Price, H.; Kearney, D.; Herrmann, U. New Heat Transfer and Storage Fluids for Parabolic Trough Solar Thermal Electric Plants. In Proceedings of the 11th Solar PACES International Symposium on concentrating Solar Power and Chemical Energy Technologies, Zurich, Switzerland, 4–6 September 2002.

19. Anthro Power. *Parabolic Trough Based Solar System—Operations & Maintenance Manual*; UNDP-GEF Project on Concentrated Solar Heat; Ministry of New & Renewable Energy, Government of India: Mumbai, India, 2014. Available online: http://mnre.gov.in/file-manager/UserFiles/CST-Manuals/PTC_E.pdf (accessed on 27 December 2016).

20. Nanjing Beite AC Equipment Co., Ltd. Available online: http://njbtkt.en.made-in-china.com/product/uBDJOxPhfCcw/China-Steel-Horizontal-Water-Storage-Tank-for-Hot-Water-Storage.html (accessed on 20 December 2016).

21. *Meteonorm Software*, version 7.1.8.29631; Global Meteorological Database for Engineers, Planners and Education; Meteotest Genossenschaft: Bern, Switzerland, 2016.

22. Kalogirou, S.A. A detailed thermal model of a parabolic through collector receiver. *Energy* **2012**, *48*, 298–306. [CrossRef]

23. Dudley, V.E.; Colb, G.J.; Sloan, M.; Kearney, D. *Test Results: SEGS LS-s Solar Collector*; SAND94-1884 Report; The Smithsonian/NASA Astrophysics Data System: Albuquerque, NM, USA, 1994.

24. McAdams, W.H. *Heat Transmission*, 3rd ed.; McGraw Hill: New York, NY, USA, 1954.

25. Stanciu, C.; Stanciu, D.; Feidt, M.; Costea, M. The available heat source influence on the operation of one stage absorption refrigeration systems. In Proceedings of the Congrès Français de Thermique, SFT'13, Gerardmer, France, 28–31 May 2013.

26. *Engineering Equation Solver*, Academic Commercial version V9.915; F-Chart Software: Madison, WI, USA, 2015.

27. Ibrahim, O.M.; Klein, S.A. Thermodynamic Properties of Ammonia-Water Mixtures. *ASHRAE Trans.* **1993**, *21*, 1495.

10

Risk Assessment Method of UHV AC/DC Power System under Serious Disasters

Rishang Long * and Jianhua Zhang

State Key Laboratory of New Energy Power System, North China Electric Power University, Beijing 102206, China; jhzhang001@163.com
* Correspondence: lrs18810667721@163.com

Academic Editor: Gianfranco Chicco

Abstract: Based on the theory of risk assessment, the risk assessment method for an ultra-high voltage (UHV) AC/DC hybrid power system under severe disaster is studied. Firstly, considering the whole process of cascading failure, a fast failure probability calculation method is proposed, and the whole process risk assessment model is established considering the loss of both fault stage and recovery stage based on Monte Carlo method and BPA software. Secondly, the comprehensive evaluation index system is proposed from the aspects of power system structure, fault state and economic loss, and the quantitative assessment of system risk is carried out by an entropy weight model. Finally, the risk assessment of two UHV planning schemes are carried out and compared, which proves the effectiveness of the research work.

Keywords: ultra-high voltage (UHV); chain failure; risk assessment; index system

1. Introduction

The construction and operation of the ultra-high voltage (UHV) power grid has played a huge role in promoting social development. However, its safety has been widely concerned by the community [1,2]. Scientific evaluation of the safety of the UHV grid is of great significance to analyze the security level of the power system, find the potential safety problems and prevent the blackouts. "N-1" or "N-2" guidelines are the conventional methods for assessing grid security [3,4], and some scholars have proposed improved methods or indicators for UHV grids [5–7]. However, these methods have two disadvantages: on the one hand, it only analyzes the consequences of failure events of single (double) elements, ignoring the probability of its occurrence. In fact, systemic risk is the combination of probability and consequence; on the other hand, it can not reflect the disaster-resistant performance under serious disasters with multi-component failure. In fact, the cascade fault characteristic is an important aspect that reflects the safety level of the power system [8].

According to the nature of fault triggering, the cascading faults can be divided into three types: overload dominant, protection dominant and structure dominant [9]. In [10,11], the transfer of post-fault flow is taken as the selection principle of N-K fault, and the influence of wind farms on power system risk is analyzed, but the model is lack of discussion on the protection hidden fault. In fact, the hidden faults cause great harm to the safety of the system. In [12,13], a protection hidden fault model is established, and the protection misoperation probability is calculated and applied to the simulation of cascading failures. However, the model lacks a discussion on the system loss during the fault recovery phase. The above studies have been focused on the overload dominant, protection dominant, but the studies on structure dominant faults caused by natural disasters are few. In [14], a cloud model and fuzzy comprehensive evaluation (FCE) method is proposed for collecting and monitoring the space charge density in an ultra-high-voltage direct-current (UHVDC) environment. In [15], the risk assessment model of an earthquake disaster on the power system is established, and the influence of the network structure and operation conditions on the seismic performance of the power system is studied. The risk index in [14,15] is single, which contains system state or loss index, but lack of a set of evaluation index systems for a UHV AC/DC power grid.

The development process of system cascading failure under severe disaster is shown in Figure 1. In severe disaster, the system structure is destroyed, and then cause line overload and protection malfunction, often contain three dominant types, which need comprehensive analysis; the process of cascading failure has not been adequately simulated in above references, in fact, the system still has risk loss in the process of fault recover; at the same time, the risk index is not comprehensive enough, and it needs to analyze the system structure, fault state and failure loss comprehensively. Finally, the simulation examples are based on the simulation system, which can not truly reflect the nature of the real AC/DC power system.

Based on the above research foundation and analysis, this paper presents a whole process risk analysis model under severe disaster. Based on the BPA stabilization procedure, this paper analyzes the cascading faults of UHV AC/DC power grid, and provides reference for the evaluation of system risk level.

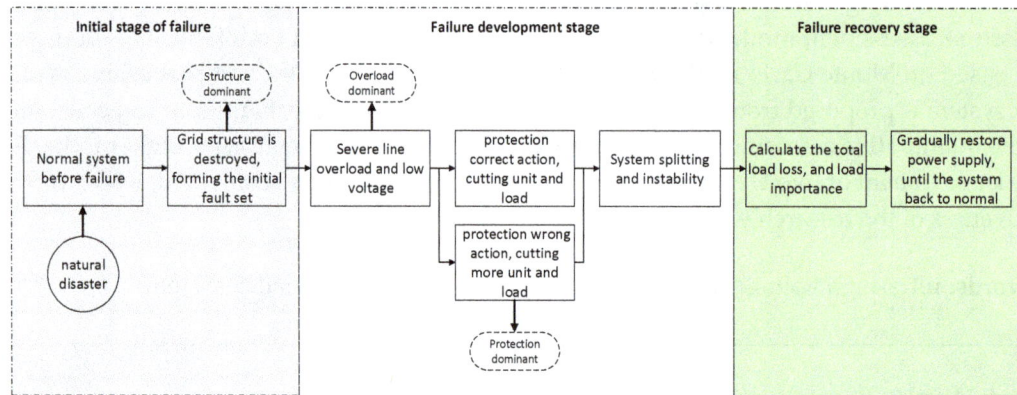

Figure 1. Development of chain failure.

2. Chain Failure Model

2.1. Chain Failure Probability Calculation

In the traditional method, a sequential Monte Carlo method and Bayesian model are usually adopted to develop the fault and calculate probability of failure. Under the large network, the computation and time-consuming are huge. In order to improve the computational efficiency, this paper combines the non-sequential Monte-Carlo method and BPA to calculate the probability of chain failure. First, make the following assumptions:

(1)   Outage of the components can usually be divided into two types: independent outages and related outages. Due to the short duration of severe disasters (such as earthquakes and typhoons), the independent outages consider the forcible outage models in which the failure can be repaired. The related outages consider the environmental interdependence and interlocking models, and do not consider the plan outage and aging failure.

(2)   The failure rate of the component is constant during the fault and no new fault occurs during system recovery.

(3)   All initial faulty components fail simultaneously under the hazard (taking into account the most severe three-phase permanent short-circuit) and are independent of each other. The subsequent failure caused by the initial failure must occur under the conditions of the simulation, which can be considered a conditional probability of 1, so we only need to calculate the probability of occurrence of the initial set of failures.

(4)   Since the probability of protection fault is low, only the primary protection faults developed by the initial failures set are taken into account.

The calculation steps are as follows:

(1)   First of all, the disaster history database is established to predict the area where the disaster may occur in the future. Combining with the data of component damage under the disaster of a domestic power system, the fault probability value of electrical equipment is calculated and the data is provided for the risk assessment.

(2)   According to the prediction result of disaster area, the initial fault set Ω (including the protection hidden fault) is set up, and the simulation is carried out in BPA to get the chain fault.

(3)   The system state is sampled by non-sequential Monte Carlo method, and the component failure frequency and average repair time are calculated. The relevant calculation is as follows [16]:

$$f_{to} = f_{ad}P_{ad} + f_{no}(1 - P_{ad}) \tag{1}$$

$$r_{to} = r_{ad}P_{ad} + r_{no}(1 - P_{ad}) \tag{2}$$

where f_{to} is the average failure frequency; r_{to} is the average repair time; f_{ad} and f_{no} are the failure frequencies under severe and normal climatic conditions, respectively; r_{ad} and r_{no} are the mean repair times under the severe and normal climatic conditions, respectively; and P_{ad} and $(1 - P_{ad})$ are probabilities of severe and normal climatic conditions, respectively.

Since data acquisition systems usually do not distinguish between failure events in normal and severe climates, only the average failure frequency and mean repair time in the past years are available. Therefore, two parameters, F and M, can be introduced, where F is the percentage of failures that occur under adverse climates and should be between 0 and 1. M is a multiple of the repair time under severe weather conditions compared to normal climatic conditions and must be equal to or greater than 1. Normally F and M can be estimated from engineering judgment. From the definitions of F and M, the variables relationship shown in Equation (3) can be derived from Equations (1) and (2):

$$\begin{cases} f_{ad} = \frac{f_{to} \cdot F}{P_{ad}} \\ f_{no} = \frac{f_{to}(1-F)}{1-P_{ad}} \\ r_{ad} = \frac{r_{to} \cdot M}{1+(M-1)P_{ad}} \\ r_{no} = \frac{r_{to}}{1+(M-1)P_{ad}} \end{cases} \tag{3}$$

(4)   After the average failure frequency and average repair time are obtained in step 3, the component failure rate can be calculated and the probability of the initial fault set can be calculated. Under simulation conditions, that is, chain failure rate, as shown in Equations (4)–(6):

$$\frac{\lambda}{\lambda + \mu} = \frac{f_{to}r_{to}}{8760} \tag{4}$$

$$\mu = \frac{1}{r_{to}} \tag{5}$$

$$E = \prod_{t \in \Omega} \lambda_t \tag{6}$$

where μ is repair rate, λ is failure rate, t is the faulty component, E is the initial failure rate of the system, and 8760 is the number of hours of one year.

2.2. The Control Strategy after Fault and the Termination Condition of Fault

A reasonable voltage and frequency control scheme is very important to ensure the safe operation of the system with UHV transmission lines. In this paper, according to "Power System Safety and

Stability Guidelines", we configure three lines of defense in the BPA for critical failures: (1) the first line of defense: protection; (2) the second line of defense: security and stability control system, including automatic reclosing, automatic bus transfer equipment, chain cutting machine, load shedding, etc.; and (3) the third line of defense: deserialization, low-frequency and low-voltage load shedding.

Based on BPA simulation, we set the following fault stopping conditions: (1) after a certain calculation, there is no new fault, that is, no new component overload and no voltage violations, and the system reaches a new equilibrium; (2) the power flow is not convergent after a fault, or the power flow also can not converge after adding the control measures, , then the grid is considered to collapse.

2.3. Fault Recovery Method

After the chain failure, fault components will be thoroughly repaired, so that there is no new hidden fault before repair, and then the power supply to the loss of load will be gradually restored. Restoration needs to first emulate the maximum power flow that the line allows to ascend:

$$P_k^{\max} = \begin{cases} \mu_k P_k^{\max} & \text{the } k\text{th line power failure and overload} \\ P_k^{\max} & \text{other conditions} \end{cases} \tag{7}$$

where $P_k^{\max}$ is the maximum power that the kth branch can pass; μ_k is the branch recovery factor, and the value depends on the degree of damage to the line.

We take the minimum load power loss and switching operations as the goal and gradually restore power supply. In the recovery process, we need to simulate the line voltage in order to prevent the occurrence of new failures. When the bus voltage rises to V_1, the load is restored to a%. When the bus voltage rises to V_2, the load is restored to b%. When the bus voltage rises to V_3 ($V_3 > V_2 > V_1$), the load is restored to c%. The risk assessment process is complete until all loads are restored.

2.4. Calculation of System Risk Indicators

On the basis of the risk assessment of the whole process, the system's loss load is calculated and the risk index of the system is completed.

3. System Risk Indicators

The UHV power grid has the following advantages: large transmission capacity, wide coverage, saving the transmission line corridor, and a decrease in the ratio of active power loss. However, it also has certain unhealthy characteristics, mainly embodied in:

(1) In the UHV power grid, with more load placement, the transmission distance is far. Therefore, the grid structure has a great influence on the grid operation and stability.

(2) In the AC/DC hybrid transmission structure, the sending and receiving ends still keep running synchronously. The stability problem of the transmission line is mainly caused by the short circuit fault of the AC system and the commutation failure of the DC system, and the large-scale power flow caused by the latch-up fault, resulting in long-distance transmission channel station voltage stability problems and transient power angle stability problems.

Therefore, the chain failure risk assessment index should not only reflect the seriousness of the accident, but also highlight the weakness of the grid structure. At the same time, special indicators should be included to meet the characteristics of UHV AC/DC power grids. Therefore, this paper establishes the risk assessment index system from the power system structure, fault state, and economic losses, as shown in Table 1.

Table 1. Risk assessment indicator system.

Classification	Indicators
System Structure	Average length of line (ALL) Network average medians (NAM) Network average degrees (NAD) Multi-infeed short-circuit ratio (MISCR)
Fault State	Low voltage severity (LVS) Power angle difference severity (PADS) System frequency severity (SFS)
Economic Losses	Power loss (PL)

3.1. Power System Structure Index

The structure of the grid has important influence on the stability of the system, so the static risk assessment of the grid structure is needed. The basic parameters describing the statistical properties of the complex network topology are called the structural feature quantities. The most important parameters include average length of line (ALL), node medians and node degrees [17], while multi-infeed short-circuit ratio (MISCR) [5] is used as the indicator of the strength of the multi-infeed AC/DC system.

(1) Average length of line (ALL)

Define ALL as the average of the shortest path length between any two nodes, that is

$$ALL = \frac{1}{N(N-1)} \sum_{i \neq j}^{K} d_{ij} \tag{8}$$

where d_{ij} is defined as the length of the shortest path connecting node i and node j; N is the number of nodes in the network, and K is the number of network edges. Generally, the higher the ALL value is, the lower the system reliability is.

(2) Network average medians (NAM) and Network average degrees (NAD)

The degree of node i is the number of edges connected to the node. The median of node i refers to the number of the shortest path through the node in the network. Degrees and medians can reflect the importance of nodes in the network, that is, the larger value of the nodes in the network are more important. The expected value of the degrees and medians are network average degrees (NAD) and network average medians (NAM), respectively. The higher the value is, the stronger the system coupling is.

(3) Multi-infeed short-circuit ratio (MISCR)

Because of the close electrical connection between the DC converter stations of the multi-DC infeed system, if the simple short-circuit ratio is used to evaluate the multi-DC infeed system, there will be a large deviation, and the results are obviously optimistic. Therefore, for a multi-DC infeed system, MISCR can better evaluate the strength of an AC/DC link system, and a strong short-ratio system is more reliable. The MISCR is defined as:

$$MISCR_m = \frac{1}{Z_{eqmm} P_{dm} + \sum_{n=1, n \neq m}^{M} Z_{eqmn} P_{dn}} \tag{9}$$

where $MISCR_m$ is the MISCR value for the mth DC line; Z_{eqmm} is the self-impedance of the mth DC line in an equivalent impedance matrix; Z_{eqmn} is the mutual impedance between the mth DC line and the

nth DC line in an equivalent impedance matrix; P_{dm} is the rated power of the mth DC line; P_{dn} is the rated power of the nth DC line; and M is the total number of DC lines.

Take the expected value of $MISCR_m$ as the system MISCR value.

3.2. Fault State Index

Voltage risk and frequency risk are two basic risk indicators that reflect the state of the power system. At the same time, the synchronous operation is an important characteristic of the stability of the system in the AC/DC mixed gird with more DC lines. Therefore, the power angle difference between units is taken as one of the risk indicators.

(1) Low voltage severity (LVS)

The two low voltage severity (LVS) functions in Figure 2 show the effect on the system when the bus voltage amplitude decreases. The discrete function is defined as: the value of the function is 1 when the bus voltage falls below the defined voltage threshold of 0.95 p.u.; otherwise, the value is 0. The continuous function is defined as: when the voltage value of the bus is 0.95 p.u., the function value is 1. When the voltage level drops further, its function value increases linearly. It overcomes the shortcoming of the discrete function. The value of the function can reflect the voltage level.

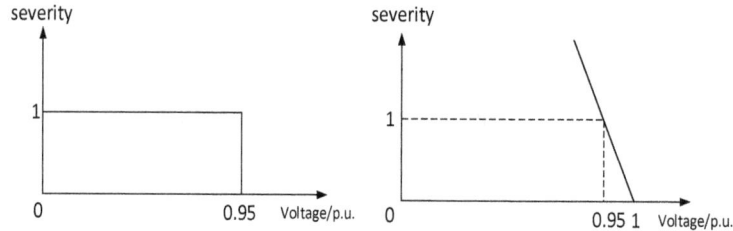

Figure 2. Low voltage severity (LVS) function.

In this paper, the continuous function is used, as shown in Equation (10):

$$s(U_k) = -20U_k + 20 \tag{10}$$

where U_k is the voltage value of the kth bus and $0 \leq U_k \leq 1$, $s(U_k)$ is the severity.

In the BPA stabilization procedure, the simulation results will give the bus voltage with the lower bus voltage. The system LVS is defined as:

$$S(U) = \omega_{u1}\overline{S}(U) + \omega_{u2}S_{\max}(U) \tag{11}$$

where ω_{u1} and ω_{u2} are the LVS weights (this paper takes equal weight), $\overline{S}(U)$ is the mean LVS, and $S_{\max}(U)$ is the maximum LVS.

(2) Power angle difference severity (PADS)

Similar to the LVS function, the power angle difference severity (PADS) function of the units also includes two forms of functions, as shown in Figure 3. In this paper, the continuous function is used, as shown in Equation (12):

$$s(\delta_p) = \begin{cases} 0 & 0 \leq \delta_p \leq 85 \\ 0.20\delta_p - 17 & 85 < \delta_p \leq 360 \end{cases} \tag{12}$$

where δ_p is the pth power angel difference, and $s(\delta_p)$ is the severity.

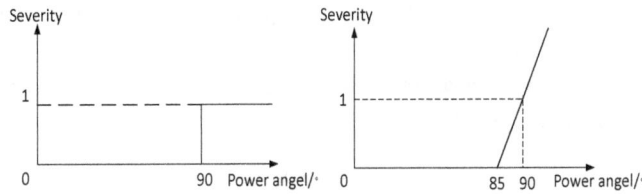

Figure 3. Power angle difference severity (PADS) function.

The system PADS is defined as:

$$S(\delta) = \omega_{\delta 1}\overline{S}(\delta) + \omega_{\delta 2}S_{\max}(\delta) \tag{13}$$

where $\omega_{\delta 1}$ and $\omega_{\delta 2}$ are the PADS weights (this paper takes equal weight), $\overline{S}(\delta)$ is the mean PADS, and $S_{\max}(\delta)$ is the maximum PADS.

(3) System frequency severity (SFS)

In the system frequency severity (SFS) function, the risk is 0 for system frequency of 50 Hz and 1 for a deviation of 0.5 Hz, as shown in Figure 4. In this paper, a continuous function is adopted, as shown in Equation (14):

$$S(f) = -2x + 100 \tag{14}$$

where f is the system frequency and $0 \le f \le 50$, $S(f)$ is SFS.

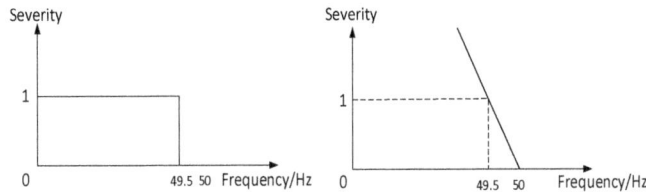

Figure 4. System frequency severity (SFS) function.

3.3. Economic Losses Index

The load of each bus can be divided into two parts, including which can be power cuts or not. For each fault condition (including fault phase and recovery phase), which may lead to load shedding or forced load reduction, calculate its power loss (PL), and then accumulate the PL value in each stage, resulting in the whole process PL function:

$$S(load) = \sum_{nd \in ND} \sum_{k \in K} W_k(d_s B_{k,nd} + d_c C_{k,nd}) \tag{15}$$

where $S(load)$ is the total PL cost (Yuan); $B_{k,nd}$ and $C_{k,nd}$ are the load reduction variables (MW·h) for users, which can be power cuts or not in the kth bus during the nd accident phase; d_s and d_c are the unit cost of power outages (Yuan/MW·h) for users, which can be power cuts or not; W_k is the weighting factor reflecting the importance of bus load, since the BPA procedure does not distinguish the severity of the load. Therefore, we take $W_k = 1$.

3.4. Comprehensive Risk Assessment

Entropy is a method of calculating the information difference between random vectors. It can determine the degree of support by judging the degree of intersection between different information sources (this paper refers to all kinds of risk indicators) and determining the weight of the information source according to the support degree. The higher the degree of support, the greater the weight.

It overcomes the subjective arbitrariness of index weights. In this paper, the entropy method [18] is used for comprehensive risk assessment.

Because of the inconsistency between the dimensions of the indicators, they need to be standardized and normalized, as follows:

$$X'_q = \frac{x_q - x_{\min}}{x_{\max} - x_{\min}} \tag{16}$$

$$Y_q = \frac{1 + x'_q}{\sum\limits_{q}^{Q} (1 + x'_q)} \tag{17}$$

where X'_q is the standardized matrix of indicators, x_q is the qth indicator, $x_{\max}$ and $x_{\min}$ are its maximum value and minimum value, Y_q is normalized index matrix, and Q is the total number of indicators.

For the indicators for which the smaller the value, the greater the risk, such as MISCR, NAM and NAD, first take its reciprocal, and then use Equations (16) and (17) to calculate.

According to the concept of entropy, the entropy of each index is defined as

$$H_q = \frac{\sum\limits_{q=1}^{Q} Y_q \ln B_q}{\ln Q} \tag{18}$$

The entropy weight of the qth index is

$$\omega_q = \frac{1 - H_q}{Q - \sum\limits_{q=1}^{Q} H_q} \tag{19}$$

The final risk value of each index is the product of its entropy weight, index value and system failure probability, as follows:

$$S_q = \omega_q X'_q E \tag{20}$$

The calculation flow of this paper is shown in Figure 5:

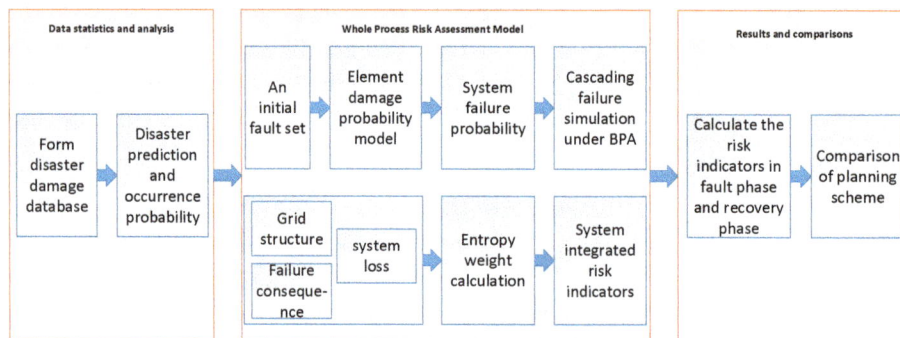

Figure 5. The calculation flow.

4. Case Study

4.1. Case Introduction

In order to verify the correctness of the proposed model in this paper, we adopt a regional power grid. The area takes 500 kV lines as the main grid structure, with the total installed capacity of 10.2589 million kilowatts and transmission load of 7.47 million kilowatts (winter big way). North and south regions form two independent power grids, and there are security and stability issues including a power surplus and a weak grid structure. In order to improve the grid structure, two kinds of target UHV grid planning schemes are proposed, as shown in Figure 6. The green line represents AC 500 kV, red represents AC 1000 kV, blue represents DC ± 800 kV, S represents source, and B represents the node (converter station, substation and load). The simulation is performed on PSD-BPA version 2.0 (China Electric Power Research Institute, Beijing, China).

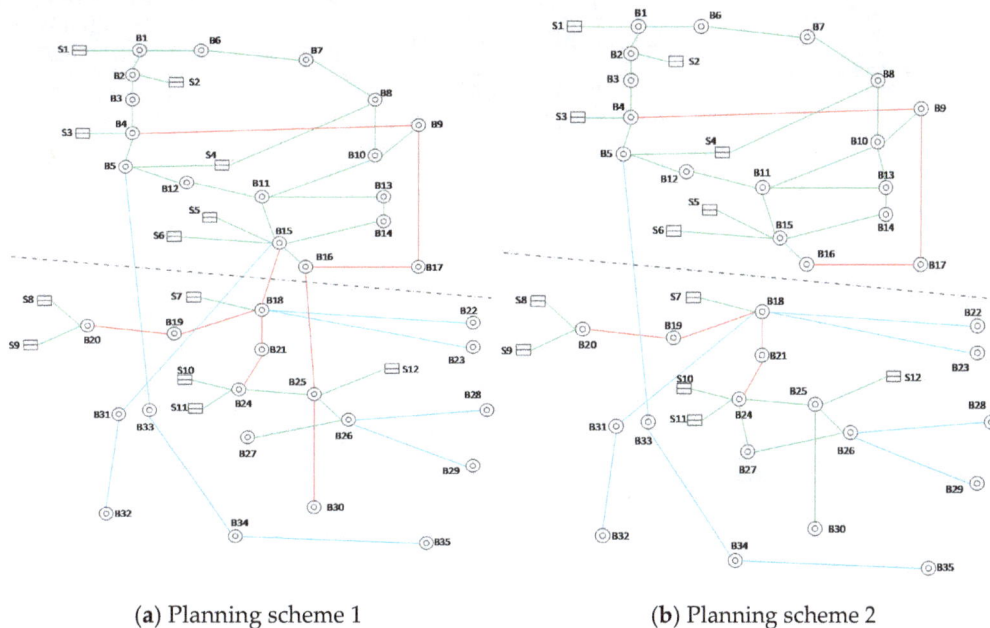

(a) Planning scheme 1   (b) Planning scheme 2

Figure 6. Planning scheme for a regional grid. (**a**) Planning scheme 1; (**b**) Planning scheme 2.

Planning scheme 1 (P1): based on the 1000 kV UHV AC power grid (B4–B30, B15–B24, B18–B20), to achieve the north–south regional system connecting and the total regional synchronous operation, with the construction of ±800 kV DC (B5–B35, B18–B22, B18–B23, B26–B28, B26–B29, B15–B32) external transmission, to solve the problem of power surplus.

Planning scheme 2 (P2): The construction of regional 1000 kV UHV (B4–B16, B18–B24, B18–B20) and 500 kV (B10–B13, B25–B30, B24–B27) lines to strengthen the regional grid structure, considering the possible spread of accidents in the disaster state, the north and south still keep running asynchronously. A DC high voltage scheme will convert B15–B32 to B18–B32.

The investment of the two schemes is basically the same and meets the requirements of "N-1" and "N-2", and the risk evaluation model proposed in this paper is used to compare and analyze the two schemes.

4.2. The Initial Failure Set and the Results of Chain Failure

From the analysis of the historical disaster database of the area, it is easy to be affected by earthquake disasters south of the dividing line. In this paper, we consider the common mainshock–aftershock type earthquakes with short duration (usually lasting several tens of minutes), but the main shock is very harmful. The initial failure set is studied in two cases.

Case 1: Do not consider the protection hidden fault. The initial failure set is: S7, S10, S11, S7–B18, S10–B24, S11–B24, B18–B19, B18–B21, and B21–B24.

Case 2: Consider protection hidden faults. Based on Case 1, B19 and B24 export main protection malfunctions.

A non-sequential Monte-Carlo method is used for sampling, with a confidence level of 95%, and a total of 2400 states are extracted. The probability of system failure and the development of cascading faults are shown in Table 2 and Figure 7 in both cases.

Table 2. Simulation results.

		Case 1	Case 2
Probability of Failure	P1	0.3914	0.0979
	P2	0.3909	0.0978
Chain Failure Process	P1	B18 → B15 → B14 (B16), B18 → B22 (B23), B15 → B31 → B32	B19 → B20, B18 → B15 → B14 (B16), B18 → B22 (B23), B15 → B31 → B32, B24 → B25 → B26 → B28 (B27, B29)
	P2	B18 → B22 (B23), B25 → B26 → B28 (B27, B29), B24 → B27, B24 → B25 → B30, B18 → B31 → B32	B19 → B20, B18 → B22 (B23), B25 → B26 → B28 (B27, B29), B24 → B27, B24 → B25 → B30, B18 → B31 → B32
Loss of Load (MW)	P1	783.76	1524.91
	P2	1683.20	1746.18

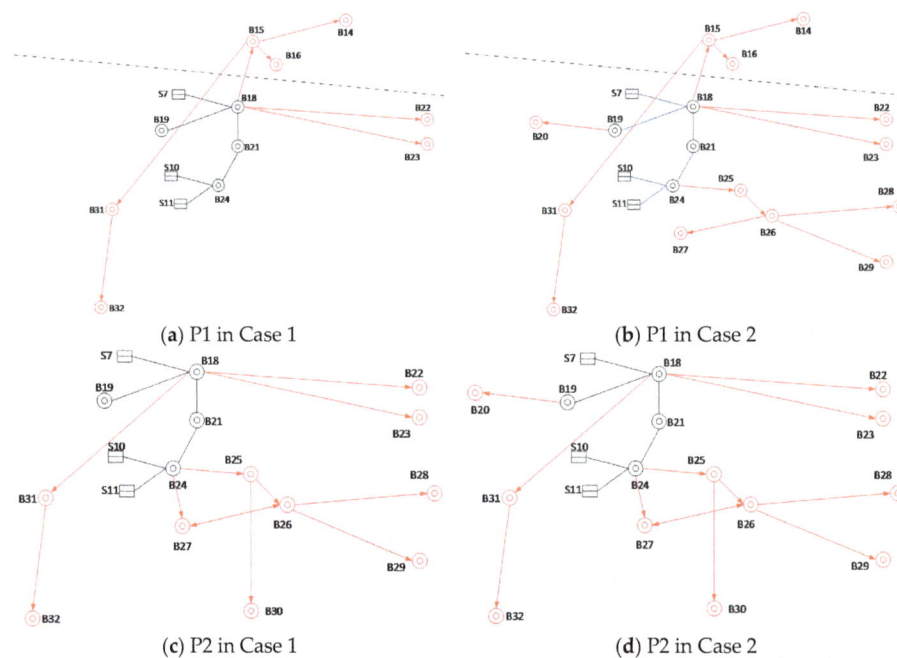

Figure 7. The development of cascading faults. (**a**) P1 in Case 1; (**b**) P1 in Case 2; (**c**) P2 in Case 1; (**d**) P2 in Case 2.

The results of the analysis are as follows:

(1) The system failure probability is low considering the protection malfunction, but the system interlocking fault is more serious, resulting in more serious load loss (more than 741,150 kW at P1 and 62,900 kW at P2). This shows that the protection hidden faults cause great harm to the system, which needs to be guarded against.

(2) In P2, even in case 1, the source S12 can not complete the power supply in the remaining area because of the large-scale sources and line failure, which leads to large-scale line overload and large-scale transfer of power flow, so that the power system lost the heavy load, The entire southern region falls into a state of collapse; In P1, although the cascading fault spread to the northern region (B18 → B15 → B14 (B16)), which results in a partial loss. However, the tie line of B16–B25 is not affected, which ensures network operation and supports the southern power grid. The tie-line and S12 together provide power to the remaining grid and reduce losses.

(3) From the development of chain failure, the stability of P1 is superior to that of P2. Quantitative evaluation results are shown in the following section.

4.3. Risk Assessment Results

As can be seen from the previous Section 3.2, the chain failure is more serious considering the protection hidden faults, so this section considers Case 2 for analysis.

Set the number of chain failure stages to day 1, after calculating the single risk index, the structural comprehensive risk, the failure comprehensive risk and economic loss risk are calculated by Formulas (16)–(20). The various types of risk indicators after the chain failure are shown in Table 3.

Table 3. System risk after chain failure.

Index	P1	P2
ALL (km)	742.14	730.28
NAD	2.54	2.41
NAM	1.33	1.30
MISCR	5.05	4.78
structural comprehensive risk	0.307	0.343
LVS	1.57	1.98
PADS	29.20	34.51
SFS	1.80	2.02
failure comprehensive risk	0.694	0.866
PL (ten thousand yuan)	914.96	1047.08
economic loss risk	89.30	102.40

The results of the analysis are as follows:

(1) From the structural risk value point of view: as there are more UHV lines in P1, resulting in an ALL value higher than P2. The MISCR of the two schemes are both more than 3.0, which belong to the strong short-circuit ratio level, but the value of P1 is higher than the value of P2, which indicates that the B15–B33 DC access scheme is better.

(2) From the fault indicators and power loss point of view, P1 is obviously superior to P2, which indicates that, in this kind of accident state, due to the strong inter-regional AC/DC link, the large-scale power flow transfer caused by the power gap can be reasonably shared, so that the southern power grid gets strong support. Compared with P2, in the event of the same serious failure, P1 can lose less load.

The risk indicators for the recovery phase are calculated by days, until the system returns to full power. Since it is assumed that there is no new fault in the recovery phase, that is, the present probability is 1 and the new failure probability is 0. Thus, the comprehensive risk value is not calculated. Finally, the system structure indicators and failure indicators take the average value, and the power loss takes the cumulative value. The various indicators for the recovery phase are shown in Table 4.

Table 4. System risk in recovery phase.

Index	P1	P2
ALL (km)	749.25	732.77
NAD	2.58	2.53
NAM	1.48	1.48
MISCR	6.25	6.01
LVS	1.36	1.58
PADS	10.96	15.91
SFS	0.51	0.52
PL (ten thousand yuan)	1862.15	2635.58

Comparing Tables 3 and 4, it can be seen that, during the recovery phase, all kinds of risk indicators of the system are still not optimistic, and the outage loss is more serious than the chain failure stage. Therefore, the loss of the recovery stage should be fully considered. As the chain fault stage in P2 suffers more serious damage, resulting in a longer recovery period, the power loss is much higher than P1.

To sum up, the ability to withstand natural disasters under synchronous operation of the regional power grid is better than asynchronous operation, so P1 is recommended as the better planning scheme.

5. Conclusions

This paper presents a whole process risk analysis method applicable to UHV power grids. The following work is completed: (1) based on the BPA stabilization procedure and non-sequential Monte Carlo method, a fast fault probability calculation method is proposed, and the whole process risk assessment model is established; (2) we have established a complete set of risk assessment indicators to effectively evaluate the structural and operational performance of AC/DC hybrid power grids; and (3) we carry on a simulation of the chain failure of an actual UHV AC/DC power grid, which provides a reference for evaluating the risk level of grid fault and guaranteeing safe operation of grids.

Acknowledgments: This work was financially supported by the National Natural Science Foundation (51277067) and the Central University Foundation (2015XS03).

Author Contributions: Rishang Long and Jianhua Zhang designed the study, Rishang Long performed the experiments and wrote the paper, and Jianhua Zhang reviewed and edited the manuscript. All authors read and approved the manuscript.

Conflicts of Interest: The authors have no conflict of interest to declare.

References

1. Xin, E.; Ju, Y.; Yuan, H. Development and application of a wireless sensor for space charge density measurement in an ultra-high-voltage, direct-current environment. *Sensors* **2016**, *16*, 1743. [CrossRef] [PubMed]

2. Dobson, I.; Chen, J.; Thorp, J.S.; Newman, D.E. Examining criticality of blackouts in power system models with cascading events. In Proceedings of the 35th IEEE Hawaii International Conference on System Sciences, Maui, HI, USA, 5–8 January 2002.

3. Kirschen, D.S.; Jayaweera, D.; Nedic, D.P.; Allan, R.N. A probabilistic indicator of system stress. *IEEE Trans. Power Syst.* **2004**, *19*, 1650–1657. [CrossRef]

4. Kirschen, D.S.; Jayaweera, D. Comparision of risk-based and deterministic security assessments. *IET Gener. Trans. Distrib.* **2007**, *1*, 527–533. [CrossRef]

5. Lin, W.-F.; Tang, Y.; Bu, G.-Q. Definition and application of short circuit ratio for multi-infeed AC/DC power systems. *Proc. CSEE* **2008**, *28*, 1–8.

6. Kwon, K.-B.; Park, H.; Lyu, J.-K.; Park, J.-K. Cost analysis method for estimating dynamic reserve considering uncertainties in supply and demand. *Energies* **2016**, *9*, 845. [CrossRef]

7. Li, J.; Yan, B.; Zhang, A.; Wu, Q.; Hao, J. Reliability research for UHVDC bipolar area DC protection system. *Power Syst. Prot. Control* **2016**, *44*, 130–136.

8. Ye, G.H.; Zhang, Y.; Zhang, Z.Q. A modified self-organized criticality method for power system risk assessment concerning influences of ultra-high-voltage transmission lines. *Autom. Electr. Power Syst.* **2015**, *39*, 44–52.

9. Vaiman, M.; Bell, K.; Chen, Y.; Chowdhury, B.; Dobson, I.; Hines, P.; Zhang, P. Risk assessment of cascading outages: Methodologies and challenges. *IEEE Trans. Power Syst.* **2012**, *27*, 631–641. [CrossRef]

10. Guo, L.; Guo, C.X.; Tang, W.H.; Wu, Q.H. Evidence-based approach to power transmission risk assessment with component failure risk analysis. *IET Gener. Transm. Distrib.* **2012**, *7*, 665–672. [CrossRef]

11. Liu, Z.; Jia, H.J.; Xu, T.; Zeng, Y. Study on N-k risk based grid-connection and capacity allocation of wind farm and energy storage system. *Power Syst. Technol.* **2014**, *38*, 889–894.

12. Nur, A.S.; Muhammad, M.O.; Mohd, S.S. Risk assessment of cascading collapse considering the effect of hidden failure. In Proceedings of the IEEE International Conference on Power and Energy, Kota Kinabalu, Malaysia, 2–5 December 2012.

13. Tamronglak, S. *Analysis of Power System Disturbances due to Relay Hidden Failures*; Virginia Polytechnic and State University: Blacksburg, VA, USA, 1994.
14. Zhao, H.; Li, N. Risk evaluation of a UHV power transmission construction project based on a cloud model and FCE method for sustainability. *Sustainability* **2015**, *7*, 2885–2914. [CrossRef]
15. He, H.; Guo, J. Seismic disaster risk evaluation for power systems considering common cause failure. *Proc. CSEE* **2012**, *28*, 44–56.
16. Li, W. *Risk Assessment of Power Systems*; IEEE Press & John Wiley & Sons, Inc.: New York, NY, USA, 2005.
17. Cuadra, L.; Salcedo-Sanz, S.; Del Ser, J.; Jiménez-Fernández, S.; Geem, Z.W. A critical review of robustness in power grids using complex networks concepts. *Energies* **2015**, *8*, 9211–9265. [CrossRef]
18. De Boer, P.-T.; Kroese, D.P.; Mannor, S.; Rubinstein, R.Y. A tutorial on the cross-entropy method. *Ann. Oper. Res.* **2005**, *134*, 19–67. [CrossRef]

11

Power Control of Low Frequency AC Transmission Systems Using Cycloconverters with Virtual Synchronous Generator Control

Achara Pichetjamroen * and Toshifumi Ise

Graduate School of Engineering, Osaka University, 2-1 Yamada-oka, Suita, Osaka 565-0871, Japan; ise@eei.eng.osaka-u.ac.jp
* Correspondence: achara@pe.eei.eng.osaka-u.ac.

Academic Editor: Paolo Mercorelli

Abstract: This paper is focused on the application of a multi-terminal line-commutated converter-type low frequency AC transmission system (MTLF) using a cycloconverter by applying a new power control scheme for multi-terminal operation. With the virtual synchronous generator (VSG) control scheme, the transmitting power among the multi-terminal system can be accomplished without a communication link for frequency synchronization in each terminal. The details of the proposed control scheme are explained in order to understand the advantages of this method. The configuration of a two-phase low frequency AC transmission system (LFAC) is adopted to examine with the proposed control scheme. Simulation results are provided to illustrate the proposed control scheme with respect to the LFAC system's performance.

Keywords: multi-terminal; low frequency AC transmission system (LFAC); power control; cycloconverters; virtual synchronous generator (VSG)

1. Introduction

The low frequency AC transmission system (LFAC) can be considered as promising for a long distance, large scale transmission system using line-commutated converters. One of the promising applications of this system is the operation on a multi-terminal system [1]. The integration of other types of power plants such as offshore wind power, High Voltage Direct Current (HVDC) or multi-terminal HVDC transmission with the main power grid are subjects of ongoing research [2,3]. In the case of LFAC, it could be easier to integrate the existing AC power plant because it can be built with commercially available power system components such as transformers and cables designed for regular frequency [4]. The transformer could be derated by a factor of transmitting frequency, with the same rated current, but partially containing the original rated voltage [5]. Given this information, the installation of LFAC can be easily paralleled with the existing transmission line.

For the application on the multi-terminal system with line-commutated converters, the operation of multi-terminal HVDC is complicated [6]. It is difficult to control power flow directions among terminals without an extra DC power flow controller, which requires greater complexity, a greater number of devices and extra costs.

In terms of the power control scheme for this LFAC transmission system, to date only one reference can be found. The concept of the power control method for the two-terminal system has been briefly mentioned in [7]. However, no details on the power control scheme were presented. In addition, the cycloconverters do not have synchronizing power with respect to the low frequency side. Therefore, absolute phase reference on the low frequency side is required to synchronize power for the low frequency output assisted by the reference time signal from the global positioning system

(GPS) or synchronous digital hierarchy (SDH), which is difficult to operate in the multi-terminal system application [8–10].

The goal of this paper is to present a new power control method for application on the multi-terminal LFAC transmission system. The controller is designed by using the advantages of a visual synchronous generator (VSG) with a governor control [11]. The voltage control is introduced to generate firing angles to the cycloconverters. The current limiter and extinction angle control are included for overcurrent and commutation failure protection of the converters. VSG with a governor control is applied to control the amount of power transmission in the LFAC multi-terminal system application. Furthermore, the proposed control scheme is applied in a two-phase multi-terminal system of LFAC to perform and verify the control operation by PSCAD/EMTDC Software (version 4.2.1, Winnipeg, MB, Canada).

2. Control Strategies for LFAC for the Two-Terminal System

The basic operation of the LFAC transmission system can be described by considering two terminals: sending and receiving ends. A cycloconverter is applied to lower the line frequency from 50/60 Hz to a smaller amount. At the receiving end, another cycloconverter is operated to convert the fraction frequency back to the grid frequency. These two conversions obtain firing angle signals from the designed control scheme, which is achieved by the different phase angles of different terminals. This is one of the common control methods of the LFAC transmitting power control. The active transmitting power over transmission lines can be expressed by Equation (1).

$$P = \frac{V_s V_R}{X} \sin \delta \qquad (1)$$

where V_S is the sending end voltage, V_R is the receiving end voltage, X is line reactance, and δ is the transmitting angle.

As indicated by Equation (1), the voltage control scheme concept with the LFAC transmission system can be described in Figure 1.

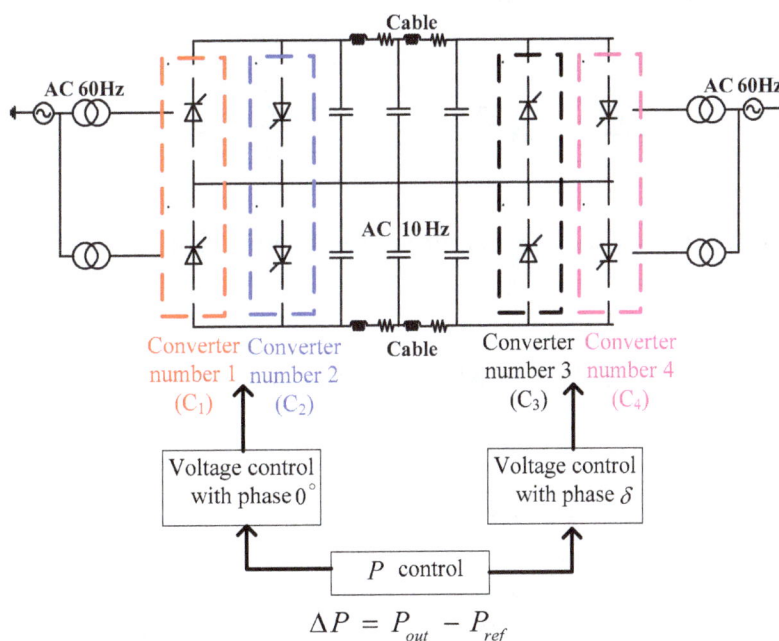

Figure 1. Concept of low frequency AC transmission system (LFAC) control scheme.

As shown in Figure 1, at the sending end, the phase is set to the constant value of 0° and the amount of transmitting power is determined by the phase difference from the receiving end.

A cycloconverter operates by following the modes in Table 1. On the low frequency side, a positive current waveform can be obtained from converter number 1 and number 4 operations. During the zero crossing point, all converters are blocked by the blocking signal from the controller. The negative current waveform can be received from converter number 3 and number 2 by reversing the current direction.

Table 1. Mode operation of the cycloconverter.

Operating Converter	Positive Current	Zero Cross Point	Negative Current
Rectifier No. of operating converter	C_1	Block	C_3
Inverter No. of operating converter	C_4	Block	C_2

2.1. The Proposed Voltage Control Scheme

Considering the mode control of the cycloconverter, this converter works as a two-set of HVDC anti-parallel connections and their inputs are a sinusoidal waveform instead of a constant value. From this point of view, the voltage control scheme can be designed by using HVDC concepts, in which the operation is much easier to understand.

$$V_d = \frac{3\sqrt{2}}{\pi} V_{LL} \cos \alpha$$
$$\frac{3\sqrt{2}}{\pi} V_{LL} = V_{d0} \tag{2}$$
$$V_d = V_{d0} \cos \alpha$$

From Figure 2, the Equation (2) is obtained to design the voltage control scheme of the LFAC by substituting V_d (DC voltage) with V_{dLF} (voltage on low frequency side) in Figure 3, and V_{LL} is the root mean square (RMS) value of 60 Hz-side AC line to line voltage. The designed control block diagram is shown in Figure 3, in which the voltage reference value of the low frequency side $v_{LF}^* = E_{max} \sin(\omega t + 0°)$ is for reference angle side, $E_{max} \sin(\omega t + \delta)$ is on another side and v_{LF} is the measured instantaneous voltage on low frequency side.

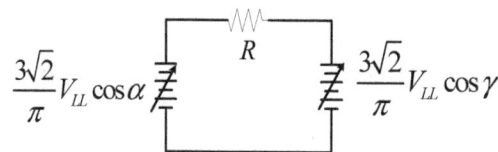

Figure 2. Equivalent circuit of High Voltage Direct Current (HVDC) link.

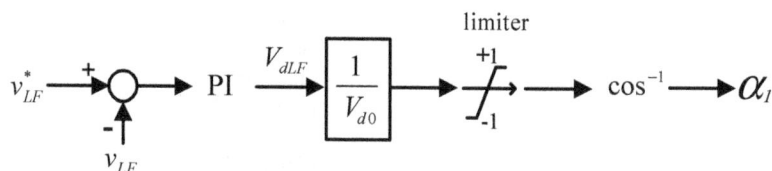

Figure 3. Voltage control of LFAC.

To obtain the phase reference (δ), the power control scheme needs to be considered. Equation (1) shows the amount of transmitting power between two terminals following the reference power. Furthermore, power on the low frequency side is not constant, thus the moving average control method is adopted to average the amount of power on the low frequency side. The power control block diagram can be achieved as shown in Figure 4.

Moving average control

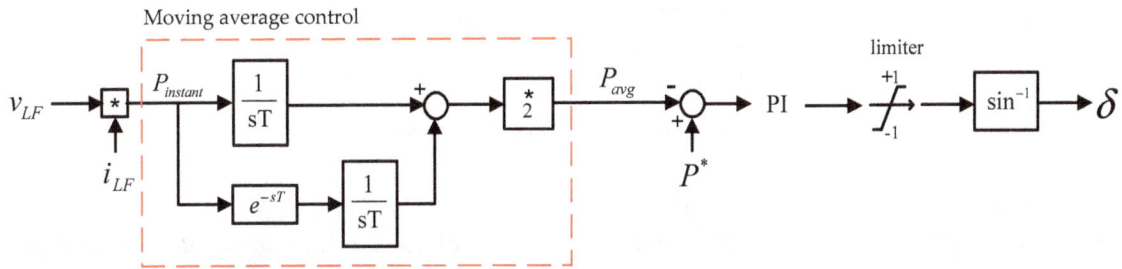

Figure 4. Power control of LFAC.

To control the amount of transmitting power between each terminal, the power control scheme is chosen for application to the cycloconverters in the LFAC transmission system. According to the system circuit mode and transmitting power in (1), power reference (P^*) is compared with average power (P_{avg}) to obtain the phase difference (δ) for the voltage reference.

2.2. Proposed Frameworks for Current Limiter and Extinction Angle Control

For protection from overcurrent and commutation failure, a current limiter and extinction angle control are applied in this system. The current limiter is used to limit overcurrent during the transmission of power where i_{LF} = the measured instantaneous current on the low frequency side. The current control scheme is shown in Figure 5.

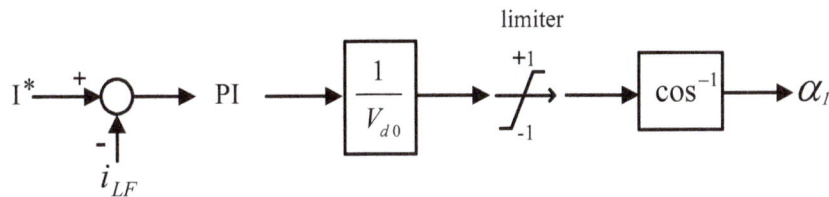

Figure 5. Current limiter of LFAC.

For line-commutated converters, in practice, there are some overlapping angles due to the commutation reactance. The constant extinction angle control (γ) is used to avoid commutation failure. As shown in Figure 6, this extinction angle control works by comparing gamma reference with gamma measure, then the minimum firing angle is chosen to be the gating pulse for cycloconverters. The typical value of extinction angle (γ) is set at 18° for 60 Hz of the AC system.

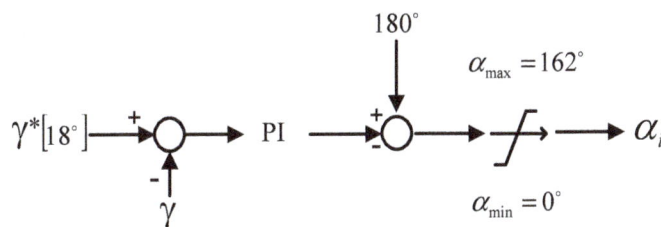

Figure 6. Extinction angle control of LFAC.

During the zero cross point current, a gate block control scheme is applied to every converter as shown in Figure 7.

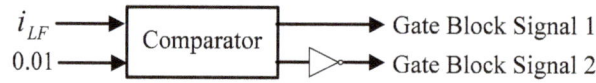

Figure 7. Gate block control scheme of LFAC.

In the complete developed control scheme of LFAC, as shown in Figure 8, the same control system is applied for every converter to control the amount of transmitting power between terminals. Following the converter operation mode, δ is set at $0°$ for one side of the converter. On another side of the converter, δ depends on the reference of transmitting power to control the amount and direction.

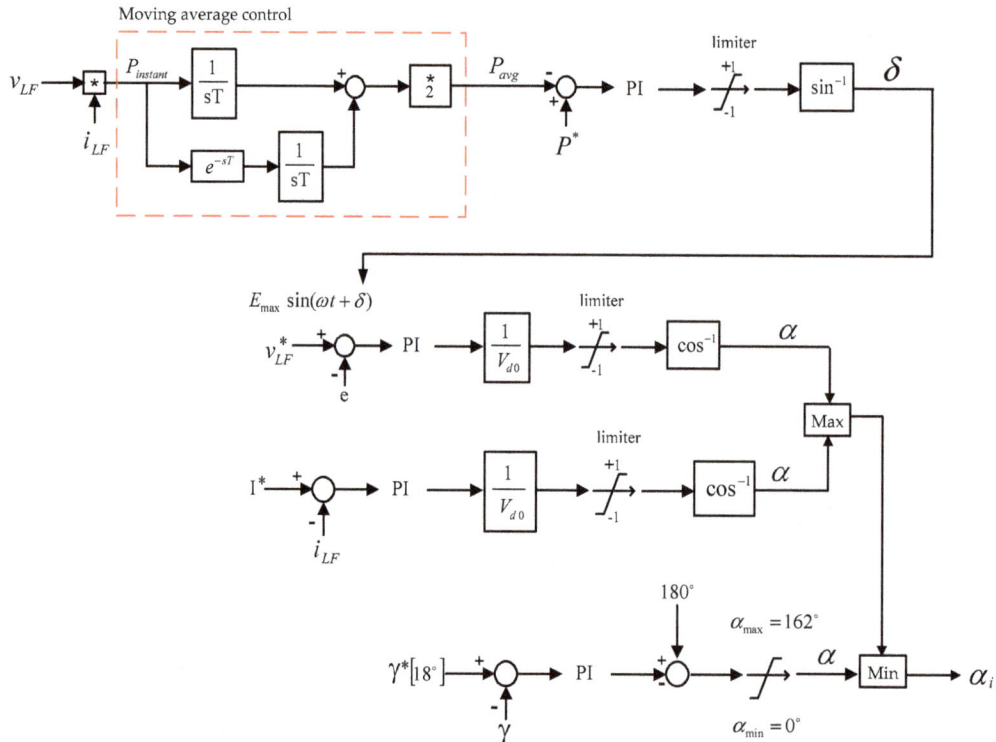

Figure 8. The proposed overall control scheme of LFAC.

3. Control Strategies for Multi-Terminal LFAC

The proposed control scheme to control the amount of transmitting power and direction among several terminals is explained in this section. In a multi-terminal application of LFAC (LCC-MTLF), a VSG is introduced to synchronize the phase and frequency among terminals instead of a communication link or GPS. The VSG model is based on the swing equation of synchronous generators. Based on this equation, the relation between the output power P_{out} and the input power P_{in} is shown in Equation (3).

$$P_{in} - P_{out} = J\omega_m \frac{d}{dt}\omega_m + D(\omega_m - \omega_g) \tag{3}$$

where ω_m is the virtual rotating angular frequency, ω_g is the angular frequency, J is the moment of inertia of rotating mass, and D is the damping factor of the damping power introduced by the damp winding.

The concept of VSG control is shown in Figure 9. For the power balance between P_{in} and P_{out}, according to the concept of VSG control, if the network frequency is lower than the reference frequency (f^*), then the generator increases its power into the network.

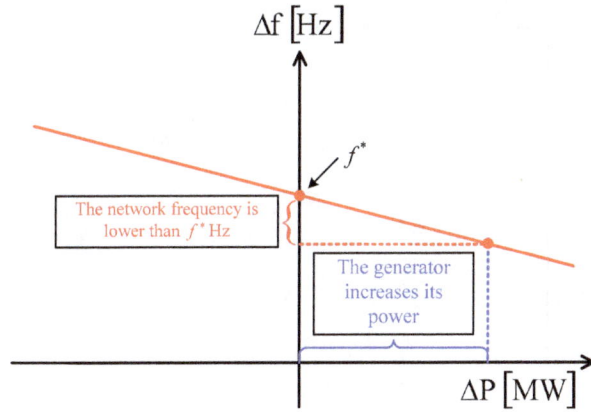

Figure 9. Concept of virtual synchronous generator (VSG) control.

Figure 10 shows the VSG control scheme applied to the LFAC transmission system. The VSG control is emulated to the swing equation in the VSG block.

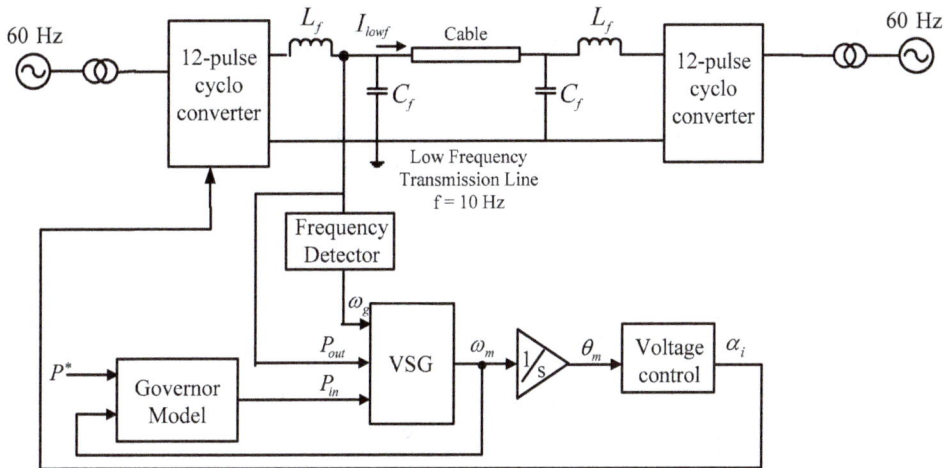

Figure 10. Block diagram of the VSG control scheme for LFAC.

For simplification, the damping factor D is considered as a small constant parameter. As the swing equation is a differential equation, an algorithm based on the Runge–Kutta iterative method is used to solve the angular frequency ω_m following the flow chart as shown in Figure 11. The calculated ω_m provides the frequency reference for the governor model and phase reference θ_m through an integrator. P_{base} is the rated capacity of transmission power.

The block of the governor model is shown in Figure 12. The governor model is implemented to tune the input power command based on the frequency deviation. The transfer function of the governor control can be expressed as Equation (4), where $K_p = 5$, $T_d = 0.1$.

$$P_{in} = P^* + \frac{K_p}{1 + T_d s}\left(f^* - f_{VSG}\right) \tag{4}$$

In the VSG control part as shown in Figure 13, P_{in}, P_{out} and low frequency side frequency $\left(f_g\right)$ are inputs of the VSG control unit. In each control cycle, the momentary ω_m is calculated via the application of the fourth-order Runge–Kutta iterative method inside the VSG block to generate the virtual phase angle θ_i ($i = 1, 2, 3$) sent through the voltage control unit.

$$\omega_g = 2\pi f_g$$

$$k_1 = \frac{P_{in} - P_{out} + P_{base} D \dfrac{\omega_g - \omega_m}{\omega_g}}{J\omega_m}$$

$$k_2 = \frac{P_{in} - P_{out} + P_{base} D \dfrac{\omega_g - (\omega_m - k_1 \Delta t / 2)}{\omega_g}}{J(\omega_m - k_1 \Delta t / 2)}$$

$$k_3 = \frac{P_{in} - P_{out} + P_{base} D \dfrac{\omega_g - (\omega_m - k_2 \Delta t / 2)}{\omega_g}}{J(\omega_m - k_2 \Delta t / 2)}$$

$$k_4 = \frac{P_{in} - P_{out} + P_{base} D \dfrac{\omega_g - (\omega_m - k_3 \Delta t / 2)}{\omega_g}}{J(\omega_m - k_3 \Delta t / 2)}$$

$$\omega_m = \omega_m + \frac{\Delta t}{6}(k_1 + 2k_2 + 2k_3 + k_4)$$

$$\omega_m \ \text{output}$$

$$f_g, P_{in}, P_{out}$$

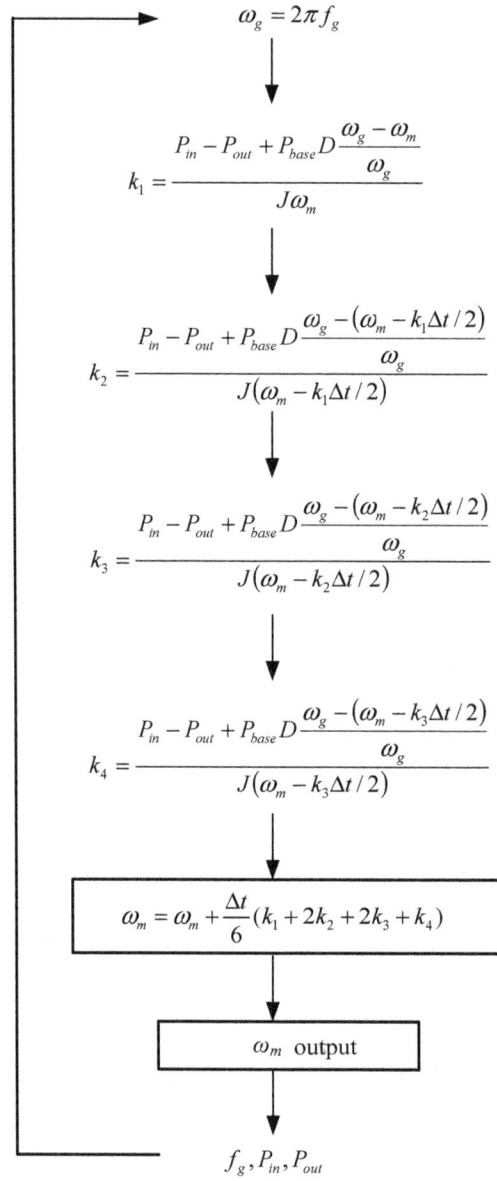

Figure 11. Flow chart of solving swing equation by Runge–Kutta method.

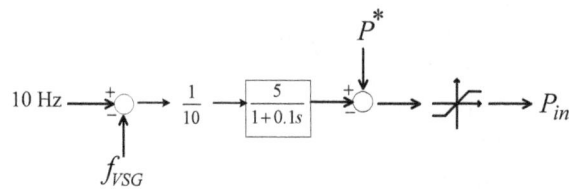

Figure 12. Governor control scheme.

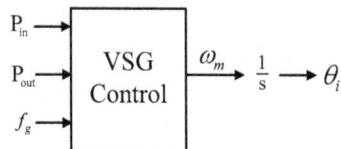

Figure 13. VSG control scheme.

As for the voltage control scheme, the concept of the control operation is the same as the voltage control scheme as described in Section 2.1. When a disturbance occurs, or more load shares are added to the system, the system can stabilize itself by regulating voltage using this control.

The current limiter works by comparing the rating current with measured current from the low frequency side to generate the firing angle to the cycloconverter, which can limit overcurrent by limiting the firing angle to the cycloconverter. Also, extinction angle control (γ) is included to avoid commutation failure in a converter.

By combining each part of the control scheme, the complete control scheme of the multi-terminal LFAC transmission system is shown in Figure 14. The voltage control receives the virtual phase angle θ_i from the VSG block to be the voltage reference, then the firing angle α_i is generated and is sent to the cycloconverters.

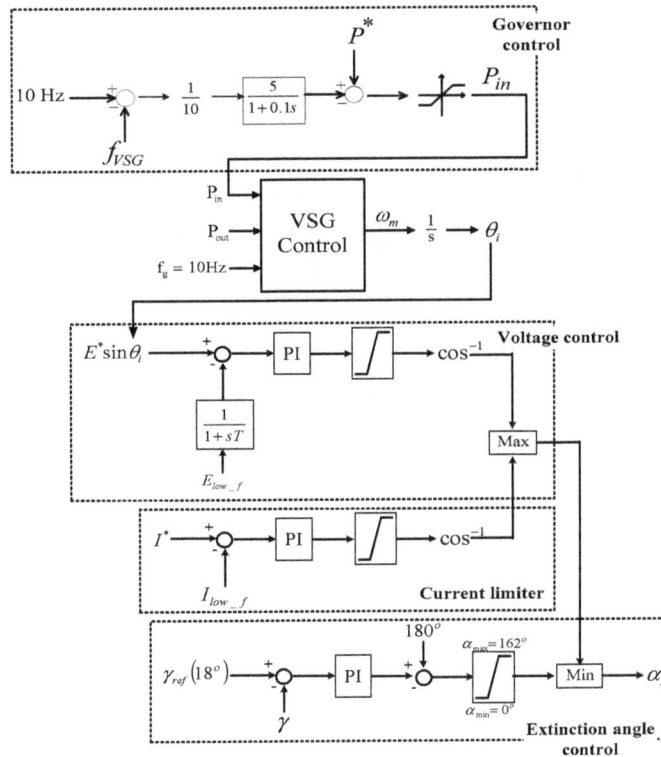

Figure 14. Complete control scheme of multi-terminal LFAC.

4. Simulation Results

To demonstrate the validity of the proposed power control scheme for the multi-terminal LFAC application, the simulations have been carried out using the tool of the PSCAD/EMTDC simulation software. The transmitting power in this system is rated at 1400 MW, and the transmission distance is 500 km. The transmission power cable is modeled by cascading 10 km of cable; its model is shown in Figure 15.

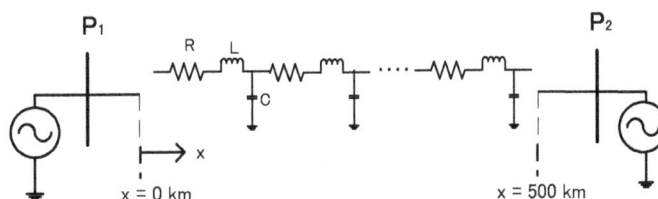

Figure 15. Cable model for the simulation.

For the simulation of the multi-terminal LFAC application, a three-terminal two-phase LFAC configuration [1] is applied with a proposed power control scheme as shown in Figure 16.

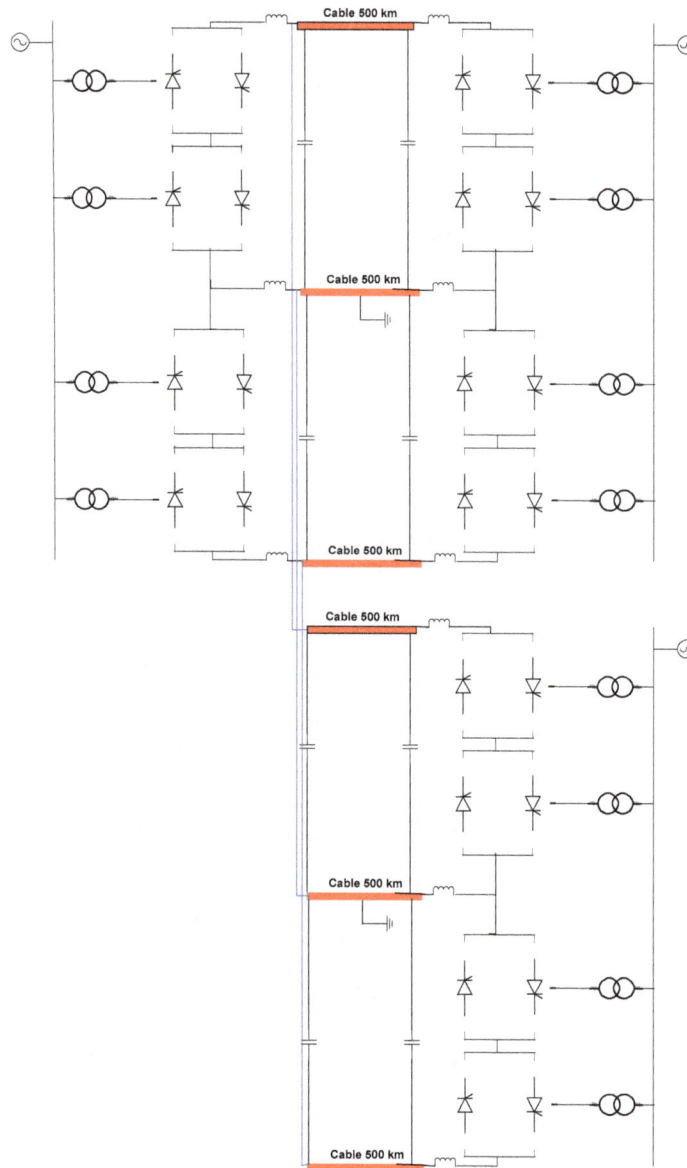

Figure 16. Three-terminal two-phase LFAC.

4.1. Current Limiter and Extinction Angle Control Operation

From the proposed control scheme in Section 2.2, some simulation results are performed to verify the control operation. Figure 17 shows the simulation result of the extinction angle control that cannot exceed the limit at the reference $\gamma = 18°$. An extinction angle γ is calculated from the equation $\gamma = 180° - (\alpha + u)$ where $u = 12°$ as shown in Figure 17b. The overlapping angle is around 12° for 60 Hz and γ is controlled at 18°. This can confirm the extinction angle control scheme works well.

During the period of transition, the mode of the cycloconverter may cause a spike current. This can malfunction with devices in the system so that the current limiter is used for protection. From the control scheme shown in Figure 5, the current limiter works by comparing the rating current with the measure current from the low frequency side to generate the firing angle to the cycloconverter. The current limiter can limit the overcurrent by limiting the firing angle to the cycloconverter. Figure 18 shows the result of the current limiter control that limits current at 20 kA on the low frequency side.

The *E_lowf* waveform is transmitting low frequency AC voltage, following the reference value from the control scheme. The *I_lowf* waveform shows a transmitting current on the low frequency side, which is limited at 20 kA due to the current limiter. From the result, it can be confirmed that the current limiter control can work properly.

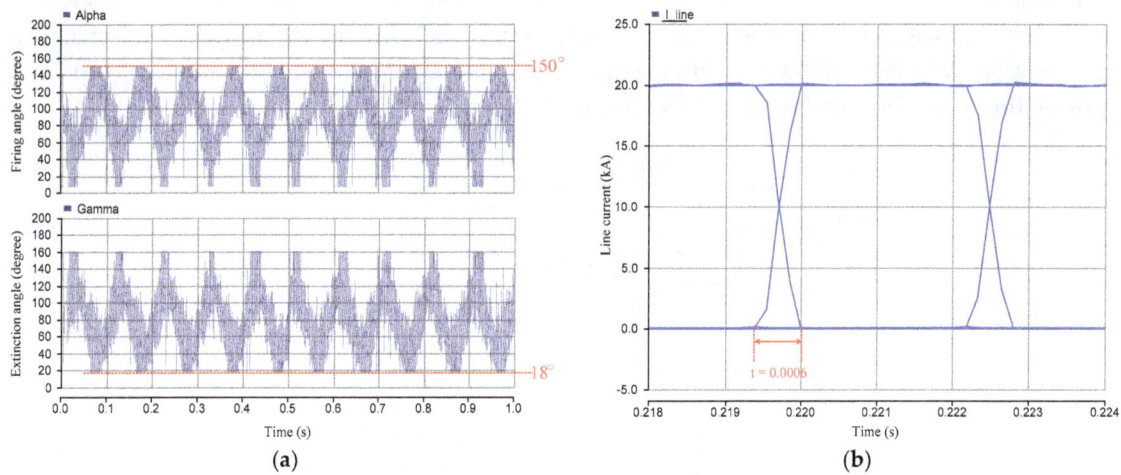

Figure 17. Simulation result of extinction control. (**a**) Firing angle and extinction angle during operation at 20 kA; and (**b**) line current waveforms.

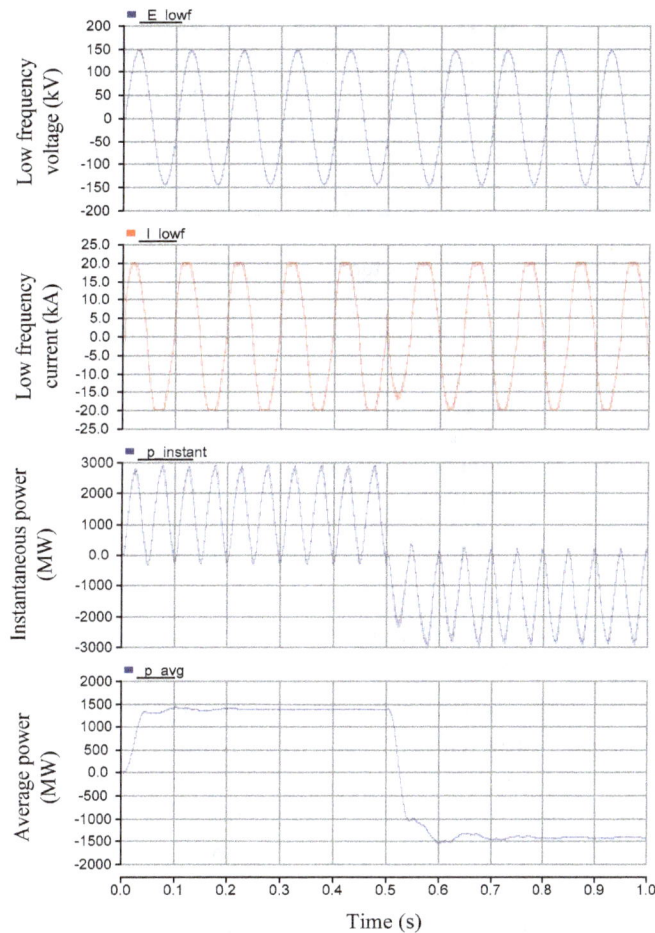

Figure 18. Simulation result for current limiter control.

4.2. Multi-Terminal LFAC Response

The simulation results of the three-terminal two-phase LFAC system are explained in this section. In the performed multi-terminal configuration, as mentioned in Figure 16, the multi-terminal LFAC demonstrates the effectiveness of the proposed system and control method. The total amount of power in this system is 1400 MW. The power flow pattern is shown Figure 19 and the details are shown in Table 2. In the power flow schedule given in Table 2, terminals 1, 2 and 3 are desired to inject a power of 800 MW, 500 MW and -1300 MW into the grid. The minus signs show that the power flow is in the opposite direction. The simulation parameters are shown in Table 3.

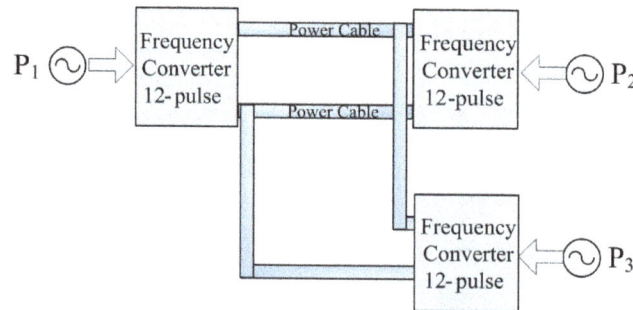

Figure 19. Power flow patterns for simulation.

Table 2. Power flow patterns.

Terminal No.	P* (Power Reference, MW)	
	Sending	Receiving
1	800	-600
2	500	-300
3	-1300	900

Table 3. Parameters of simulated system.

LFAC System	500 kV, 10 Hz
Line frequency	60 Hz
Transmitting frequency	10 Hz
Transformer	500 kV/110 kV
Maximum power transfer	1400 MW
Maximum cable length	500 km
Number of terminals	3
Power reversal	After 0.5 s

Figure 20 depicts the results of a simulation where the transmitting power flows from terminals 1 at 800 MW, and 2 and 3 at 500 MW and 1300 MW, respectively. After 0.5 s, the power direction is reversed following power references (terminal 1 at 600 MW, terminal 2 at 300 MW and terminal 3 at 900 MW). The voltage (E_lowf), current (I_lowf), and average power (p_avg) are the results on the low frequency side. As can be seen from the simulation results, the amount of power in each terminal can be transferred following the power flow patterns given in Table 2. The transmitting power on the low frequency side is 10 Hz following the reference. Figure 19 shows the power that flows from terminal 1 at 800 MW and in the reverse direction at 0.5 s following the power control scheme command.

Figure 20b,c shows the result of the power transmission at terminals 2 and 3. The results show that the amount and direction of power can follow the power flow patterns. It can be confirmed that the proposed power control for the multi-terminal LFAC using the VSG control works properly.

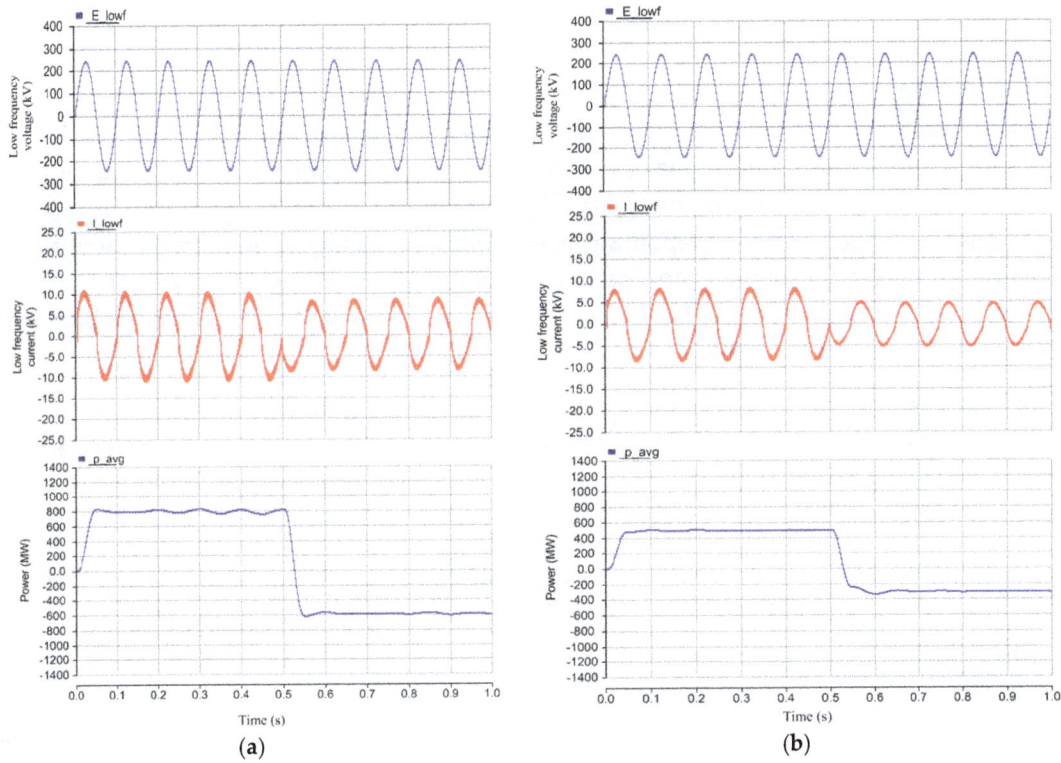

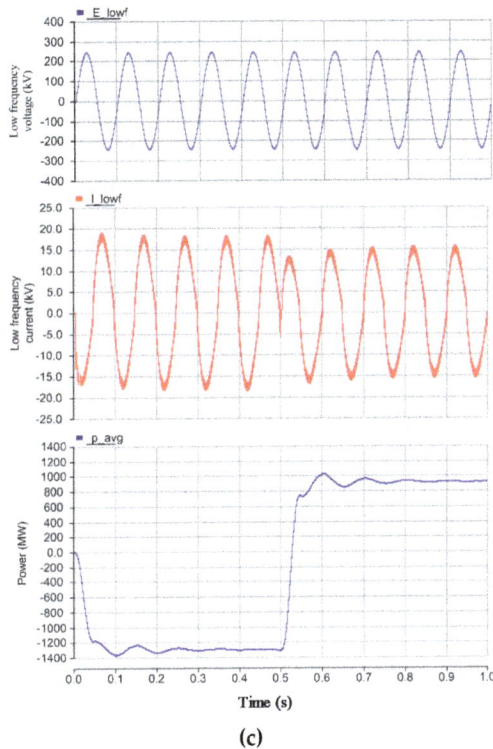

Figure 20. Low frequency side waveforms (voltage, current and power) on (**a**) terminal 1; (**b**) terminal 2; and (**c**) terminal 3.

5. Conclusions

This paper explains the proposed power control scheme of the multi-terminal LFAC application. A three-terminal two-phase LFAC cycloconverter configuration is applied to operate with the proposed control scheme. There are three main important points to be concluded as follows:

(1)   The proposed power control scheme which was explained in detail is feasible for synchronized frequency and regulated voltage by integrating VSG and the governor control to the control scheme;

(2)   A two-phase LFAC configuration was adopted to operate as a multi-terminal system LFAC. The simulation results confirm the operation of the current limiter and extinction angle control, which operated using this configuration;

(3)   The multi-terminal LFAC application was operated by using the proposed power control scheme to transmit power among three terminals following the reference power patterns. The simulation results verified the operation of the proposed control scheme with the given transmission system without a communication link.

This study can lead us to consider more details such as adding additional terminals to operate with the proposed control scheme and configuration. Comparing merits and drawbacks of the system with other existing multi-terminal transmission systems in the presence of various disturbance scenarios could be considered as the future steps for the present work.

Author Contributions: Achara Pichetjamroen conceived the proposed control strategy, performed calculation, simulation of the system and wrote the paper. Toshifumi Ise as research supervisor provided guidance and key suggestions, participated in proposing the ideas of the system configurations and control scheme, and helped in writing the paper.

Conflicts of Interest: The authors declare no conflict of interest.

References

1.   Achara, P.; Ise, T. A proposal on low frequency AC transmission as a multi-terminal transmission system. *Energies* **2016**, *9*, 687.

2.   Bozhko, S.; Asher, G.; Li, R.; Clare, J.; Yao, L. Large offshore DFIG based wind farm with line-commutated HVDC connection to the main grid: Engineering studies. *IEEE Trans. Energy Convers.* **2008**, *23*, 119–127. [CrossRef]

3.   Mau, C.N.; Rudion, K.; Orths, A.; Eriksen, P.B.; Abildgaard, H.; Styczynski, Z.A. Grid connection of offshore wind farm based DFIG with low frequency AC transmission system. In Proceedings of the Power and Energy Society General Meeting, San Diego, CA, USA, 22–26 July 2012.

4.   Tuan, N.; Min, L.; Surya, S. Steady-state analysis and performance of low frequency AC transmission lines. *IEEE Trans. Power Syst.* **2016**, *31*, 3873–3880.

5.   Slade, P.G.; Smith, R.K. A comparison of the short circuit interruption performance using transverse magnetic field contacts and axial magnetic field contacts in low frequency circuits with long arcing times. In Proceedings of the Discharges and Electrical Insulation in Vacuum International Symposium, Yalta, Crimea, 27 September–1 October 2004.

6.   Etienne, V.; Boon, T.O. Multiterminal HVDC with thyristor power-flow controller. *IEEE Trans. Power Deliv.* **2012**, *27*, 1205–1212.

7.   Funaki, T.; Matsuura, K. Feasibility of the low frequency AC transmission. In Proceedings of the Power Engineering Society Winter Meeting, Singapore, 23–27 January 2000.

8.   Cho, Y.; Cokkinides, G.J.; Meliopoulos, A.P. Time domain simulation of a three-phase cycloconverter for LFAC transmission systems. In Proceedings of the Transmission and Distribution Conference and Exposition, Orlando, FL, USA, 7–10 May 2012.

9.   Nakagawa, R.; Funaki, T.; Matsuura, K. Installation and control of cycloconverter to low frequency AC power cable transmission. In Proceedings of the Power Conversion Conference, Osaka, Japan, 2–5 April 2002.

10.   Chen, H.; Johnson, M.H.; Aliprantis, D.C. Low frequency AC transmission for offshore wind power. *IEEE Trans. Power Deliv.* **2013**, *28*, 2236–2244. [CrossRef]

11.   Bevrani, H.; Ise, T.; Miura, Y. Virtual synchronous generators: A survey and new perspectives. *Int. J. Electr. Power Energy Syst.* **2014**, *54*, 244–254. [CrossRef]

12

The Effect of the Angle of Inclination on the Efficiency in a Medium-Temperature Flat Plate Solar Collector

Orlando Montoya-Marquez and José Jasson Flores-Prieto *

National Center of Research and Develop of Technology-TecNM-SEP, Interior Internado Palmira s/n, Cuernavaca 62490, Morelos, Mexico; orlando_m_marquez@hotmail.com
* Correspondence: jasson@cenidet.edu.mx

Academic Editor: José Antonio Sánchez Pérez

Abstract: In this experimental work, the effects of the inclination angle β and the $(T_i - T_a)/G$ on the efficiency and the U_L-value were investigated on a medium-temperature flat plate solar collector. The experiments were based on steady-state energy balance, by heat flow calorimetry at indoor conditions and considering the standard American National Standard Institute/American Society of Heating Refrigerating and Air Conditioning Engineers (ANSI/ASHRAE) 93-2010. The solar radiation was emulated by the Joule effect using a proportional integral derivative (PID) control considering two conditions of the absorber temperature, Case 1: $(T_o - T_i) > 0$, and Case 2: $(T_o - T_i) = 0$. The inclination angles were 0°–90° and the $(T_i - T_a)/G$ were 0.044–0.083 m^2·°C/W and 0.124–0.235 for Case 1 and Case 2, respectively. The variations of β and $(T_i - T_a)/G$ cause efficiency changes up to 0.37–0.45 (21.6%) and 0.31–0.45 (45.0%), respectively, for Case 1. Also, the $U_L(\beta)$ reached changes up to 10.1–12.0 W/m^2·°C (19.2%) and 8.4–12.0 W/m^2·°C (41.7%), respectively, for Case 1. The most significant changes of $U_L(\beta)/U_L(90°)$ vs. β were 8.0% at the horizontal position for Case 1, while for Case 2, the maximum change was 1.8% only. Therefore, the changes of the inclination angle cause significant variations of the convective flow patterns within the collector, which leads to considerable variation of the collector efficiency and its U_L value.

Keywords: covered solar collectors; tilt solar collector; inclined solar collector; overall heat loss coefficient

1. Introduction

The solar heating systems for industrial applications have a great potential to reduce the demand for conventional energy. This technology could supply 406 GW$_{th}$ [1]. Currently, the world's solar heating by flat plate collectors is about 83.9 GW$_{th}$ [1]. The glazing flat plate solar collectors are one of the most used devices for solar heating, due to being able to reach more than the desired temperatures, collecting direct and diffuse radiation at a lower cost due to their construction and operation simplicity [2]. The characterization and simulation have improved the commercialization and their applications in most heating processes. Therefore, anything focusing on better characterization of solar collectors has been welcomed for the solar heating industry because that allows reducing estimation uncertainty, in order to have a more accurate picture of their performance.

Most of the previous reports show theoretical and experimental models to determine the thermal efficiency of solar collectors, in terms of the transmittance-absorbance factor $\tau\alpha$, the overall heat transfer U_L and the heat removal factor F_R [3]. These models usually consider the collector tilt angle with respect to the normal axis of the surface, mainly taking into account the variations with respect to the east-west axis, as a common daily variation. The ANSI/ASHRAE 93-2010 [4] and International Standard Organization (ISO) 9806:2013 [5] show methods to determine the influence of the incidence angle which only amend the first coefficient of the efficiency curve ($F_R \tau \alpha$). Then the tilt angle with respect to the North-South (N-S) horizontal axis is currently taken into account to determine a constant value of U_L, as this varies only during the few days of the collector's test [4,5]. Currently, the effect of the change of the inclination angle is included in the third coefficient of the collector efficiency curve along with many other effects, as well as in the Incident Angle Modifier (IAM) to fit the first coefficient [4].

Among the research models, previous works have also proposed an overall heat loss coefficient and experimental generalized correlations as complements to determine the solar collector's efficiency [6–9]. However, these proposed models generally use a fixed inclination with respect to the N-S horizontal axis, according to the latitude of the evaluation place [10,11], thereby disregarding the fact that the second and third coefficients of the efficiency curve are functions of the U_L. This latter is, in turn, dependent on the collector's inclination, due to the changes of the convective flow pattern of the confined fluid between the absorber and the cover of the collector. To date, the determinations of U_L as a function of the collector inclination β have been weakly studied, separating the effect of the confined fluid and the incidence solar radiation, as well as the effect of working at higher differences of temperature ($T_p - T_a$), as is the case for medium-temperature solar collectors for industrial processes.

Ozoe et al. [12,13] and Alvarado et al. [14], among others, have shown the changes of the flow pattern with the collector inclination with respect to the horizontal position. However, lower ($T_p - T_a$) and the collector's inclination as used at characterization place do not make a big difference. Bava and Furbo [15] carried out a parametric experimental study varying the collector type, the solar collector fluid, the volume flow rate and the collector inclination to determine the performance in terms of efficiency. They found that the flat plate collector is most commonly used at inclination angles of 30°, 45° and 60° of, in a range of 20–100 °C. On the other hand, Sabatelli et al. [16] developed a test method to evaluate the efficiency uncertainty for the ISO 9806/1 standard. Beikircher et al. [17] developed a procedure to determine the collector thermal efficiency based on heat loss measurements without insulation. Bava and Furbo [15] found that the most important parameter in the sensibility of the U_L is the inclination angle, which can represent 5%–8% of U_L change. Thus, the use of a higher temperature between the absorber plate and the glazing cover, as well as the increase of the movement of products around the world, causes the effect of inclination on the collector's efficiency, and more careful studies are required.

Therefore, a study of the effects of the inclination angle on the efficiency and the overall heat transfer coefficient was proposed and investigated in a glazing flat plate collector under inclinations from horizontal to vertical, varying ($T_i - T_a$)/G for working conditions of some of the medium-temperature solar collectors.

2. Materials and Methods

2.1. Sampling

Figures 1 and 2 show a sketch and a picture of the flat plate solar collector respectively. The sample is 2.00 m^2 (1.00 m × 2.00 m) of gross collector area, with an aspect ratio of 40. The solar collector has a couple header tubes and five raising finned tubes, all of them of cooper, joined by tin-lead solder. The working fluid was water-glycol 10%–90%, considering variable heat capacity as function of the temperature [18]. The surface of absorber plate was covered with black matte paint with absorptance of 0.94. The solar transmittance of the glazing was 0.86. The spectral absorptance and transmittance were obtained according to CIE 130-1998 [19], using a spectrophotometer Shimadzu UV-3100, (Shimadzu Corpotation, Nakagyo-ku, Kyoto, Japan), in the range of 300–2500 nm, every 2.0 nm with ±0.1% of photometric uncertainty and 1.0% of wavelength uncertainty. The absorptance was obtained by normalizing the measured spectral absorptance according ISO 9050-2003 [20]. Table 1 shows the solar collector construction characteristics.

Table 1. Construction specifications of the solar collector.

Parameter	Dimension	Units
Aspect ratio of enclosure	40	-
Copper absorber area	1.95×0.95	m
Fin width	0.19	m
Diameter of heaters	0.0381	m
Diameter of reassign tubes	0.0127	m
Fin thickness	0.5	mm
Absorptance	0.94	-
Fiber glass insulation	0.0254	m
Cover of glass shit	3.0	mm

2.2. Experimental Design

The experimental design allows determining the U_L and collector efficiency η, under inclinations from horizontal to vertical, at a range of $(T_i - T_a)/G$. All of the rest variables involved in the experiment were considered without significant variations. Once the experiment works emulating the solar heating, part of the supplied energy heats the working fluid Q_u, and the rest is transferred to the ambient as heat losses $Q_l(\beta)$, both of them also dependent from the β and $(T_i - T_a)/G$. The $U_L(\beta,(T_i - T_a)/G)$ and $\eta(\beta,(T_i - T_a)/G)$ are determined by flow calorimetry, based on steady-state energy balance according Figure 1, at indoor controlled conditions. Constant solar heating at $(T_o - T_i) > 0$ and isothermal absorber plate at $(T_o - T_i) = 0$ are considered in the study for comparison, looking for achieve suitable experimental uncertainty.

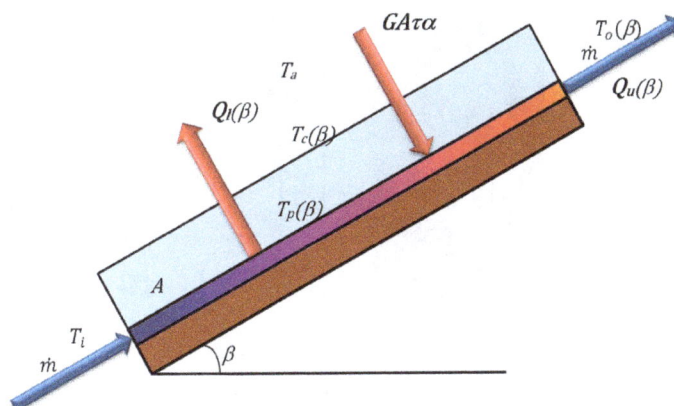

Figure 1. Physical model.

The incoming heat flux is GA, which is fixed and is not considered function of β. The outlet heat is the sum of the $Q_u(\beta)$ plus the heat loss flux $Q_l(\beta)$. The following considerations are taken in to account in the experiment: (a) steady state; (b) constant surrounding temperature and emissivity; (c) constant radiative exchange; (d) linear variation of C_p with the temperature; and (e) the mean plate temperature, $T_p(\beta)$ is considered as the average temperature between outlet and inlet temperature $(T_o(\beta) - T_i)/2$.

Throughout each test once the absorber plate works, it heats the working fluid that flows through the raising tubes. Part of supplied energy heats the working fluid and the rest is transferred to the ambient as heat losses, both of them dependent from the β and $(T_i - T_a)/G$. At indoor conditions, the solar heating is emulated by Joule effect and proportional integral derivative (PID) control, using electrical heater, making possible to replace $(T_i - T_a)/G$ by $(T_i - T_a)/(VI/A\tau\alpha)$, to achieve better experimental uncertainty, thus the solar heating is given by Equation (1).

$$GA = \frac{VI}{\tau\alpha} \tag{1}$$

where τ and α are the glazing solar transmittance and the solar absorptance of the absorber respectively, and V and I are the electrical voltage and current respectively. Considering the above, due to the convective flow pattern varies with the tilt angle then the overall heat transfer coefficient in terms of β is calculated by:

$$U_L(\beta) = \frac{VI - Q_u(\beta)}{A\left[T_p(\beta) - T_a\right]} \tag{2}$$

where the useful energy is $Q_u(\beta) = \int_{t_1}^{t_2} \dot{m}C_p(T_o(\beta) - T_i)\mathrm{d}t$ and the $Q_l(\beta) = AU_L(\beta)\left[T_p(\beta) - T_a\right]$. And then, the collector efficiency, according ANSI/ASHRAE 93-2010, is given by:

$$\eta(\beta) = \frac{Q_u(\beta)}{\left(\frac{VI}{A\tau\alpha}\right)} \tag{3}$$

2.3. Experimental Setup

The heating of the absorber is homogenously distributed by means of the electrical heater and remains almost constant over each test. The experimental setup is sketched in Figure 2.

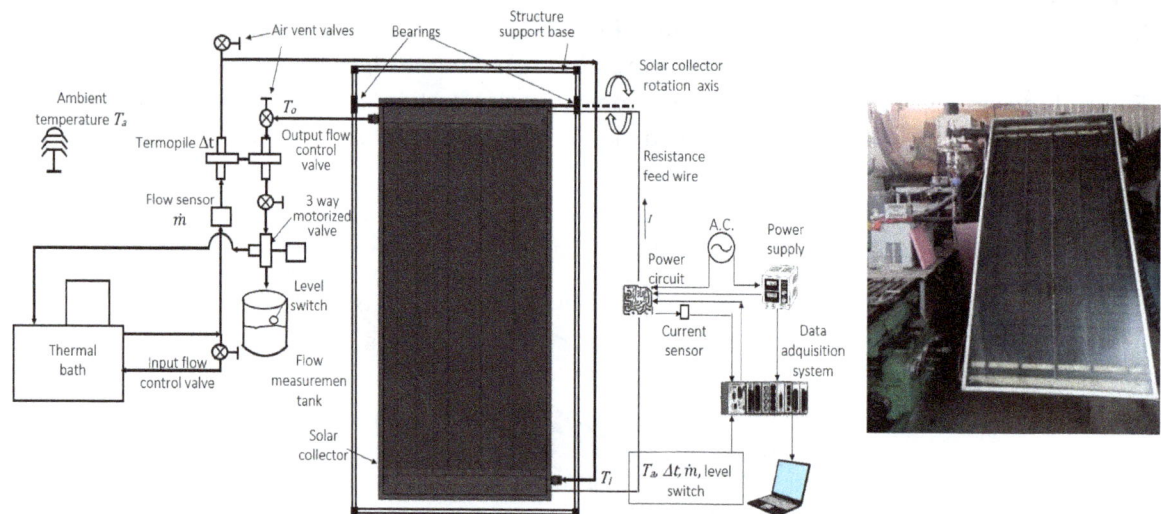

Figure 2. Experimental Setup.

According the heating of the absorber two cases were considered. The Case 1 ($(T_o - T_i) > 0$, $Q_u(\beta) \geq 0$): with considerable profile temperature in the raising tubes, also called non-isothermal absorber plate, which allows to achieve the U_L and the collector efficiency. The Case 2 ($(T_o - T_i) = 0$, $Q_u(\beta) = 0$): also called isothermal absorber plate, which allows to achieve the U_L only. In the Case 1, the GA is set to a specified value, therefore $(T_o - T_i)$ and $(T_p - T_a)$ are the output variables. In the Case 2, the $(T_o - T_i)$ are controlled close to zero by adjusting the VI, thus $(T_p - T_a)$ is the output variables only.

As is shown in Figure 2, the experimental set up allows mounting the sample with variable angle β at $0°$, $30°$, $45°$, $60°$ and $90°$, with uncertainty of $\pm0.1°$. The $(T_i - T_a)/G$ are fixed at series of specified steps, by adjusting T_i and VI in the range of part of low and medium temperature solar collectors (0.044–0.084). The electrical heater supplied up to a maximum of 2000 W, with an uncertainty of ±5 W.

The temperature differentials $(T_o - T_i)$ and $(T_i - T_a)$ were measured using a type T thermocouple, 32 gauge wires, with an uncertainty of ±0.1 °C. A thermal bath supplied 10%–90% water-glycol mixture as working fluid, with an uncertainty of ±0.01 °C. The mass flow rate was 0.016 kg/s [15,21]; it was monitored with a turbine-flowmeter, with an uncertainty of 3%, and was also verified by weighing the water-glycol mixture, at specified time steps during the experimental campaign.

The experimental indoor condition allows to keep uniform surrounding temperature, surrounding emissivity, as well as to run the experiments with non-considerable changes of solar heating, ambient temperature, surrounding temperature and wind velocity. The 10%–90% water-glycol mixture allows minimizing adverse boiling effects.

A programmable Field Programmable Gate Array, NI-CompactRIO, 9022 (National Instruments, Austin, TX, USA), 32 bits data acquisition and Lab-VIEW software 2012 (National Instruments, Austin, TX, USA) were used to monitor, recorded and calculate the experimental variables at time steps of one second. During experiments, the steady state was verified by monitoring experimental data without considerable changes at each 30 min. Each data point corresponds to an average of over 1800 measurements, taken during a 30 min period. The test was made by triplicate for comparison.

3. Effect of the Parameters

The experiment focused on the behavior of the overall heat transfer and its effect on the collector's efficiency as a function of the inclination angle and $(T_i - T_a)/G$, emphasizing that $G \neq f(\beta)$ and minimizing the effect of the changes of the rest of the variables. Three sets of 20 experiments were carried out for this purpose. The U_L was studied at β ($0°$, $30°$, $45°$, $60°$ and $90°$) and $(T_i - T_a)/G$ at (0.044, 0.056, 0.070, 0.083, 0.124, 0.140, 0.160, 0.195 and 0.235). The efficiency was the same for range β and at $(T_i - T_a)/G$ of (0.044, 0.056, 0.069, 0.083). Case 1 allows the study of the U_L and efficiency at $(T_o - T_i) > 0$, $(Q_u(\beta) > 0)$; the $(T_i - T_a)/G$ and T_p were 0.044, 66.8:0.056, 76.1:0.69, 85.3:0.82, 94.4, respectively. Case 2 allows the study of the U_L only at $(T_o - T_i) = 0$, $(Q_u(\beta) = 0)$; the $(T_i - T_a)/G$ and T_p were 0.235, 60:0.195, 70.0:0.160, 80:0.140, 90:0.124, 100, respectively. Table 2 shows the experimental conditions for the two cases.

Table 2. Experimental conditions for the two study cases.

Parameter	Units	Case 1	Case 2
β	grades	0, 30, 45, 60, 90	0, 30, 45, 60, 90
$(T_o - T_i)$	°C	>0	=0
$Q_u(\beta)$	W	>0	=0
VI	W	constant	Adjustable
$(T_i - T_a)/G$	(m^2·°C/W)	0.044, 0.056, 0.070, 0.083	0.124, 0.140, 0.160, 0.195, 0.235
T_i	°C	60.0, 70.0, 80.0, 90.0	60, 70, 80, 90, 100
T_p	°C	66.8, 76.1, 85.3, 94.4	60, 70, 80, 90, 100
$\dot{m}$	kg/s	0.016	0.016

3.1. U_L vs. β and $(T_i - T_a)/G$

3.1.1. Case 1: $(T_o - T_i) > 0$, $(Q_u(\beta) > 0)$

Figure 3 shows U_L vs. β and $(T_i - T_a)/G$ with its corresponding average of T_p. The experimental uncertainty of U_L was ±0.25 W/m^2·°C [22]. The T_i was increased to increase the value of T_p and then $(T_i - T_a)/G$ increased as well. It can be observed that the Q_u decreased slower than the increase of $(T_p - T_a)$, causing a U_L value reduction with the increase of T_p. As expected, according Equation (3), the results show that the energy loss $(VI - Q_u)$ increases as $(T_p - T_a)$ increases, while Q_u slowly decreases and VI remains constant. Thus, the U_L decrease due to the Q_u decrease is slower than the increase of $(T_p - T_a)$, causing the U_L value reduction with the increase of T_p.

As seen, $U_L(\beta)$ can achieve a reduction up to 12–10.1 W/m^2·°C (19.2%), increasing the inclination angle from $0°$ to $90°$ at a constant $(T_i - T_a)/G$ of 0.44. At a constant β, the decreasing rate of the U_L was accentuated at lower values of $(T_i - T_a)/G$. The U_L can change up to 12.0–8.4 W/m^2·°C (41.7%), at $0.0°$ of inclination, into the range we studied of the $(T_i - T_a)/G$. A critical angle was found between $30°$ and $45°$ of inclination for each case of $(T_i - T_a)/G$.

The $U_L(\beta)/U_L(90°)$ vs. β correlation for the four different values of $(T_i - T_a)/G$ at $Q_u(\beta) > 0$ is shown in Figure 4. As seen, changing β from 90° to 0°, a significant growth of $U_L(\beta)/U_L(90°)$ ranging from 1.12 to 1.21 (8.0%) was reached at $\beta = 0$. The critical angle was also found between 30° and 45° of inclination, which is similar for all the values of the $(T_i - T_a)/G$, while the variation between 0°–30° and 45°–90° seems to be almost linear behavior.

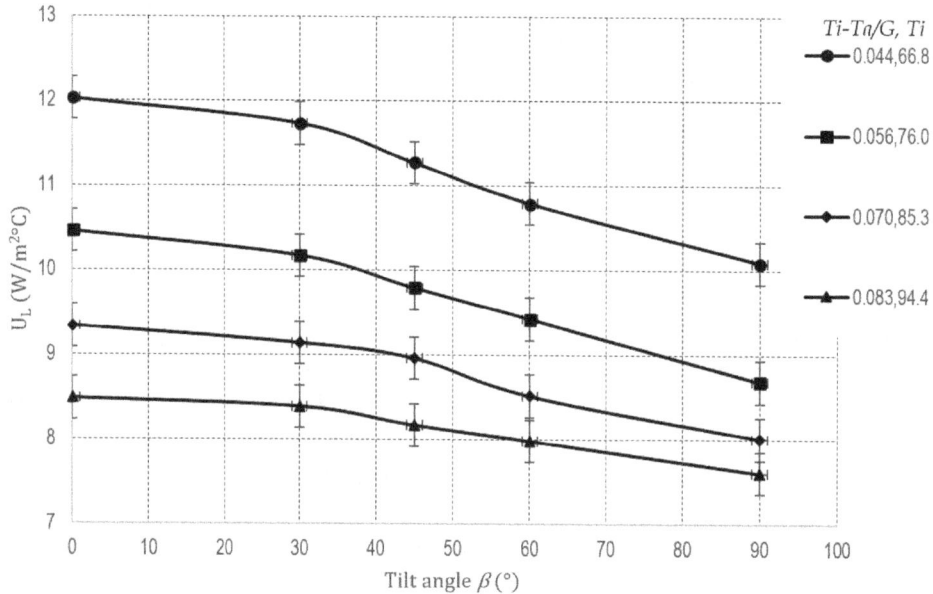

Figure 3. U_L vs. β, for variable $(T_i - T_a)/G$ at its corresponding T_p, at $Q_u(\beta) > 0$.

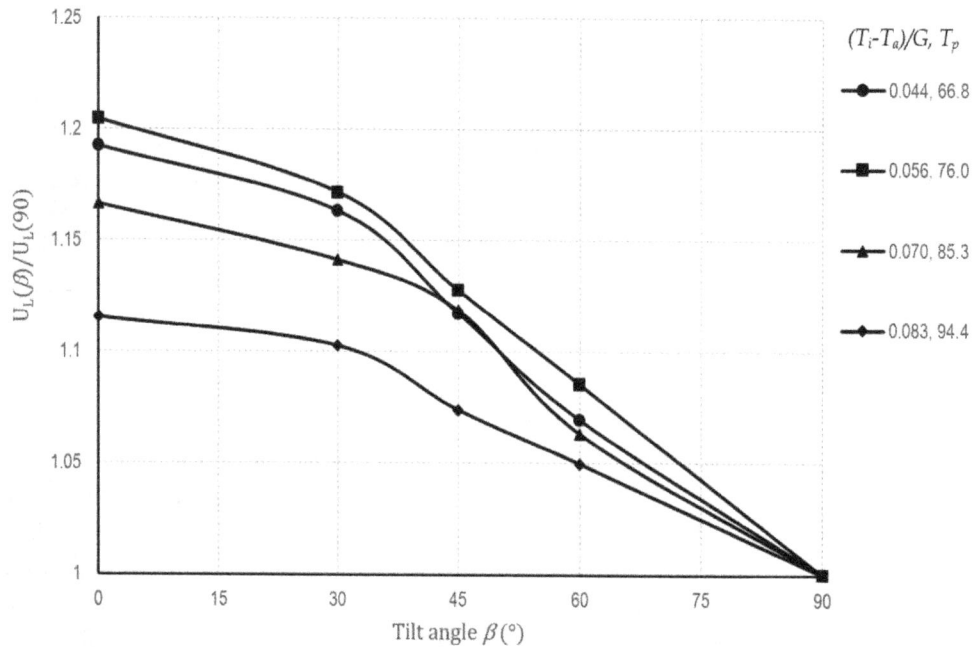

Figure 4. $U_L(\beta)/U_L(90°)$ vs. β and $(T_i - T_a)/G$, for Case 1, at $Q_u(\beta) > 0$.

3.1.2. Case 2: $(T_o - T_i) = 0$, $(Q_u(\beta) = 0)$

Figure 5 shows U_L vs. β and $(T_i - T_a)/G$ at its corresponding average of T_p. The experimental uncertainty was only ± 0.07 W/m²·°C on average, 247% less than in Case 1. The most important reduction of $U_L(\beta)$ was for lower values of $(T_i - T_a)/G$, from 8.42 to 7.63 W/m²·°C (10.4%), increasing the inclination angle from 0° to 90° at a constant $(T_i - T_a)/G$ of 0.124. As seen in Figures 3 and 5, there is higher linearity at $(T_o - T_i) = 0$ than when $(T_o - T_i) > 0$, which can be due to the even and uneven surface temperature, which causes a transition flow within the cavity.

Figure 6 shows $U_L(\beta)/U_L(90°)$ vs. β, which ranges from 1.09 to 1.11 only. The $U_L(0)$ could only be increased up to 1.8% of the value of $U_L(90)$, from the highest to lowest values of $(T_i - T_a)/G$ at $\beta = 0$.

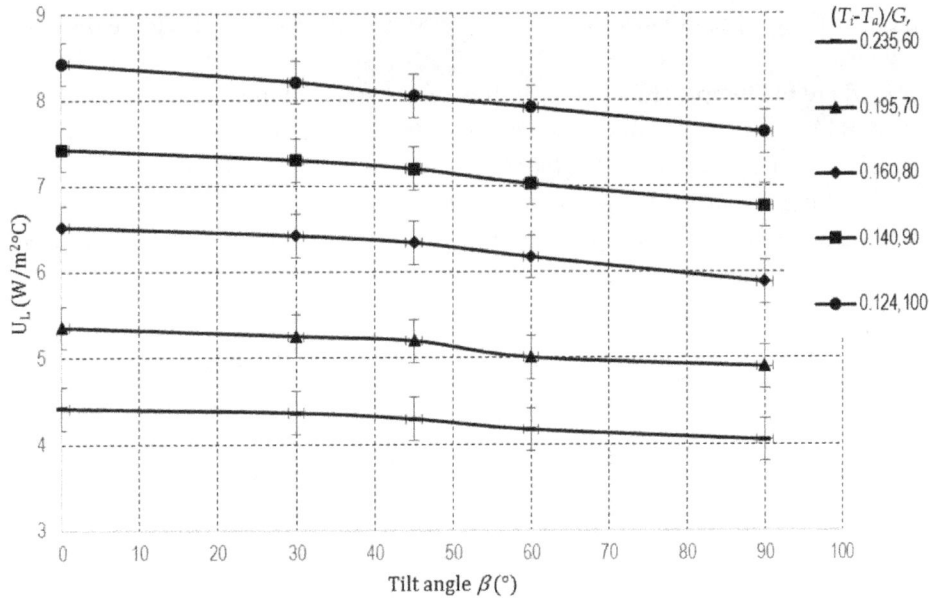

Figure 5. U_L vs. β, for variable $(T_i - T_a)/G$ at its corresponding T_p, at $Q_u(\beta) = 0$.

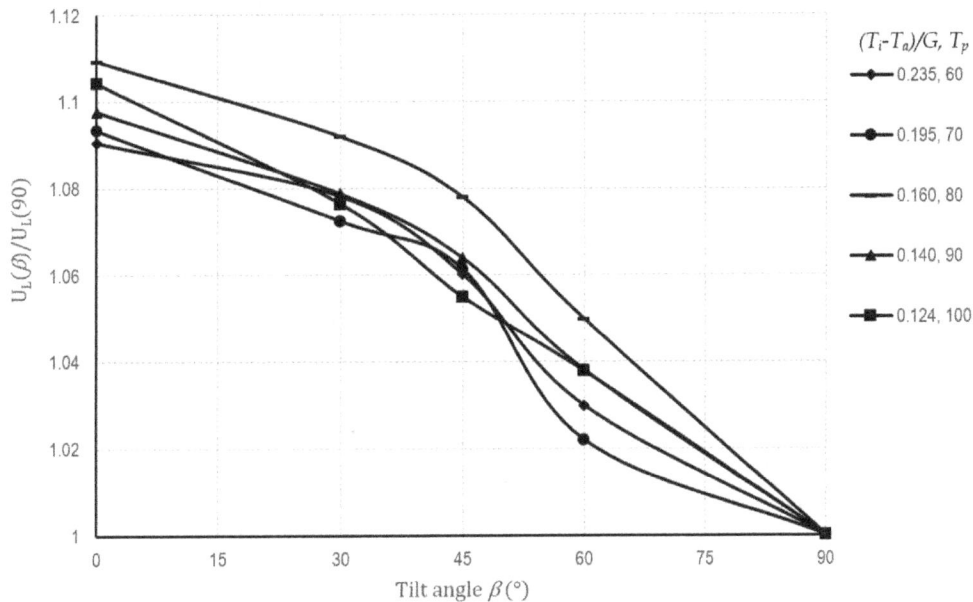

Figure 6. $U_L(\beta)/U_L(90°)$ vs. β and $(T_i - T_a)/G$, at $(T_o - T_i) = 0$.

3.1.3. Comparison with Previous Works

Figure 7 shows a comparison of $U_L(\beta)/U_L(90°)$ vs. β for Case 1 ($Q_u(\beta) > 0$), considering that $U_t \approx U_L$, according Duffie and Beckman [3] and Cooper et al. [8], to calculate U_t. The Hollands correlation [23] was used to calculate the Nussetl number in both methods. Significant changes of $U_L(\beta)/U_L(90°)$ are evident up to 10.0%, comparing different results with different types of collectors and experimental conditions.

According to the uncertainty propagation method [23], in Case 2 the experimental uncertainty of U_L was 0.07 W/m²·°C, which was 247% lower than the value obtained in Case 1. The reduction in uncertainty was reached due to the elimination of $Qu(\beta)$ in Equation (2). The elimination of $Qu(\beta)$ implied that $(T_o - T_i) = 0$, which caused the absorber plate not to have significant temperature changes through the raising tubes. Also, the Joule effect controlled by PID allows us to achieve lower uncertainty, but in Case 1 the greatest contribution of uncertainty continues to be the uncertainty of the mass flow rate.

In Case 1, where the temperature profile of the raising tubes of the collector changes significantly, the U_L is considerably sensitivity to the β and $(T_i - T_a)/G$, reaching up to 19.2% and 41.7%, respectively. The most significant change of $U_L(\beta)/U_L(90°)$ vs. β was 8.0%, while for Case 2, where the raising tubes do not have a significant temperature profile, the maximum change of $U_L(0)/U_L(90°)$ vs. β was only 1.8%. On the other hand, the comparison of the results with different authors showed that there are still considerable differences.

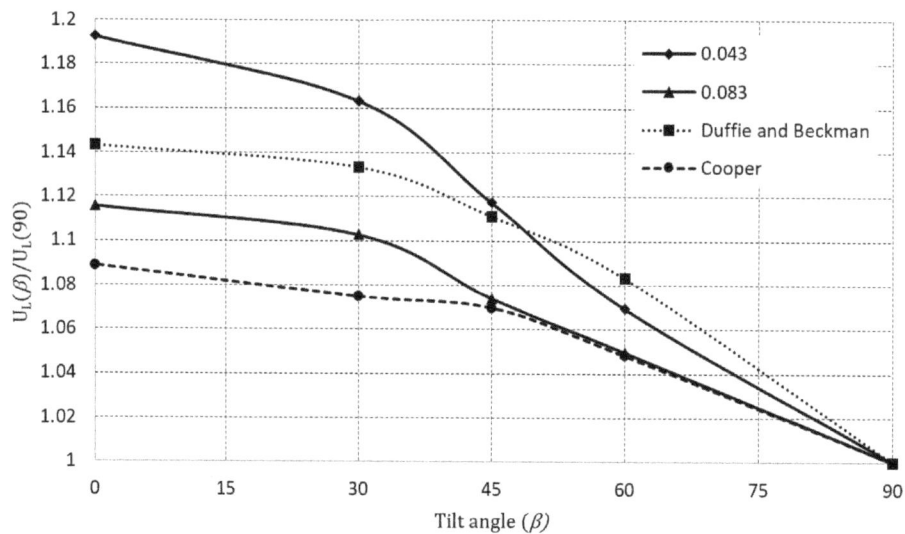

Figure 7. Comparison of $U_L(\beta)/U_L(90°)$ vs. β and $(T_i - T_a)/G$ at $(T_o - T_i) > 0$.

3.2. Efficiency vs. β and $(T_i - T_a)/G$

Figure 8 shows the thermal efficiency η vs. $(T_i - T_a)/G$ at β (0°, 30°, 45°, 60° and 90°). Within the range we studied, the variation of the efficiency with respect to $(T_i - T_a)/G$ reached up to 0.31–0.45 (45%). Similarly, the variation of the efficiency with respect to β reached up to 0.37–0.45 (21.6%). According to the slope of each linear approach, the $F_R U_L(\beta)$ at each inclination angle was 3.12, 3.27, 3.38, 3.52 and 3.65, respectively. These variations behave in almost a linear manner over the inclination range, as in Equation (4), changing 17.6%.

$$F_R U_L(\beta) = 3.112 + 0.0061\beta \qquad (4)$$

As expected, the efficiency slowed down as β and $(T_i - T_a)/G$ increased, showing average variations of 21.6% and 45.0%, respectively, over the studied range. According to the least squares fitting method, considering double fitting, the efficiency curve in terms of β and $(T_i - T_a)/G$ is equated in Equation (5). The *RMSE* and R^2 were 0.008 and 0.982, respectively, taking into account the experimental data.

$$\eta\left(\beta, \left[\frac{T_i - T_a}{G}\right]\right) = 0.5282 - 3.386\left[\frac{T_i - T_a}{G}\right] + 0.0008145\beta \qquad (5)$$

Equation (5) allows us to observe that the heat transfer convection between the absorber plate and glazing cover is significantly affected by changes of β and $(T_i - T_a)/G$, which in turn significantly affects the accuracy of the overall heat transfer coefficient and the thermal efficiency of the flat plate collectors.

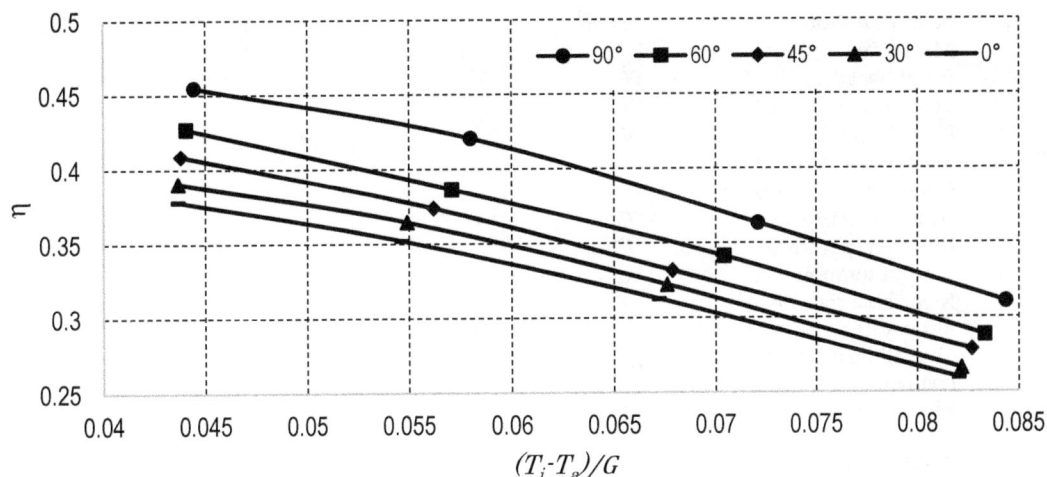

Figure 8. η vs. β and $(T_i - T_a)/G$.

4. Conclusions

The effects of varying the angle of inclination are reflected directly in the overall loss coefficient which in turn significantly affects the thermal efficiency. Within the range we studied, the average variation of the efficiency with respect to β and $(T_i - T_a)/G$ was 21.6% and 45.0%, respectively. At the same time, the U_L is also considerably sensitivity to the β and $(T_i - T_a)/G$, reaching up to 19.2% and 41.7%, respectively. A two-fitting variable correlation was obtained from the efficiency with respect to the $(T_i - T_a)/G$ and β, with *RMSE* and R^2 values of 0.008 and 0.982, respectively.

The most significant changes of $U_L(β)/U_L(90°)$ vs. β were found for Case 1: $(T_o - T_i) > 0$, 8.0%, where the temperature profile of the raising tubes of the collector changes significantly. On the other hand, for Case 2: $(T_o - T_i) = 0$, where the raising tubes do not have a significant temperature profile, the maximum change of $U_L(β)/U_L(90°)$ vs. β was 1.8% only, which is negligible. A critical angle was found between 30° and 45° of the collector's inclination, which is similar for all the values of the $(T_i - T_a)/G$, while the variation between 0°–30° and 45°–90° seems to be almost linear in its behavior.

The $F_R U_L(β)$ varied up to 17.6% from the horizontal to vertical position. These variations behave almost linearly over the inclination range with respect to the average value. Therefore, the effects of the changes of β considerably influence the efficiency calculations of solar collectors if changes of the convective flow patterns within collector are not considered.

Acknowledgments: The authors are grateful to Consejo Nacional de Ciencia y Tecnolgía, CONACyT and Tecnológico Nacional de México, TecNM, whose financial support made this work possible.

Author Contributions: Orlando Montoya-Marquez and José Jasson Flores-Prieto conceived and designed the experiments; Orlando Montoya-Marquez performed the experiments; both authors analyzed the data and contributed reagents/materials/analysis tools; also, Orlando Montoya-Marquez and José Jasson Flores-Prieto wrote the paper.

Conflicts of Interest: The authors declare that they have no conflict of interest.

Nomenclature

Variables	Description	Units
A	Collector area	m^2
Cp	Specific heat	$J/kg°C$
F_R	Removal factor	-
G	Solar radiation	W/m^2
$\dot{m}$	Mass flow	kg/s
Q_i	Input heat	W
Q_l	Loss heat	W
Qu	Useful heat	W
$RMSE$	Root mean square error	-
R^2	Coefficient of determination	-
T_a	Ambient Temperature	°C
T_i	Input temperature	°C
T_o	Output temperature	°C
T_p	Mean absorber plate temperature	°C
U_L	Overall heat transfer coefficient	$W/m^2·°C$
VI	Electric power	W
Symbols		
α	Absortance	-
β	Tilt angle	°
τ	Transmittance	-
η	Efficiency	-
Acronyms		
ASHRAE	American Society of Heating Refrigerating and Air Conditioning Engineers	
ANSI	American National Standard Institute	
IAM	Incident Angle Modifier	
IEA	International Energy Agency	
CIE	Commission Internationale de L'Eclairage	
ISO	International organization for standardization	

References

1. Mauthner, F.; Weiss, W.; Spörk-Dür, M. *Solar Heat Worldwide*; IEA Solar Heating & Cooling Programme: Gleisdorf, Austria, 2016.

2. Kalogirou, S.A. Solar thermal collectors and applications. *Prog. Energy Combust. Sci.* **2004**, *30*, 231–295. [CrossRef]

3. Duffie, J.; Beckman, W. *Solar Engineering of Thermal Processes*, 4th ed.; John Wiley & Sons, Inc.: Hoboken, NJ, USA, 2013.

4. American Society of Heating, Refrigerating, and Air-Conditioning Engineers, Inc. *ANSI/ASHRAE 93-2010 (RA 2014), Methods of Testing to Determine the Thermal Performance of Solar Collectors*; ASHRAE: Atlanta, GA, USA, 2014.

5. *ISO 9806-1 International Standard, Test Methods for Solar Collectors—Part 1: Thermal Performance of Glazed Liquid Heating Collectors Including Pressure Drop*; ISO: Vernier, Switzerland, 1994.

6. Whillier, A. Prediction of performance of Solar Collectors. In *Applications of Solar Energy for Heating and Cooling of Buildings*; ASHRAE GRP 170; ASHRAE: New York, NY, USA, 1977.

7. Cooper, P.I. The effect of inclination on the heat loss from flat-plate solar collectors. *Sol. Energy* **1981**, *27*, 413–420. [CrossRef]

8. Cooper, P.I.; Dunkle, R.V. A non-linear flat plate collector model. *Sol. Energy* **1981**, *26*, 133140. [CrossRef]

9. Rodríguez-Hidalgo, M.C.; Rodríguez-Aumente, P.A.; Lecuona, A.; Gutiérrez-Urueta, G.L.; Ventas, R. Flat plate thermal solar collector efficiency: Transient behavior under working conditions. Part I: Model description and experimental validation. *Appl. Therm. Eng.* **2011**, *31*, 2394–2404. [CrossRef]

10. Chang, T.P. Study on the optimal tilt angle of solar collector according to different radiation types. *Int. J. Appl. Sci. Eng.* **2008**, *6*, 151–161.

11. Shariah, A.; Al-Akhras, M.A.; Al-Omari, I.A. Optimizing the tilt angle of solar collectors. *Renew. Energy* **2002**, *26*, 587–598. [CrossRef]

12. Ozoe, H.; Sayama, H.; Churchill, S.W. Natural convection in an inclined rectangular channel at various aspect ratios and angles—Experimental measurements. *Int. J. Heat Mass Transf.* **1975**, *18*, 1425–1431. [CrossRef]

13. Ozoe, H.; Yamamoto, K.; Sayama, H.; Churchil, S.W. Natural convection in an inclined rectangular channel heated on one side and cooled on the opposing side. *Int. J. Heat Mass Transf.* **1974**, *17*, 1209–1217.

14. Alvarado, R.; Xamán, J.; Hinojosa, J.; Álvarez, G. Interaction between natural convection and surface thermal radiation in tilted slender cavities. *Int. J. Therm. Sci.* **2008**, *47*, 355–368. [CrossRef]

15. Bava, F.; Furbo, S. Correction of Collector Efficiency Depending on Variations of Collector Type, Solar Collector Fluid, Volume Flow Rate and Collector Tilt. IEA-SHC Info Sheet 45.A.1. 22 December 2014. Available online: http://task45.iea-shc.org/data/sites/1/publications/IEA-SHC-T45.A.1-INFO-Correction-of-collector-efficiency.pdf (assessed on 21 December 2016).

16. Sabatelli, V.; Marano, D.; Braccio, G.; Sharma, V.K. Efficiency test of solar collectors: Uncertainty in the estimation of regression parameters and sensitivity analysis. *Energy Convers. Manag.* **2002**, *43*, 2287–2295. [CrossRef]

17. Beikircher, T.; Osgyan, P.; Fischer, S.; Drück, H. Short-term efficiency test procedure for solar thermal collectors based on heat loss measurements without insolation and a novel conversion towards daytime conditions. *Sol. Energy* **2014**, *107*, 653–659. [CrossRef]

18. *Glycols*; Curme, G.O.; Johnston, F., Eds.; Reinhold Publishing Corp.: New York, NY, USA, 1952.

19. *CIE 130-1998, Practical Methods for the Measurement of Reflectance and Transmittance*; Technical Report No. 130; Commission Internationale de L'Eclairage: Vienna, Austria, 1998.

20. *ISO9050-2003 Glass in Building—Determination of Light Transmittance, Solar Direct Transmittance, Total Solar Energy Transmittance, Ultraviolet Transmittance and Related Glazing Factors*; ISO: Vernier, Switzerland, 2003.

21. Zauner, C.; Hengstberger, F.; Hohenauer, W.; Reichl, C.; Simetzberger, A.; Gleiss, G. Methods for medium temperature collector development applied to a CPC collector. *Energy Procedia* **2012**, *30*, 187–197. [CrossRef]

22. Holman, J.P. *Experimental Methods for Engineers*, 8th ed.; McGraw-Hill Book Company: New York, NY, USA, 2012.

23. Hollands, K.G.T.; Unny, T.E.; Raithby, G.D.; Konicek, L. Free convection heat transfer across inclined air layers. *J. Heat Transf.* **1976**, *98*, 189–193. [CrossRef]

Permissions

All chapters in this book were first published in ENERGIES, by MDPI; hereby published with permission under the Creative Commons Attribution License or equivalent. Every chapter published in this book has been scrutinized by our experts. Their significance has been extensively debated. The topics covered herein carry significant findings which will fuel the growth of the discipline. They may even be implemented as practical applications or may be referred to as a beginning point for another development.

The contributors of this book come from diverse backgrounds, making this book a truly international effort. This book will bring forth new frontiers with its revolutionizing research information and detailed analysis of the nascent developments around the world.

We would like to thank all the contributing authors for lending their expertise to make the book truly unique. They have played a crucial role in the development of this book. Without their invaluable contributions this book wouldn't have been possible. They have made vital efforts to compile up to date information on the varied aspects of this subject to make this book a valuable addition to the collection of many professionals and students.

This book was conceptualized with the vision of imparting up-to-date information and advanced data in this field. To ensure the same, a matchless editorial board was set up. Every individual on the board went through rigorous rounds of assessment to prove their worth. After which they invested a large part of their time researching and compiling the most relevant data for our readers.

The editorial board has been involved in producing this book since its inception. They have spent rigorous hours researching and exploring the diverse topics which have resulted in the successful publishing of this book. They have passed on their knowledge of decades through this book. To expedite this challenging task, the publisher supported the team at every step. A small team of assistant editors was also appointed to further simplify the editing procedure and attain best results for the readers.

Apart from the editorial board, the designing team has also invested a significant amount of their time in understanding the subject and creating the most relevant covers. They scrutinized every image to scout for the most suitable representation of the subject and create an appropriate cover for the book.

The publishing team has been an ardent support to the editorial, designing and production team. Their endless efforts to recruit the best for this project, has resulted in the accomplishment of this book. They are a veteran in the field of academics and their pool of knowledge is as vast as their experience in printing. Their expertise and guidance has proved useful at every step. Their uncompromising quality standards have made this book an exceptional effort. Their encouragement from time to time has been an inspiration for everyone.

The publisher and the editorial board hope that this book will prove to be a valuable piece of knowledge for researchers, students, practitioners and scholars across the globe.

List of Contributors

Norhisam Misron
Department of Electrical & Electronic, Faculty of Engineering, Universiti Putra Malaysia, 43400 Serdang, Selangor, Malaysi
Institute of Advanced Technology, Faculty of Engineering, Universiti Putra Malaysia, 43400 Serdang, Selangor, Malaysia

Tsuyoshi Hanamoto
Department of Biological Functions Engineering, Graduate School of Life Science and Systems Engineering, Kyushu Institute of Technology, Kitakyushu 808-0916, Japan

Shehu Salihu Mustafa, Norman Mariun and Mohammad Lutfi Othman
Department of Electrical & Electronic, Faculty of Engineering, Universiti Putra Malaysia, 43400 Serdang, Selangor, Malaysia

Ming-Tse Kuo
Department of Electrical Engineering, National Taiwan University of Science and Technology, No. 43, Section 4, Keelung Road, Da'an District, Taipei 106, Taiwan

Ming-Chang Tsou
Leadtrend Technology Corporation, No. 1, Taiyuan 2nd St., Zhubei City, Hsinchu County 30288, Taiwan

Mohammed H. Alsharif and Jeong Kim
Department of Electrical Engineering, College of Electronics and Information Engineering, Sejong University, 209 Neungdong-ro, Gwangjin-gu, Seoul 05006, Korea

Feifan Ji, Ji Xiang and Wuhua Li
College of Electrical Engineering, Zhejiang University, Hangzhou 310027, China;

Quanming Yue
State Grid Zhejiang Electric Power Company, Hangzhou 310007, China

Wenqiang Sun
Department of Thermal Engineering, School of Metallurgy, Northeastern University, Shenyang 110819, China
State Environmental Protection Key Laboratory of Eco-Industry, Northeastern University, Shenyang 110819, China

Yuhao Hong
Department of Thermal Engineering, School of Metallurgy, Northeastern University, Shenyang 110819, China
Department of Technology, Hangzhou Boiler Group Co., Ltd., Hangzhou 310021, China

Yanhui Wang
Department of Thermal Engineering, School of Metallurgy, Northeastern University, Shenyang 110819, China

Guoqiang Li, YutingWu, Ruiping Zhi, JingfuWang and Chongfang Ma
Key Laboratory of Enhanced Heat Transfer and Energy Conservation, Ministry of Education and Key Laboratory of Heat Transfer and Energy Conversion, Beijing Municipality, College of Environmental and Energy Engineering, Beijing University of Technology, Beijing 100124, China

Yeqiang Zhang
School of Energy and Power Engineering, Zhengzhou University of Light Industry, No. 5 Dongfeng Road, Zhengzhou 450002, China

Mahmoud A. Eissa
Department of Mathematics, Harbin Institute of Technology, Harbin 150001, China
Department of Mathematics, Faculty of Science, Menoufia University, Menoufia 32511, Egypt

Boping Tian
Department of Mathematics, Harbin Institute of Technology, Harbin 150001, China

Ruixiong Li, Huanran Wang, Erren Yao and Shuyu Zhang
School of Energy and Power Engineering, Xi'an Jiaotong University, Xi'an 710049, Shaanxi, China

Camelia Stanciu, Dorin Stanciu and Adina-Teodora Gheorghian
Department of Engineering Thermodynamics, University Politehnica of Bucharest, Splaiul Independentei 313, 060042 Bucharest, Romania

Rishang Long and Jianhua Zhang
State Key Laboratory of New Energy Power System, North China Electric Power University, Beijing 102206, China

Seunghyoung wRyu and Hongseok Kim
Department of Electronic Engineering, Sogang University, 35 Baekbeom-ro, Mapo-gu, Seoul 121-742, Korea

Jaekoo Noh
Software Center, Korea Electric Power Corporation (KEPCO), 105 Munji Road, Yuseong-Gu, Daejeon 305-760, Korea

Index